AF686196

Communications and Control Engineering

Springer
London
Berlin
Heidelberg
New York
Barcelona
Hong Kong
Milan
Paris
Singapore
Tokyo

Published titles include:

Optimal Sampled-Data Control Systems
Tongwen Chen and Bruce Francis

Nonlinear Control Systems (3rd edition)
Alberto Isidori

Theory of Robot Control
C. Canudas de Wit, B. Siciliano and G. Bastin (Eds)

Fundamental Limitations in Filtering and Control
María M. Seron, Julio Braslavsky and Graham C. Goodwin

Constructive Nonlinear Control
R. Sepulchre, M. Janković and P.V. Kokotović

A Theory of Learning and Generalization
M. Vidyasagar

Adaptive Control
I.D. Landau, R. Lozano and M.M'Saad

Stabilization of Nonlinear Uncertain Systems
Miroslav Krstić and Hua Deng

Passivity-based Control of Euler-Lagrange Systems
Romeo Ortega, Antonio Loría, Per Johan Nicklasson and Hebertt Sira-Ramírez

Stability and Stabilization of Infinite Dimensional Systems with Applications
Zheng-Hua Luo, Bao-Zhu Guo and Omer Morgul

Nonsmooth Mechanics (2nd edition)
Bernard Brogliato

Nonlinear Control Systems II
Alberto Isidori

L_2-Gain and Passivity Techniques in Nonlinear Control
Arjan van der Schaft

Control of Linear Systems with Regulation and Input Constraints
Ali Saberi, Anton A. Stoorvogel and Peddapullaiah Sannuti

Robust and H∞ Control
Ben M. Chen

Computer Controlled Systems
Efim N. Rosenwasser and Bernhard P. Lampe

Dissipative Systems Analysis and Control
Rogelio Lozano, Bernard Brogliato, Olav Egeland and Bernhard Maschke

Stanislav V. Emelyanov and Sergey K. Korovin

Control of Complex and Uncertain Systems

New Types of Feedback

With 336 Figures

Springer

Stanislav V. Emelyanov, PhD

Sergey K. Korovin, PhD
Office 733b, Faculty Computational Mathematics and Cybernetics, Building #2, Moscow State University, Vorob'evy Gory, Moscow, 119899, Russia

Series Editors

E.D. Sontag • M. Thoma

ISSN 0178-5354

ISBN 1-85233-302-2 Springer-Verlag London Berlin Heidelberg

British Library Cataloguing in Publication Data
Emelyanov, Stanislav V.
Control of complex and uncertain systems : new types of feed back. - (Communications and control engineering)
1.Control theory 2.Automatic control 3.Cybernetics
ITitle II.Korovin, Sergei K.
629.8'312
ISBN 185233302

Library of Congress Cataloging-in-Publication Data
Emel'ianov, Stanislav Vasil'evich.
Control of complex and uncertain systems : new types of feed back / Stanislav V. Emelyanov and Sergey K. Korovin.
p. cm. -- (Communication and control engineering)
Includes bibliographical references and index.
ISBN 1-85233-302-2 (alk.paper)
1. Feedback control systems. 2. Control theory. I. Korovin, S. K., 1945- II. Title. III. Series.
TJ216.E415 2000
629.8'3--dc21 00-037372

Apart from any fair dealing for the purposes of research or private study, or criticism or review, as permitted under the Copyright, Designs and Patents Act 1988, this publication may only be reproduced, stored or transmitted, in any form or by any means, with the prior permission in writing of the publishers, or in the case of reprographic reproduction in accordance with the terms of licences issued by the Copyright Licensing Agency. Enquiries concerning reproduction outside those terms should be sent to the publishers.

© Springer-Verlag London Limited 2000
Printed in Great Britain

The use of registered names, trademarks, etc. in this publication does not imply, even in the absence of a specific statement, that such names are exempt from the relevant laws and regulations and therefore free for general use.

The publisher makes no representation, express or implied, with regard to the accuracy of the information contained in this book and cannot accept any legal responsibility or liability for any errors or omissions that may be made.

Typesetting: Camera ready by authors
Printed and bound by Athenæum Press Ltd., Gateshead, Tyne & Wear
69/3830-543210 Printed on acid-free paper SPIN 10761412

The authors want to express their gratitude to Mikhail Yuryevich Zhivilo whose sincere wish to contribute to the development of Russian science deserves deep esteem.

Preface

In this monograph we consider one of the main problems of the automatic control theory, namely, the stabilization problem and the methods of its solution in their evolution, i.e., beginning with the simplest statement of this problem and gradually making it more complicated, we analyze in detail the possibilities of different methods of its solution. The complication begins with an increase in the uncertainty factors in the statement of the problem, and the methods of solution become more complicated correspondingly. This approach makes it possible to consider the general trends in the development of the principles and methods of the automatic control theory. The last fact is obviously very important and urgent since, in the new situation, the mastering of the general mechanisms of formation of control is more important than the knowledge of specific methods of its synthesis.

It should be pointed out that we do not consider the proposed point of view concerning the development of the automatic control theory to be the only possible approach since the problem of the mechanism of generation of feedback is far from being trivial and different approaches to this mechanism are possible. The larger the number of these approaches the better since they bring us closer to the understanding of the fundamental mechanisms of functioning of feedback. This is very significant both theoretically and practically since the modern methods of stabilization are oriented, in the main, to an "intensive" solution of the problem whereas nature demonstrates remarkable examples of solving stabilization problems with the use of very limited means and in rather strained circumstances. This essential difference testifies that a genuine feedback theory has not yet been worked out, that many things are not yet clear, and that the principal discoveries in this sphere are yet to come.

Investigating this complicated and delicate problem, we are far from laying claim to grasping the crux of the matter, but we are sure that the theory that we propose directly concerns the matter and seems to be quite natural and convincing.

Some words are due about the structure of the monograph. As was already pointed out, we try to go from simple things to more complicated ones and begin, naturally, from linear objects and the methods of the theory of linear control systems. Since we place special emphasis on the principles of problem solving and on the conceptual interpretation of the results, we avoided mathematically strict statements and proofs. It stands to reason that all facts and statements presented in the book can be strictly substantiated and many of them are well known from literature.

In the monograph we compare the applications of different methods of control for solving stabilization problems under different conditions, namely, external forces, parameters, the structure and order of an object. For this purpose, sample models of objects are, obviously, especially useful, and therefore only models of this kind are considered in this monograph. However, when carrying out the analysis, we use different forms of description of control objects, namely, structural, operator, and differential forms, since some facts seem to be more convincing in a certain description and other facts are more convincing in a different description.

It can be seen from the book that the nonlinearity becomes more and more important as the stabilization problem becomes more complicated. In addition, it becomes clear that there cannot be good stabilization without nonlinear feedback and it is precisely the nonlinear feedback that makes a control system capable of demonstrating the needed behavior in complicated and constantly varying external and internal conditions.

It turns out that beginning with a certain level of complexity of the problem, a "good" control is necessarily nonlinear. It is known that in the nonlinear world there are no regular ways or universal methods typical of local theories since the specific features of nonlinearity impose certain constraints. For the theory developed in this monograph the structural methods of analysis and synthesis of systems turn out to be very useful, and therefore we pay so much attention to the description of these methods.

The purposeful use of nonlinearities in control makes it possible to operate with principally new "nonintensive" or "compensational" mechanisms of suppressing uncertainty factors, in particular, the techniques which are based on the use of positive feedback and unstable motions and which allow the system to gather momentum by itself and work until conditions are created for suppressing the disturbances and uncertainty factors. It is precisely the positive feedback and the instability that play a key role in some problems.

It should finally be pointed out that the stabilization problem should not be considered in the restricted sense since many important problems of control theory can be reduced to the stabilization problem, say, the problems of differentiation and optimization. However, since problems of this class are important and rich in content, we devote special sections to their analysis.

The authors express their deepest gratitude to many people who made possible the appearance and development of the binary control theory: to some people for their benevolent reaction and mild criticism when we first appeared in public with our reports, to other people for their selfless and creative work on the topical problems of the theory, to our opponents for their severe, may be not always justified but, in the end, useful criticism.

We express our special gratitude to Academicians A.A. Krasovskii, E.P. Popov, and Ya.Z. Tsypkin whose remarks were always to the point, concerned the essence of the matter, and made for a correct development of the theory.

We are also grateful to our disciples and followers who, for many years, worked fruitfully and with enthusiasm in this field and made a significant contribution to the new theory. First of all, we want to acknowledge the contribution made by I.G. Mamedov, A.L. Nersisyan, V.I. Sizikov, A.P. Nosov, and L.V. Levantovskii.

We want to point out the atmosphere of well wishing, creative activity, and self-support that prevailed among the researchers of the Institute of Systems Analysis of the Russian Academy of Sciences. This is mainly due to Academician D.M. Gvishiani, the first director of the Institute of Systems Analysis of the Russian Academy of Sciences.

We want to express our special gratitude to V.B. Betelin, Corresponding Member of the Russian Academy of Sciences. Without his support and real help the publication of this book would be delayed for many years.

We are especially grateful to Irene Aleksanova, the translator of this work. She did an outstanding job which lay beyond the scope of a simple translation and improved the book considerably. Finally, we want to point out the laborious work of the proof-reader B.N. Rusakova who frequently "exceeded her commission" and made useful suggestions, and the computer operator K.E. Pankratiev who put the manuscript of the book into shape and prepared the camera-ready copy. We are sincerely grateful to them for their highly professional work.

When writing the book, we received financial support as grants of the Russian Foundation for Basic Research and the European Economic Association.

Table of Contents

Introduction

It is believed now that the elaboration of the basic themes of automatic control theory is almost completed and, correspondingly, the center of investigation is now in the domain of applications, working out efficient methods of analysis, and designing control systems. It is also believed that the appearance of new ideas and principles is possible only upon a transition to objects of a new nature.

It should be pointed out that there are, indeed, certain reasons for this viewpoint, and these reasons are rather weighty. The automatic control theory has made an impressive progress, and today it can propose a wide spectrum of solution methods for various problems of applied automatics. The field of practical applications of the automatic control theory is very wide, and one cannot think of the contemporary technology without means of automation and, hence, without the use of recommendations of control theory. This is one aspect. Another aspect is that the most modern mathematical apparatus is used more and more actively in control theory whereas the books and articles in scientific journals concerning control theory are, in the main, of generalizing, summing up character. It may seem that there is a clear evidence that control theory is close to perfection and completes its development.

Is this really so and are there ready solutions in control theory for every specific case? In certain situations this is really the fact, but most often control theory does not give recipes but only gives recommendations which must be subjected to experimental verification as to the adequacy to the situation under consideration. Therefore it is no accident that there exists a generally accepted sequence of stages in the development of automatic control systems, namely, the elaboration of a mathematical model of an object, the investigation and identification of the model, the formulation of requirements to the properties of the system, the choice of the law of control and performance of immitation experiment, the technological realization of the system and the conduction of a natural or seminatural experiment, and the adjustment of the system. In this sequence, the whole chain constituting the elaboration of a system or some of its links may be used repeatedly. If we also take into account that the realization of each stage requires certain creative effort, it becomes clear that the creation and exploitation of control system is a complicated process which requires the enlisting of services of highly qualified experts. This requirement obviously contradicts the mass character of automation.

Consequently, the elaboration, designing, and keeping a control system in working order is a “bottleneck” which retards the progress in the technology of automation. This is a challenge to control theory which must produce methods and tools that would make it possible to work out and exploit control systems with the use of small effort and without resort to highly qualified specialists.

To a certain degree, this problem can be solved with the use of systems of computer aid design (CAD), but not only of these systems. The CAD-system is a tool which is efficient only when it is well "equipped" theoretically. Otherwise the CAD-system can cope with the routine phase of the development but will not promote the solution of the creative problems of automation, and, strictly speaking, it is precisely these problems that require high qualification. Only a developed theory that not only gives strict recommendations for a certain class of situations but also suggests rules for reasonable actions in nonstandard situations and the techniques of obtaining an adequate solution for every specific case can become a foundation which will allow a further qualitative progress in automatic control. The CAD-system with elements of "intellect" in the theoretical basis is what we need today.

Does the modern control theory satisfy this requirement (the requirement of high "intellect")? We must state that it does not. There are many reasons for this.

The designer who works out a control system has to resolve an objective contradiction between the elaboration of an object in detail and the possibility of a further analytic investigation of the system, namely, the identification of its parameters and the synthesis of the controller. Very likely, this is the most difficult stage whose formalization is hardly possible. Although a certain real process lies, as a rule, at the basis of elaboration of a system, the researchers who work in automatics strive to design not an exact model but only an immitating model of the process which reflects "the most important properties" with respect to the preassigned input and output variables. This is the main thing that distinguishes the models of control theory from the models that are exploited in such fundamental disciplines as physics, chemistry, etc. And it should be pointed out that the concept of "the most important properties" often has an intuitive sense which poorly gives in to formalization. It is, perhaps, the reason why, when designing a control system, we have to return iteratively to this stage and make necessary corrections.

Because of this circumstance, the elaboration of "as simple models as possible" seemed to be the most natural way. This way led to the formation of a collection of standard models which are, in the main, exploited in control theory. At present, this arsenal is very poor and is based on linear models or models close to them. In this way, often to the detriment of the real circumstances, but in order to oblige the theory, a bank of simplified models has been formed with which we deal, in the main, in control theory and which is, in essence, one of the obstacles against which control theory "stumbles" in practice.

Thus, the priority in the development of the theory was given to analytics, and this, in turn, led to a hypertrophied development of analytical methods which are often similar as concerns their final results but differ in the means of their achievement and the conditions of their application, namely, in transfer functions, differential equations, input-output representations, frequency and time characteristics, etc. However, we cannot wring out much from simple models even when we have a powerful apparatus. This especially concerns automatic control systems (ACS) since the attention given to the solution of problems of synthesis of controllers is inadequate. In fact, this branch of control theory remains almost virginal. There exists a rather restricted collection of techniques of synthesis for a small number of standard situations.

We can say without exaggeration that today the processes of origination of controlling mechanisms are not clear. In all cases, the appearance of a new method of synthesis is due to invention rather than to theory. Therefore, in our view, the problem of finding the general principles of synthesis which would make it possible to obtain the required law of control in concrete circumstances as if automatically is very attractive. The elaboration of these general principles will predetermine, in our opinion, the development of control theory in the near future.

We can try to guess certain features of their development. It is clear, first of all, that nonlinearity must become an inalienable element of the theory. In the first place, this is the requirement of practice: constraints, nonlinearity of elements, etc. But this is not the only reason, examples from other branches of science (and control theory as well) clearly demonstrate that due account of nonlinear phenomena substantially enriches the theory many times over: the nonlinear "world" is incommensurably richer than the linear one, and precisely on this way new phenomena, principles and laws originate.

As an illustration, we can cite an example when the automatic control theory was considerably enriched due to the solution of problems on absolute stability, to the investigation of self-excited oscillation processes, and adaptive control. Examples from other branches of science, say, physics and chemistry, are even more expressive. However, this statement is almost obvious, and it is much more difficult to indicate some constructive way that would lead to the nonlinear "world". Does it exist?

In our opinion, it does, and this way lies in the direction of systematic use of the most important principle of cybernetics, namely, the principle of feedback. We have only to learn how to use it correctly in nonstandard situations. It is clear today that this principle is the basis of self-control and the development of all living things. However, only the negative feedback and, correspondingly, stable processes "work" now in full measure in automatic control theory. The "press" of linearity hinders the use of positive feedback and unstable processes. Only the transition to principally nonlinear systems will allow us to involve new effects connected with the employment of positive and alternating feedback. Our monograph serves as an illustration of this scientific paradigm.

Part I

Principles Underlying the Design of Automatic Control Systems

The first part of the monograph contains a brief description of the main principles underlying the design of automatic control systems. The main emphasis is placed on the methods of compensation for the effect produced by external disturbance on the controlled coordinate. For simplicity, we only consider scalar stationary control objects, and this allows us, when necessary, to represent the operators of their input-output correspondence by transfer functions.

We have accepted the form of exposition with a parallel use of the modes of describing dynamical systems that are most often employed in the theory of automatic control, namely, description in terms of the structural flow charts and description in terms of the state space. This ensures the possibility of analyzing comprehensively the problems in question giving different interpretations to the obtained solutions. Moreover, different examples may be used to compare the advantages and limitations of the two modes.

The exposition is carried out in an orderly sequence which makes it possible to follow the evolution of the main ideas, principles, and problems of control theory as well as the methods of their solution. This allows us to formulate some urgent problems of control theory whose solution requires new ideas.

The first part of the monograph consists of two chapters. In the first chapter we consider principles of designing linear automatic control systems and in the second chapter we study some principles of constructing nonlinear controllers. The general theoretical constructions are followed by examples which allow us to use simple situations for elucidating the strong and weak aspects of each approach.

Some important results and conclusions which are known from the automatic control theory and which are necessary for better apprehension of the material are given in the form of Examples.

These Examples constitute separate structural units of the text and are provided with a title and a number, which makes it easier for the reader to find them when they are referred to. Below we give a list of these themes (with the pages indicated in parentheses).

- Relay stabilization of an object with a relative order equal to 2 (65).
- Hierarchy of feedbacks (78).
- Inclusion method (87).
- Synthesis of an adaptive control system (95).
- An exact tracking of a reference action signal (102).
- Switching mode in variable structure systems (105).
- Variable structure systems with motion along degenerate trajectories (107).
- Sliding mode on a straight line (108).
- VSS under a singular perturbation (116).
- VSS under a functional perturbation (117).

Each chapter includes a brief review of the corresponding literature given in the form of bibliographical comments.

Chapter 1

Principles Underlying the Design of Linear Automatic Control Systems

Below we introduce the principal concepts from the automatic control theory and describe the main mechanisms of compensation for an external disturbance for a linear control object.

1.1. Statement of a Control Problem and Preliminaries

In the automatic control theory we deal with mathematical models of real processes which are, of course, always incomplete and only approximately reflect the characteristics features of a real process which are significant in the context of a specific investigation. A chosen mathematical model is called an object of control or simply an object and, for convenience, is represented graphically as

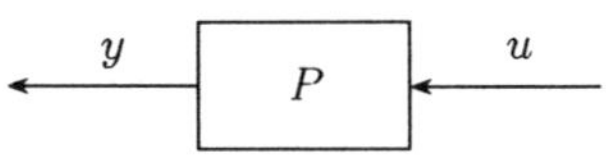

Figure 1.1.

a block diagram with input u and output y (Fig. 1.1). In this block diagram the object is characterized by an input-output correspondence operator P, i.e., an operator that establishes a relationship between the sets of input and output signals

$$y = Pu.$$

It is customary to call the input u of the object a control and its output y a controlled coordinate.

In the linear control theory the operator P is assumed to be linear, and this means that the relation

$$P(\alpha_1 u_1 + \alpha_2 u_2) = \alpha_1 P u_1 + \alpha_2 P u_2$$

is satisfied for any numbers α_1, α_2 and for arbitrary inputs u_1, u_2. It stands to reason that this assumption simplifies the matter and ensures the possibility of carrying out an analytic solution of automatic control problems, which can hardly be expected in a nonlinear case.

One more important assumption that we shall make consists in the invariance of the operator P of the object under a time shift. Objects of this kind are said to be stationary, i.e., time independent, and they are the most simple objects studied in the automatic control theory. In addition to the operator P which establishes the relationship between functions of time, we can also introduce for stationary objects an equivalent operator $W(s)$ which establishes a relationship between Laplace transformations of the input and output of the object under special, namely, zero initial conditions, i.e.,

$$Y(s) = W(s)\, U(s),$$

where $Y(s) = \mathcal{L}[y]$, $U(s) = \mathcal{L}[u]$, and for scalar objects, i.e., objects which have one input and one output, the transfer function is given, by definition, by the expression

$$W(s) = \left. \frac{\mathcal{L}[y]}{\mathcal{L}[u]} \right|_{\text{init.cond.}=0} .$$

In these relations s is a complex number, participating in the one-sided Laplace transformation, of the function of time $\xi(t)$, i.e.,

$$\mathcal{L}[\xi] = \int_0^\infty \xi(t)\, e^{-st}\, dt.$$

It is known from the theory of this integral transformation that the indicated operations have sense only for the time functions $\xi(t)$ which are originals. The assumption usually holds in linear stationary systems. Since we often use transfer functions in the sequel, it pays to note that the transfer functions of finite-dimensional linear stationary scalar objects are always linear fractional functions in the complex variable s, i.e., functions defined by the ratio of two polynomials, $\beta(s)$ and $\alpha(s)$, of degrees $m = \deg \beta(s)$ and $n = \deg \alpha(s)$, respectively, i.e.,

$$W(s) = \frac{\beta(s)}{\alpha(s)} = \frac{b_{m+1} s^m + b_m s^{m-1} + \cdots + b_1}{s^n + a_n s^{n-1} + \cdots + a_1},$$

where a_i, b_j are fixed real parameters of the object.

When the principle of causality is satisfied (the consequence does not precede the cause which leads to it), the degree m of the polynomial of the numerator $\beta(s)$ does not exceed the degree n of the polynomial of the denominator $\alpha(s)$ of the transfer function $W(s) = \beta(s)/\alpha(s)$.

These transfer functions are said to be physically realizable (feasible).

The zeros of the polynomial $\beta(s)$ are called zeros of the transfer function $W(s)$ or zeros of the object and the zeros of the polynomial $\alpha(s)$ are called poles of $W(s)$ or poles of the object. The transfer function $W(s) = \beta(s)/\alpha(s)$ is nondegenerate (nonsingular) if it does not have coincident zeros and poles. In what follows, we assume this condition to be a priori fulfilled. In this case, for a physically realizable transfer function, the number $n = \deg \alpha(s)$ is called the order of the object.

Using the nondegenerate transfer function of an object, it is easy to derive a differential equation for input-output time functions which describe the same object. Indeed, by definition we have a relation

$$Y(s) = W(s)\,U(s) = \frac{\beta(s)}{\alpha(s)}\,U(s),$$

which, after a term-by-term multiplication by $\alpha(s)$, reduces to the form

$$(s^n + a_n s^{n-1} + \cdots + a_1)\,Y(s) = (b_{m+1}s^m + \cdots + b_1)\,U(s).$$

After the application of the inverse Laplace transformation to this relation, we obtain the required differential equation of the object,

$$y^{(n)} + a_n y^{(n-1)} + \cdots + a_1 y = b_{m+1}u^{(m)} + \cdots + b_1 u,$$

where $\xi^{(k)} = d^k\xi/dt^k$ is the kth derivative of the time function $\xi(t)$. Note that the order of the leading derivative of the function of output which appears in this differential equation coincides with the order of the object.

We can pass from one nth-order differential equation to an equivalent set of n first-order differential equations which should be describe the evolution of the object under investigation in the state space.

Indeed, we introduce into consideration n functions $x_1,\ x_2,\ \ldots,\ x_n$, connected, via the input $u(t)$ and the output $y(t)$, by the relations

$$\dot{x}_i = x_{i+1}, \quad i = 1, 2, \ldots, n-1,$$

$$\dot{x}_n = -\sum_{i=1}^{n} a_i x_i + u,$$

$$y = \sum_{i=1}^{m+1} b_i x_i.$$

It is easy to verify that the transfer function of such an object from u to y coincides with the initial function $W(s) = \beta(s)/\alpha(s)$, and, hence, the given relations also describe the control object in question. In this case, $(x_1, x_2, \ldots, x_n)$ is a collection of certain dummy variables which characterize, in contrast to, say, y and u, the internal state of the object. We can assign to the last fact a useful geometrical meaning if we make use of the Cartesian system of coordinates $(x_1, x_2, \ldots, x_n)$. Then every internal state of the object in some n-dimensional space is associated with a point x with coordinates $(x_1, x_2, \ldots, x_n)$

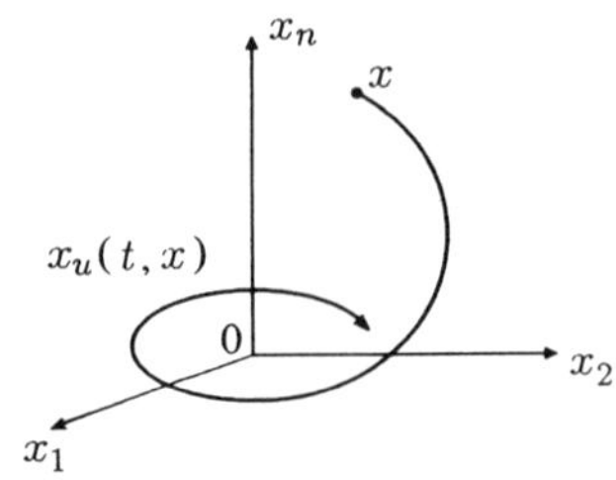

Figure 1.2.

which is also called a phase point. With a variation of time it draws, in the n-dimensional space, a curve $x_u(t,x)$ which is a geometrical image of the solution of a system of first-order differential equations for a certain input function $u(t)$ (Fig. 1.2). The orthogonal projection of the solution $x_u(t)$ onto the normal $c^\top = (b_1, b_2, \ldots, b_{m+1}, 0, \ldots, 0)^\top$ of the hyperplane $y = cx$ defines the output $y(t)$ corresponding to the input $u(t)$ of the object.

The use of vector-matrix notation

$$x = \begin{bmatrix} x_1 \\ x_2 \\ \vdots \\ x_n \end{bmatrix}, \quad A = \begin{bmatrix} 0 & 1 & 0 & \ldots & 0 \\ 0 & 0 & 1 & \ldots & 0 \\ \ldots & \ldots & \ldots & \ldots & \ldots \\ -a_1 & -a_2 & -a_3 & \ldots & -a_n \end{bmatrix}, \quad b = \begin{bmatrix} 0 \\ \vdots \\ 0 \\ 1 \end{bmatrix},$$

$$c = (b_1, b_2, \ldots, b_{m+1}, 0, \ldots, 0)$$

makes it possible to represent the model of the object in the following form which is a standard form for the contemporary control theory:

$$\dot{x} = Ax + bu,$$
$$y = cx.$$

Here the differential equation is called an equation of state and the static relation is called an equation of output. It stands to reason that all of these three techniques of mathematical description of a nondegenerate object are equivalent, and it is easy to indicate the corresponding one-to-one transformations. We omit the details of these transformations and only give a relation

$$c(sE - A)^{-1} b = \frac{\beta(s)}{\alpha(s)}$$

whose validity can be easily verified by the reader.

Finally, note that the property of the vector of state $x(t)$ to be dummy must be understood in the sense that any other vector z connected with the dummy vector by the nondegenerate transformation M ($\det M \neq 0$) can be taken as this vector, i.e.,

$$z = Mx.$$

After such a substitution we obtain equations for the object in the former form but with different parameters ($A_M = MAM^{-1}$, $b_M = Mb$, $c_M = cM^{-1}$):

$$\dot{z} = A_M z + b_M u,$$
$$y = c_M z.$$

It stands to reason that the input-output correspondence (the transfer function, in particular) remains the same. Indeed,

$$\begin{aligned} c_M\,(sE - A_m)^{-1} b_M &= cM^{-1}(sE - MAM^{-1})^{-1} Mb \\ &= cM^{-1}M(sE - A)^{-1}M^{-1}Mb = c(sE - A)^{-1}b. \end{aligned}$$

Thus, whereas the transfer function $W(s)$ does not depend on the choice of the phase vector, its "delicate" structure obviously depends on it (Fig. 1.3). In particular, this is the reason for using generalized transfer functions in the

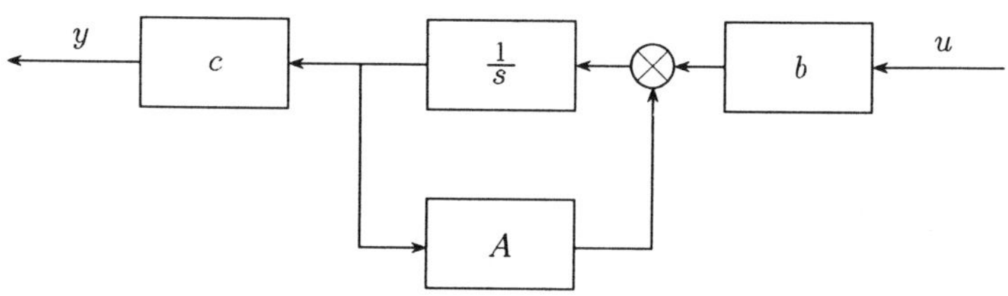

Figure 1.3.

investigation of composite systems since in this way it is especially easy to find the transfer function of the composite system.

Direct calculations make it possible to verify the following rules of connection of transfer functions:

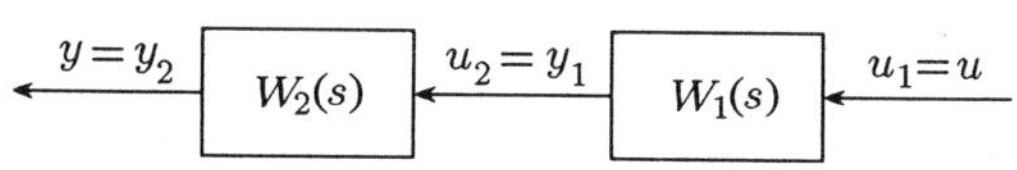

Figure 1.4.

- when systems are connected in series (Fig. 1.4), the transfer functions are multiplied together and the output of the system can be found from the relation

$$y = W(s)\,u = W_2(s)\,W_1(s)\,u;$$

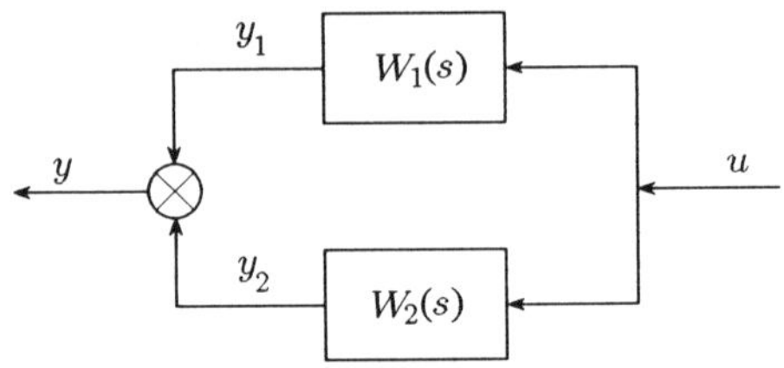

Figure 1.5.

- when systems are connected in parallel (Fig. 1.5), the transfer functions are added together and the output of the system can be found from the relation

$$y = W(s)\,u = \big[W_1(s) + W_2(s)\big]\,u;$$

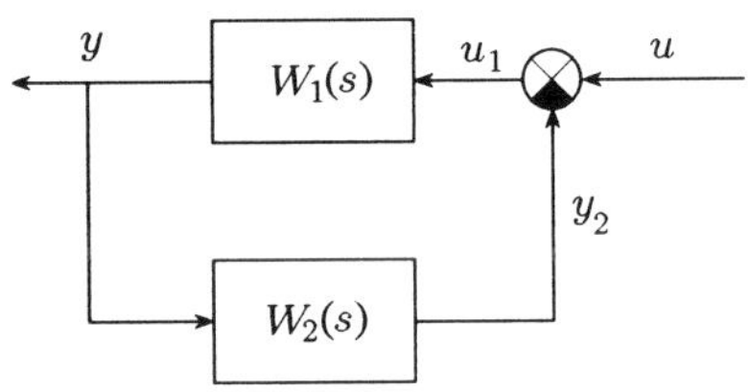

Figure 1.6.

- when transfer functions are connected by feedback (Fig. 1.6), the output of the system is transformed in accordance with the expression

$$y = W(s)\,u = \frac{W_1(s)}{1 + W_1(s)\,W_2(s)}\,u.$$

The white sector of the summator $\otimes$ shown in the figures denotes the addition and the black sector denotes the subtraction of the corresponding input signal.

We shall describe now how the control problem under consideration is formulated. Following the contemporary terminology, we shall call this problem a stabilization problem. The essence of the control problem arising in this connection is the choice of a control u such that the output y of the object coincides with the predefined function of time $y^s(t)$ which expresses the requirements placed upon the character of variation of the output of the object. The function $y^s(t)$ is called a reference action signal. The problem of stabilization, which is not very simple by itself, is made still more complicated by the action of external disturbances on the control object. These disturbances may be of two types, a coordinate type $f(t)$ and an operator type $a(t)$. Under the action of external disturbances, the information about which is often insufficient (for instance, we only know that it belongs to certain sets of functions $\mathcal{F}$ and $\mathcal{A}$, i.e., $f \in \mathcal{F}$, $a \in \mathcal{A}$), the relationship between the input and output of the object is no longer

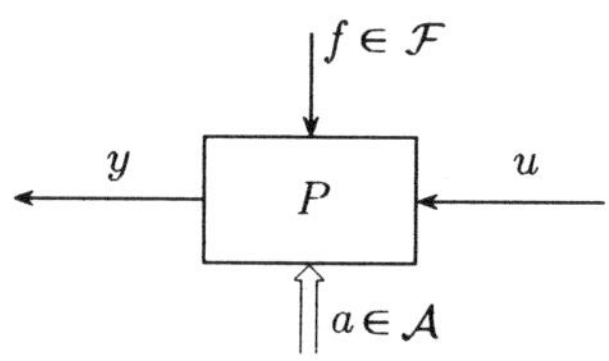

Figure 1.7.

uniquely defined and becomes uncertain, and this, of course, hinders the solution of the stabilization problem (Fig. 1.7). However, we must point out an essential difference between the effects produced by disturbances of coordinate and operator types on the object. Bearing this in mind, we shall consider in detail

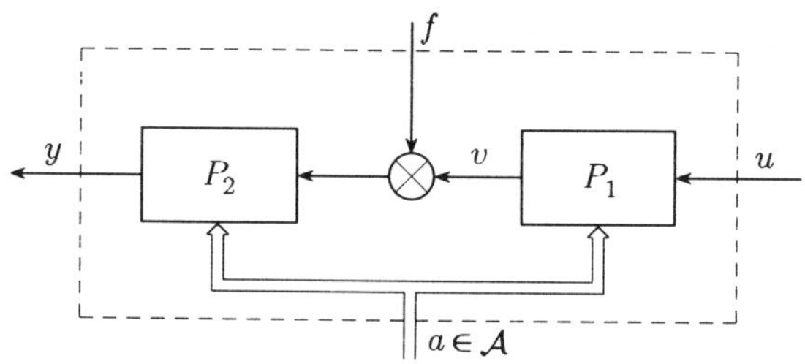

Figure 1.8.

the structure of a control object and the actions produced on it by disturbances using the block diagram given in Fig. 1.8. The equations that describe this diagram have the form

$$y = P_2[a]\,(f + v), \quad v = P_1[a]u$$

or, in greater detail,

$$y = P_2[a]\,P_1[a]u + P_2[a]f = P[a]u + P_2[a]f,$$

where the dependence of the operators P_1 and P_2 on the operator perturbation a is explicitly defined.

Now one can see the difference between the effects produced by the disturbance f and perturbation a on the output of the object which is emphasized by symbols in the block diagrams. The coordinate disturbance f makes an additive, independent of the input u, contribution to the reaction of the object equal to $P_2[a]f$. As to the operator perturbation a, it only changes the form, or parameters, of the operators $P_1[a]$, $P_2[a]$ and does not produce any effect on the output of the object independent of u and f. Thus, the disturbance f simulates the "linear" effect of the surrounding medium on the controlled coordinate and the perturbation a simulates its "nonlinear" action.

The effect produced by the operator perturbation on a control process is not studied in the linear automatic control theory, and therefore we put $a \equiv 0$ below and do not denote this perturbation in block diagrams.

Note that the reference action y^s can also be an output of a certain dynamical system which is called a reference device and is denoted by S in the block diagram given below (Fig. 1.9). It stands to reason that, just as a control

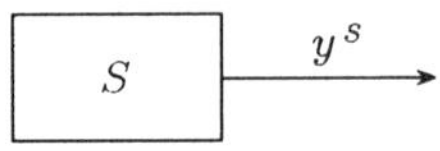

Figure 1.9.

object, a reference device can have an output and be subjected to the effect of interferences, but we do not consider these possibilities for the sake of simplicity. In this notation and terminology a stabilization problem can be associated with the block diagrams shown in Fig. 1.10. The notation $e = y^s - y$ in the figure is a control error and R is a controller that uses the available information (y^s, e,

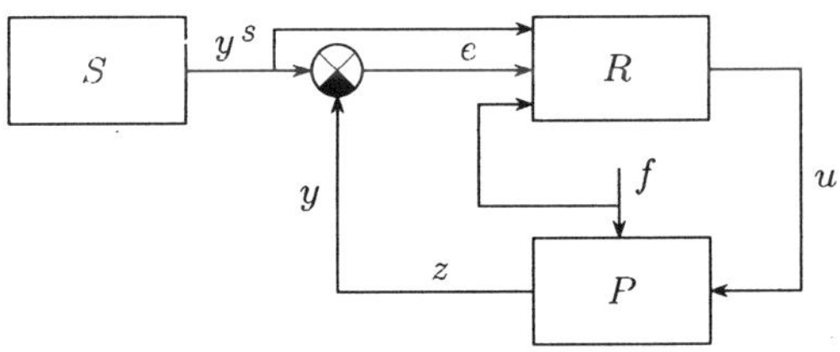

Figure 1.10.

f etc.) to form a control signal u for which the control error e is zero or lies in admissible limits.

We shall outline the fundamental possibilities that control theory possesses for achieving the objective indicated above.

In the first place, as a rule, a specialist in automation is prevented from influencing directly the internal structure of a technological process which would have led to the desired equality $y = y^s$ without any control. Most often he has to deal with an object which was designed without any account of this circumstance. Therefore the only possibility of producing a pronounced effect on the output of the process and, hence, on the possibility of solving a control problem is connected with the manipulation of the input signal u. And here, as a matter of fact, two "pure" strategies of behavior are revealed at once. The first strategy is connected with the appropriate formation of the input signal from the available signals so that its subsequent transformation by the operator of the object would lead to the desired result $y = y^s$ and the second strategy is connected with the variation of the input-output correspondence operator with the aid of feedback.

In the first case that corresponds to the use of feedforward, the input signal u is supplemented by an auxiliary signal u^s, which depends, for instance, on the reference action y^s and is transformed by an appropriate operator R (Fig. 1.11a). As a result of these transformations, the output of the object assumes the form

$$y = P\,Ry^s + Pu + P_2 f,$$

and, under certain conditions (say, $R = P^{-1}$, $P_2 f \equiv 0$, $u \equiv 0$), it may turn out that the required equality $y = y^s$ is attained.

In the second case, the input signal of the object is changed with the aid of feedback according to the block diagram given in Fig. 1.11b which shows the

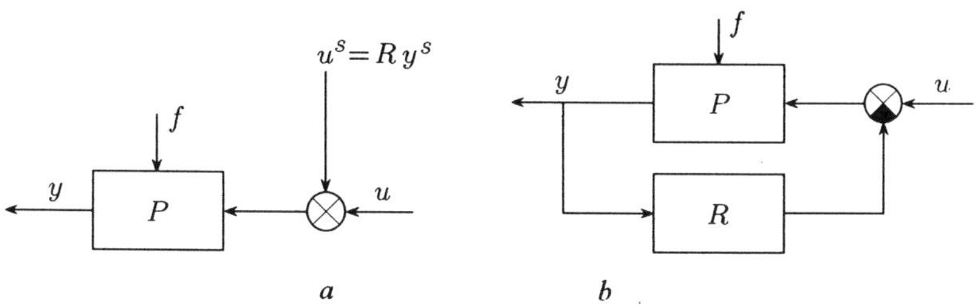

Figure 1.11.

feedback operator R. The output equation corresponding to this block diagram has the form

$$y = P(u - Ry) + P_2 f,$$

and, after a simple transformation, we find out that the output of the object covered by the feedback is connected with the input u and the interference f by the relation

$$y = \frac{P}{1 + PR} u + \frac{P}{1 + PR} f,$$

from which we can see that feedback changes the transfer operators from the inputs u and f to the output y without any intervention into the technological process, only by a skilful use of the information about the output. This principle has huge potentialities whose utilization for the needs of automation constitutes the main content of control theory.

It is also easy to realize that the combination of feedforwards and feedbacks may lead to a greater influence exerted on the object and, as a consequence, to a widening of the scope of possibilities of the automatic control system.

We shall describe now the principles indicated above using the language of differential equations, by way of the following example. Suppose that an object is described by the second-order differential equation

$$\ddot{y} + a_2 \dot{y} + a_1 y = u + f \tag{1.1}$$

with constant parameters a_1, a_2 and disturbance f. Note that in the "operator" description of the object, this example is associated with the operators P and P_2 with transfer functions

$$W(s) = W_2(s) = \frac{1}{s^2 + a_2 s + a_1}.$$

Let R be a feedforward operator with respect to the reference action y^s with a transfer function

$$W_R(s) = b_2 s + b_1,$$

where b_1, b_2 are constant parameters. Then, in accordance with Fig. 1.11a, we have

$$u^s = b_2 \dot{y}^s + b_1 y,$$

and the output of the object is now a solution of the differential equation

$$\ddot{y} + a_2 y + a_1 y = b_2 \dot{y}^s + b_1 y^s + f + u, \tag{1.2}$$

which differs from the initial equation (1.1) by an additional term on the right-hand side. Since the general solution of the linear equation (1.2) consists of an arbitrary solution of the homogeneous equation

$$\ddot{y} + a_2 \dot{y} + a_1 y = 0$$

and a particular solution of the nonhomogeneous equation (1.2), the feedforward affects only this particular or, as we say, forced solution and does not affect its free motion at all.

On the contrary, if we use feedback according to the block diagram presented in Fig. 1.11b with, say, the same operator R, then we get the differential equation for the feedback control system first in the form

$$\ddot{y} + a_2 \dot{y} + a_1 y = -b_2 \dot{y} - b_1 y + u + f,$$

and then, after a simple transformation, in the final form

$$\ddot{y} + (a_2 + b_2)\dot{y} + (a_1 + b_1) y = u + f. \tag{1.3}$$

It follows from (1.3) that the feedback does not change the right-hand side of the equation but can be used to change the parameters of the differential equation. This means that feedback not only affects the proper motions of the object but also its forced motions. It stands to reason that the use of feedforward and feedback can only strengthen the action exerted on the behavior of the object.

It is expedient to use this example to illustrate two concepts which are important for the automatic control theory, namely, the concepts of statics and dynamics of a control system.

Let us consider Eq. (1.1) and assume that the input $u \equiv 0$ and the disturbance $f = f_0 = \text{const}$. Then we get

$$\ddot{y} + a_2 \dot{y} + a_1 y = f_0.$$

As is known, the general solution of this equation contains two terms,

$$y(t) = y_{\text{free}}(t) + y_f(t),$$

where $y_{\text{free}}(t)$ is an arbitrary solution of the homogeneous equation and $y_f(t)$ is a particular solution of the nonhomogeneous equation. In the case under consideration we have

$$y_f(t) = \frac{f_0}{a_1} = y^* = \text{const}.$$

what can be easily verified by a direct substitution.

The solution is also known as a steady-state solution or the statics of the system and the difference $y(t) - f_0/a_1$ is called a transient solution, transient process, or dynamics of the system. If an object is asymptoticaly stable, and this is the fact for positive parameters a_1 and a_2, then the component $y_{\text{free}}(t)$ exponentially vanishes and the object tends to the static position y^* in an oscillatory

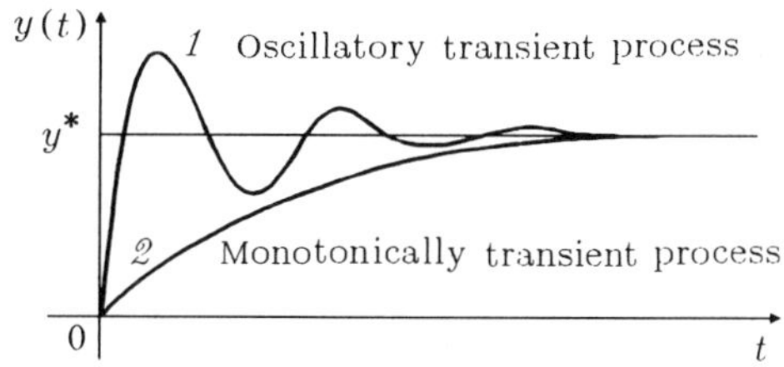

Figure 1.12.

way (curve *1* in Fig. 1.12) or monotonically (curve *2* in Fig. 1.12) since in the state of rest $\dot{y} = \ddot{y} = 0$. If we use feedforward, then we must consider Eq. (1.2). Suppose, for simplicity, that $f \equiv f_0$, $u \equiv 0$ as before and, in addition, $y^s = \text{const}$. Then we have an equation

$$\ddot{y} + a_2\dot{y} + a_1 y = b_1 y^s + f_0,$$

and it follows from the arguments given above that if the object is asymptotically stable, then all solutions of this equation exponentially tend to the static mode

$$y^* = \frac{b_1}{a_1} y^s + \frac{1}{a_1} f_0.$$

Consequently, feedforward does not affect the dynamics but only affects the statics of a control system. Since $y = y^s$ in the desired mode, the difference

$$\eta = y^s - y^* = \frac{a_1 - b_1}{a_1} y^s - \frac{1}{a_1} f_0$$

defines the stabilization error and is called a static error of a control system. Since the parameter b_1 characterizes feedforward, the latter can be used to diminish the static error; for instance, for $b_1 = a_1$ the error

$$\eta = -\frac{1}{a_1} f_0.$$

We shall consider now the feedback in Eq. (1.3) under the same assumptions ($u \equiv 0$, $f \equiv f_0$), i.e., study the equation

$$\ddot{y} + (a_2 + b_2)\dot{y} + (a_1 + b_1)y = f_0.$$

In this case, by an appropriate choice of the feedback parameters b_1, b_2 we can always make the feedback control system asymptotically stable, and, consequently, feedback changes the dynamics of a system. But not only the dynamics. Since

$$y^* = \frac{f_0}{a_1 + b_1},$$

in the steady-state operation mode, the statics also depends on feedback. In particular, by increasing the parameter b_1 we can diminish the static error. It should be emphasized that there is no such a possibility in systems with feedforward.

We shall now describe the methods of synthesis of stabilizing controls.

1.2. Load Control Principle

A simple stabilization problem arises when there is no coordinate disturbance, i.e., when $f \equiv 0$. In this case, the equation for the object has the form

$$y = P\,u, \tag{1.4}$$

and the solution of the problem is given by the so-called program control

$$u^s = P^{-1}y^s, \tag{1.5}$$

where P^{-1} is the inverse of the operator P, i.e., an inverse operator satisfying the relation $P\,P^{-1} = 1$. Indeed, substituting (1.5) into Eq. (1.4) for the object, we successively find

$$y = Pu^s = P\,P^{-1}y^s = y^s$$

and solve the problem.

This control principle leads to the block diagram of an automatic control system, see Fig. 1.13. In this figure the relationship between the reference action signal y^s and the control u is shown explicitly. Since the reference action signal y^s characterizes the "load" with which the object functions, it is more convenient to call this control principle a load control principle. This principle of control has chronic weaknesses among which we want to point out the following.

In the first place, according to the physical meaning of the problem, the operator P simulates a real process, and therefore it satisfies the condition of physical realizability. Suppose, for instance, that the transfer function of the operator P has the form

$$W(s) = 1/s.$$

Then the transfer function of its inverse P^{-1} is given by the expression

$$W^{-1}(s) = s$$

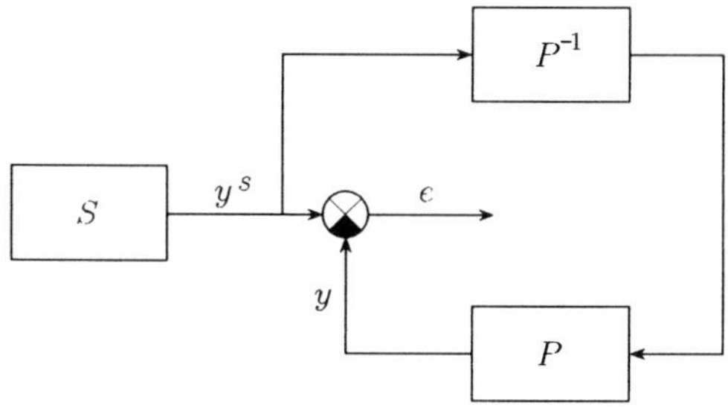

Figure 1.13.

and, consequently, the degree of the polynomial in the numerator in its expression is higher than the degree of the polynomial in the denominator, and, hence, the transfer function does not satisfy the principle of causality, i.e., the operator P^{-1} is physically nonrealizable. Therefore we can only speak about an approximate realization of the program control $u^s = P^{-1}y^s$ and, hence, about an approximate solution of the stabilization problem by a system of load control.

Second, a stabilization problem, i.e., a problem in which the object is unstable, cannot be solved only in this way, i.e., without resort to any other ideas (Example 1).

Third, it should be pointed out that, in a system of program control, the control being realized does not depend on the actual behavior of the object, and therefore operational corrections are impossible when an unexpected change occurs in the behavior of the object or the reference device. As a result, even a small perturbation may lead the output of the object away from the prescribed one.

Example 1 (A program control of an unstable object). Suppose that the control object has a transfer function

$$W(s) = \frac{1}{s^2 + a_2 s + a_1},$$

where a_1, a_2 are constant parameters, and we have to stabilize the output of the object at zero, i.e., we assume that $y^s \equiv 0$. In accordance with recommendations given in Sec. 1.1., in order to obtain a differential equation describing the object, we must first find the relationship between the Laplace transformations of the input and output,

$$(s^2 + a_2 s + a_1)Y(s) = U(s),$$

and then write out the required equation

$$\ddot{y} + a_2\dot{y} + a_1 y = u. \tag{1.6}$$

Following the above procedure of constructing a system of load control, we obtain a block diagram of the control system shown in Fig. 1.14. It should be pointed out at once that the transfer function of the control

$$W^{-1}(s) = s^2 + a_2 s + a_1$$

can be realized only approximately, and, consequently, the control system must be modified. However, even if we assume that the transfer function $W^{-1}(s)$ is exactly realized, a serious difficulty is encountered all the same when the parameter $a_1 < 0$. In fact, since $u^s = P^{-1}y^s \equiv 0$,

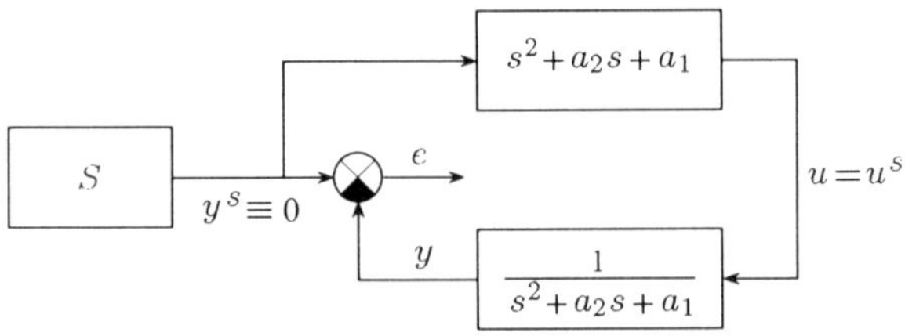

Figure 1.14.

only proper oscillations caused by the nonzero initial conditions and described by the differential equation

$$\ddot{y} + a_2\dot{y} + a_1 y = 0$$

are observed at the output of the object. It is clear that for $a_1 < 0$ its characteristic equation

$$s^2 + a_2 s + a_1 = 0$$

has a positive root, and, hence, the output y of the object exponentially increases.

In other words, if the control object is unstable, then the problem of stabilization with the use of the principle of load control cannot be solved. It is easy to understand that for $y^s \not\equiv 0$ as well the stability of an object is a necessary condition for the efficiency of a load control system. Indeed, the programmed control $u^s = P^{-1}y^s$ has the form

$$u^s = \ddot{y}^s + a_2\dot{y}^s + a_1 y,$$

and, substituting it into the equation for the object (1.6), we obtain an equation

$$\ddot{y} + a_2\dot{y} + a_1 y = \ddot{y}^s + a_2\dot{y}^s + a_1 y.$$

Writing this equation for the error of control $e = y^s - y$, we get

$$\ddot{e} + a_2\dot{e} + a_1 e = 0$$

and see that if $e(t) \to 0$ as $t \to \infty$, then the stabilization problem is solved. However, the last condition precisely signifies the asymptotic stability of the free motions of the object.

Let us now consider the case where a control object is at the boundary of stability $a_1 = 0$, $a_2 > 0$ and the programmed control is not exactly realized but is realized with an error $\varepsilon = \text{const}$, i.e., $u = u^s + \varepsilon = \varepsilon$ since $u^s = P^{-1}y^s \equiv 0$. In this case, the output of the object satisfies the differential equation $\ddot{y} + a_1\dot{y} = \varepsilon$ whose particular solution, which is caused by the existence of the right-hand side ε, increases indefinitely in time in its absolute value taking the output of the object away from the desired zero. Note that we can also make this inference when the programmed control is realized exactly but a constant interference acts as the input of a natural object.

Thus,

- the load control principle has a limited application in the synthesis of control systems and can only be applied, generally speaking, in conjunction with other principles of control.

In particular, it can be applied in conjunction with the principle of disturbance control which we shall now consider.

1.3. Principle of Disturbance Control

Suppose that the coordinate disturbance f, which is identically nonzero, is applied to a certain interior point of the object as is shown in Fig. 1.15. Suppose, in addition, that $z = P_1 u$, $v = z + f$ and

$$y = P_2 v = P_2 (z + f) = P_2 P_1 u + P_2 f = P u + P_2 f, \tag{1.7}$$

where it is taken into account that $P_2 P_1 = P$. We have here a new situation since the output of the object depends now not only on the input u but also on

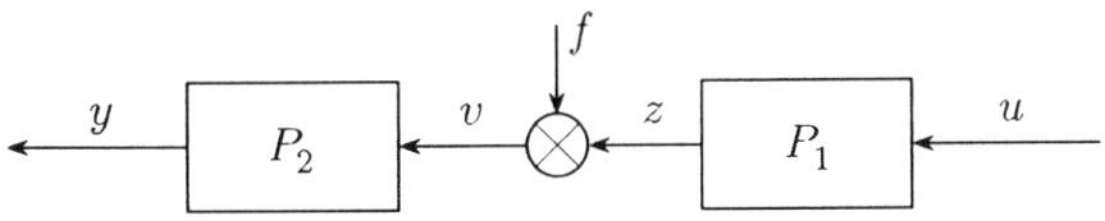

Figure 1.15.

the disturbance f. Indeed, even if we follow exactly the principle of load control and apply the corresponding programmed control

$$u = u^s = P^{-1} y^s,$$

we obtain, as a result, a control system whose output is defined by the relation

$$y = P P^{-1} y^s + P_2 f = y^s + P_2 f$$

and, as we can see, depends explicitly on the disturbance f if $P_2 f \not\equiv 0$. If the disturbance f can be measured, then we can use the disturbance control principle in order to eliminate this dependence.

In accordance with this principle, the control must contain a component proportional to the disturbance. As applied to the given case, this means that we must take the control as the sum of two components, namely, a programmed component and a compensating one, i.e.,

$$u = u^s + \mathcal{D} f,$$

where $\mathcal{D}$ is the required operator. After the substitution of this expression into (1.7), we obtain a relation

$$y = P (u^s + \mathcal{D} f) + P_2 f = y^s + (P\mathcal{D} + P_2) f,$$

analyzing which, we see that the stabilization problem can be exactly solved and the output of the object does not depend on the disturbance f if the relation

$$P\mathcal{D} + P_2 = 0,$$

known as a condition of compensation, is satisfied.

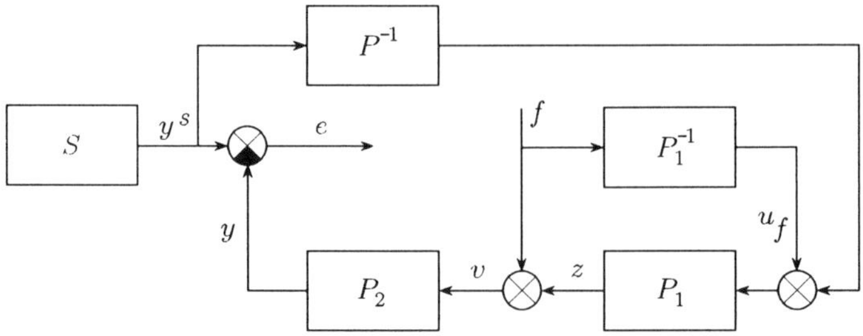

Figure 1.16.

The procedure of synthesis described above leads to the block diagram of the control system shown in Fig. 1.16. The advantages of this method of control are reduced by the following circumstances.

In the first place, as before, unstable objects cannot be stabilized in this way.

Second, in nontrivial cases (i.e., when $P_1 \neq$ const) the operator $\mathcal{D}$, which satisfies the condition of compensation, is physically unrealizable. Indeed, from the compensation condition we have a relation

$$\mathcal{D} = -P^{-1} P_2 = -(P_1 P_2)^{-1} = -P_1^{-1},$$

and since the operator P_1 simulates a real process, its inverse P_1^{-1} cannot be physically realized.

Third, the compensation condition is given by an equality, and, if the smallest error E appears on the right-hand side of the corresponding operator relation

$$P\mathcal{D} + P_2 = E,$$

the dependence of the output of the object on disturbance cannot be eliminated since, in this case, we have

$$y = y^s + (P\mathcal{D} + P_2) f = y^s + Ef,$$

and if $Ef \not\equiv 0$, then the second term in this relation can lead to an inadmissible deviation of the output y from the reference action y^s. An essential limitation of this principle is the assumption that the disturbance f can be directly measured, which is seldom realized in practice.

Example 2 (Direct compensation for a disturbance). We shall illustrate the capabilities and limitations of the principle of disturbance compensation using the object from Example 1. We assume that in the composition of operators shown in Fig. 1.15

$$P = P_2 P_1,$$

and the transfer functions corresponding to the operators P_1 and P_2 are given by the expressions

$$W_1(s) = \frac{1}{s + \lambda_1}, \quad W_2(s) = \frac{1}{s + \lambda_2},$$

where, naturally,

$$\lambda_1 + \lambda_2 = a_2, \quad \lambda_1 \lambda_2 = a_1.$$

In the state space, this operator description of the control object is associated with the system of differential equations

$$\dot{y} + \lambda_2 y = z + f, \quad \dot{z} + \lambda_1 z = u$$

or, which is more convenient in the given case, one second-order differential equation

$$\ddot{y} + a_2\dot{y} + a_1 y = u + \dot{f} + \lambda_1 f. \tag{1.8}$$

Then, following the principle of disturbance control described above, from the condition of compensation $\mathcal{D} = -P_1^{-1}$ we find the transfer function $W_\mathcal{D}(s)$ of the operator $\mathcal{D}$ in the form

$$W_\mathcal{D}(s) = -(s + \lambda_1),$$

and this (in conjunction with the load control for obtaining the programmed part u^s of the control function) leads to the structure of the control system shown in Fig. 1.17. Using this block diagram and the differential equation (1.8), we can obtain the corresponding equation in terms of the state space.

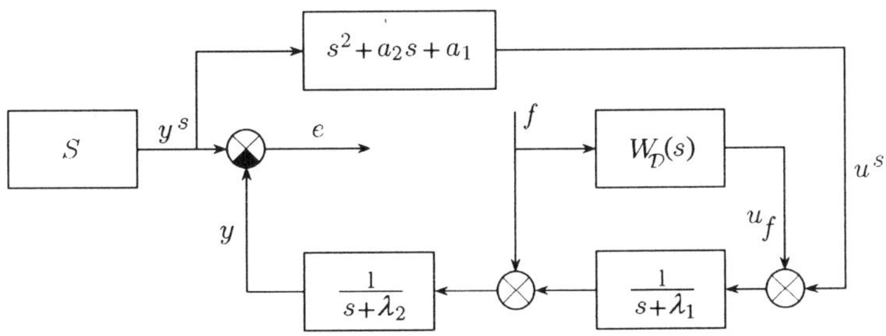

Figure 1.17.

After the substitution of the compensating control $u = -(\dot{f} + \lambda_1 f)$ into the right-hand side of Eq. (1.8), the motion of the system can be described by the homogeneous differential equation

$$\ddot{y} + a_2\dot{y} + a_1 y = 0.$$

As before, these equations imply that the asymptotic stability of the object is a necessary condition of solvability of the problem in the framework of the load control principle. However, this limitation of the method is far from being the only one since the controller of the synthesized control system uses physically unrealizable operators with transfer functions $s^2 + a_2 s + a_1$ and $s + \lambda_1$. Moreover, even if we shut our eyes to it, all the same the constructed system may turn out to be inefficient since it is very sensitive to the smallest violation of the compensation condition: $u = -(\dot{f} + \lambda_1 f)$.

Assume that we have introduced an error in the realization of the transfer function of the operator $\mathcal{D}$ and, instead of $W_\mathcal{D}(s)$, used the transfer function

$$W_\Delta(s) = -(s + \lambda_1 + \Delta\lambda),$$

where $\Delta\lambda$ is a constant. Then the error in the realization of the compensation condition is described by the expression

$$E = -\frac{\Delta\lambda}{s^2 + a_2 s + a_1}$$

and can be arbitrarily small for a small error $\Delta\lambda$. Since the control error is defined, in this case, by the expression

$$e = y^s - y = \frac{\Delta\lambda}{s^2 + a_2 s + a_1} f,$$

we again, as in Example 1, make sure that for $a_1 = 0$, $a_2 > 0$ even an arbitrarily small constant disturbance f can lead to an arbitrarily large deviation of the output y from the reference action y^s.

Thus,

- only an approximate realization of the direct compensation principle is possible, and its most considerable limitation here is the necessity of a direct measurement of the disturbance.

An attempt was made to ease off this limitation in the compensation principle with an indirect measurement of the disturbance. Let us consider this principle of control.

1.4. Compensation Principle in an Indirect Measurement of Disturbance

One of the classical ideas of an indirect measurement of disturbance can be illustrated with the aid of Fig. 1.18.

Suppose that the output y of the object and its interior coordinate z are measured. Then, as a result of transformation of the output of the object by the operator P_2^{-1} we obtain a signal $\eta = P_2^{-1} y$ which is the estimate of the signal

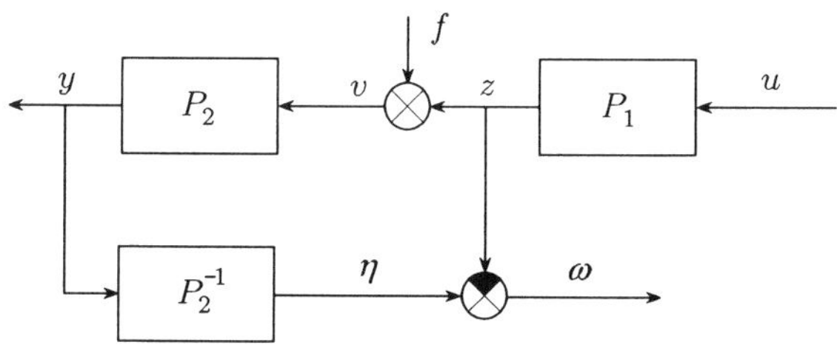

Figure 1.18.

v. Since $v = f + z$ and the signal z is known, we can obtain the estimate of the disturbance f from the formula $\omega = \eta - z$.

We are speaking about the estimates of the signals v and f because the link with the operator P_2^{-1} under consideration can have its own dynamics up to which we must understand the relations $\eta \cong v$, $\omega \cong f$. Next we must use the principle of direct compensation and formulate control in the form of a sum of two components, namely, a programmed component u^s and a compensating one $\mathcal{D}\omega$, i.e.,

$$u = u^s + \mathcal{D}\omega,$$

where the operator $\mathcal{D}$ must satisfy the condition of compensation $\mathcal{D} = P_1^{-1}$ given in the preceding subsection.

As a result, we must form the control which solves the stabilization problem in question as

$$u = P^{-1}y^s - \mathcal{D}(P_2^{-1}y - z).$$

The block diagram of the control system shown in Fig. 1.19 corresponds to this relation. A considerable difficulty is encountered in the application of this compensation method, namely, a loop of internal positive feedback (hatched in

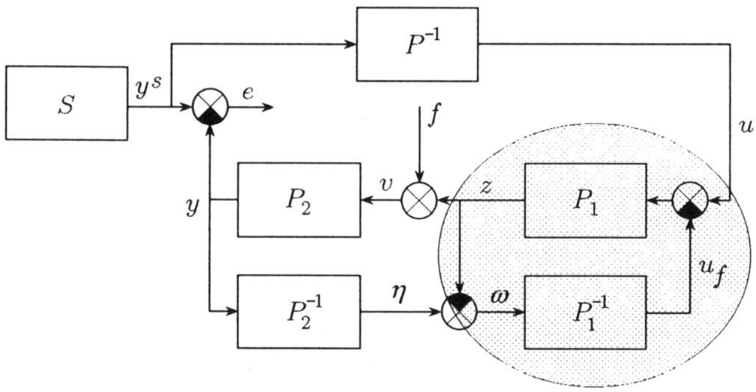

Figure 1.19.

Fig. 1.19) appears in the structure of the control system which is actually a critical one since its stability or instability is defined by hidden parameters.

Indeed, the operator $P_{u,z}$ of this loop from the input u^s to the output z is given by the expression

$$\frac{P_1}{1 - \mathcal{D}P_1} = \frac{P_1}{1 - P_1^{-1}P_1},$$

and, consequently, problems arise connected with the division by zero. We can solve these problems if we take into account the hidden parameters.

Indeed, if P_1 is a model of a real process, then the operator $\widetilde{P}_1 = P_1 Q$ actually acts in this loop, where Q is an operator which we neglected when making the model. However, in the circumstances that arise, this operator plays an essential role, and the dynamical properties of the loop in question are actually defined by this operator since the operator of the loop from u^s to z has the form $P_{u,z} = P_1/(1 - Q)$.

In other words,

- the properties of the loop are predetermined by uncontrollable factors, and this is unacceptable. As a rule, this leads to the instability of proper motions of the internal loop. This means that the signal z increases indefinitely and violates the correct functioning of the control system.

In the following example we study in detail some peculiarities of this phenomenon. It should be emphasized here that the indicated difficulty is not the only one. Of course, all problems connected with the use of the principle of direct compensation and indicated in the preceding subsections are inherited by compensation systems with indirect measurements of disturbance considered in this subsection.

Example 3 (Indirect compensation for a disturbance). We shall illustrate the method of compensation of disturbance described above for the object considered in Example 1. In accordance with the exposed theory, we have, for the indicated object, a control system

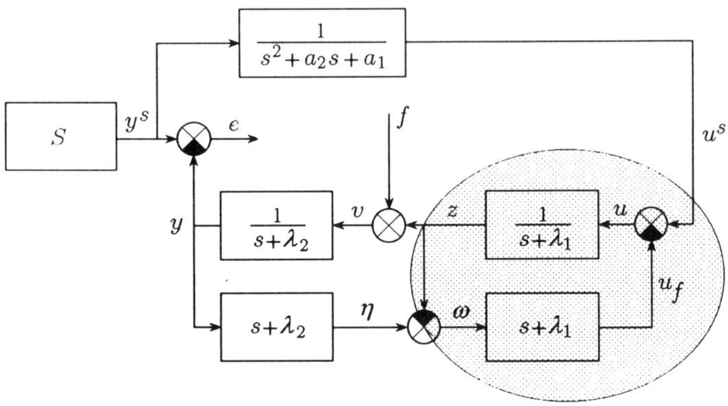

Figure 1.20.

whose block diagram is given in Fig. 1.20. We isolate from this block diagram a fragment with a local positive feedback (Fig. 1.21) and study its properties in greater detail. We directly establish from Fig. 1.21 that the transfer function $W_{u,z}(s)$ from the input u^s to the output z

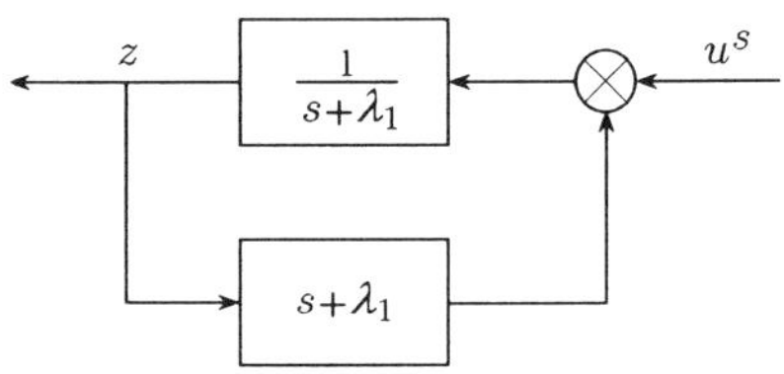

Figure 1.21.

is defined by the relation

$$W_{u,z}(s) = \frac{1}{1-1}\frac{1}{s+\lambda} = \frac{k_\infty}{s+\lambda_1},$$

and, consequently, the indicated positive feedback leads to the effect of employment of the infinite gain factor k_∞ in the direct channel. Assume that this loop includes not the operator P_1 with the transfer function $W_1(s) = 1/(s+\lambda_1)$ but an operator $\widetilde{P}_1 = P_1 Q$ with a transfer function

$$\widetilde{W}_1(s) = \frac{1}{s+\lambda_1}\frac{1}{\tau s+1},$$

where $\tau = \text{const} > 0$ is a small parameter. The second factor simulates "fast" dynamic of the operator Q which was not taken into account in the original model of the object. Then we have a block diagram shown in Fig. 1.22 and the corresponding relations

$$(s+\lambda_1)(\tau s+1)z = u, \quad u = u^s + (s+\lambda_1)z.$$

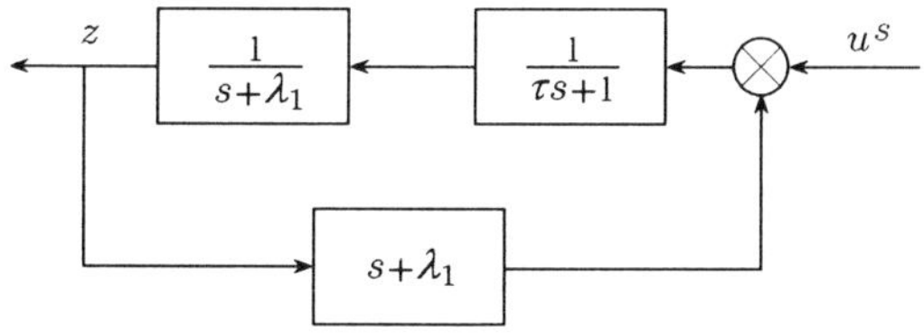

Figure 1.22.

As a result, the required transfer function $\widetilde{W}_{u,z}(s)$ from u^s to z is uniquely defined and is expressed as

$$\widetilde{W}_{u,z}(s) = \frac{1}{\tau s(s + \lambda_1)}.$$

It follows that the object is on the boundary of stability and an arbitrarily small perturbation of its parameters may lead to instability. The coordinate z increases indefinitely in absolute value and leads to an unlimited growth of the error of control.

We shall now investigate Example 3 with the use of differential equations rather than with the use of structural or operator description as it was done above. From the block diagram of a control system given in Fig. 1.20 we have differential equations describing the control object

$$\dot{y} + \lambda_2 y = f + z, \tag{1.9}$$

$$\dot{z} + \lambda_1 z = u \tag{1.10}$$

and equations of the signal that compensates for the external disturbance

$$u = u^s - (\dot{\omega} + \lambda_1 \omega), \tag{1.11}$$

$$\omega = \eta - z, \tag{1.12}$$

$$\eta = \dot{y} + \lambda_2 y. \tag{1.13}$$

From (1.9) and (1.13) we have $\eta = f + z$ and $\omega = -f$, and therefore, as a result, the control has the form

$$u = u^s - (\dot{f} + \lambda_1 f). \tag{1.14}$$

If we apply the differential operator $d/dt + \lambda_1$ to (1.9), then, with due account of (1.10), we get an equation for the object in the form

$$\ddot{y} + a_2 \dot{y} + a_1 y = u + \dot{f} + \lambda_1 f,$$

where, as before, the parameters

$$a_2 = \lambda_1 + \lambda_2, \quad a_1 = \lambda_1 \lambda_2.$$

After substituting the control from (1.14) into this equation, we find that the motion of the control system is described by the equation

$$\ddot{y} + a_2 \dot{y} + a_1 y = u^s - (\dot{f} + \lambda_1 f) + (\dot{f} + \lambda_1 f) = u^s$$

and does not depend on disturbance. Since (see. Fig. 1.20)

$$u^s = \ddot{y}^s + a_2\dot{y}^s + a_1 y^s \quad \text{and} \quad e = y^s - y,$$

the motion equation with errors has the form

$$\ddot{e} + a_2\dot{e} + a_1 e = 0, \tag{1.15}$$

and if it is stable, then the stabilization problem is solved without direct measurement of the disturbance f.

Note that if we substitute Eq. (1.12) into Eq. (1.11) and substitute the result

$$u = u^s + (\dot{z} + \lambda_1 z) - (\dot{\eta} + \lambda_1\eta)$$

into Eq. (1.10), then we obtain

$$\dot{z} + \lambda_1 z = u^s + (\dot{z} + \lambda_1 z) - (\dot{\eta} + \lambda_1\eta)$$

or, after simplification,

$$\dot{\eta} + \lambda_1\eta = u^s. \tag{1.16}$$

This means that under this control the proper motions of the coordinate z are not uniquely defined, i.e., the pole and the zero are cancelled out.

Let us assume now that an operator $\widetilde{P}_1 = P_1 Q$ acts in the system instead of the operator P_1, and the operator Q is associated, as before, with the transfer function

$$W_\tau(s) = 1/(\tau s + 1).$$

Then, as can be seen from Figs. 1.20, 1.21, instead of the differential equations (1.9), (1.10) we must consider equations of the form

$$\dot{y} + \lambda_2 y = f + z, \quad \tau\ddot{z} + (\tau + 1)\dot{z} + \lambda_1 z = u. \tag{1.17}$$

Repeating the transformations, we verify the validity of Eq. (1.15) but, instead of relation (1.16), we now have an equation

$$\tau\ddot{z} + (\tau + 1)\dot{z} + \lambda_1 z = (\dot{z} + \lambda_1 z) + u^s - (\dot{\eta} + \lambda_1\eta),$$

which, after collection of terms, assumes the form

$$\tau\ddot{z} + \tau\dot{z} = u^s - (\dot{\eta} + \lambda_1\eta).$$

It follows that the stability of free motions of the variable z is defined by the roots of the characteristic polynomial

$$\varphi(s) = \tau^2 s^2 + \tau s,$$

which has a zero root and, consequently, reflects the boundary between a stable and an unstable situation:

- a small variation of the parameters of the object is sufficient for the appearance of an unstable root and, along with it, an infinite increase in the coordinate z.

This inference completely coincides with the result obtained from the operator analysis of the control system. A subtler approach to the indirect measurement of disturbance with a subsequent compensation for its influence on the coordinate being controlled is used in Petrov's double-channel principle.

1.5. Double-Channel Principle

The double-channel principle is an heuristic technique of structural synthesis of invariant systems of automatic control or systems in which the coordinate being controlled does not depend on the uncontrolled external disturbance, i.e., disturbance which cannot be directly measured. As any heuristic technique, the double-channel principle does not give a unique solution and does not reduce to any unique sequence of actions. However, the main idea of this technique is very transparent and can be formulated as follows: in order to make the controlled coordinate of a control system independent of an external disturbance, it is necessary to organize at least one more additional channel along which the disturbance would influence the controlled coordinate and "adjust" it so that a mutual compensation of the components of the signals which are due to the action of disturbance would occur at a given point of the control system.

This principle of constructing a control system with a complete compensation for disturbance can be applied to the stabilization problem being considered in the following circumstances, for instance. Suppose that the component P_2 of the operator of the object P is represented as a composition of two operators, P_2' and P_2'', i.e.,

$$P_2 = P_2'' \, P_2',$$

and suppose that the output signal z of the subsystem with operator P_2' can be measured and, in addition, when necessary, some external signal q can be added to this signal. This composition is associated with the structure of the object shown in Fig. 1.23. We can see from this figure that the output y of the object does not depend on the disturbance f when the signal $v = z + q$ does not depend on f. Futhermore, since the signals z and r are defined by the relations

$$z = P_2' \, (f + r), \quad r = P_1 \, u^s,$$

it is obvious that $z = P_2' f + P_2' \, P_1 u^s$ and the signal v does not depend on the disturbance f if the signal q appropriately depends on this disturbance. This consideration is the informal expression of the double-channel principle.

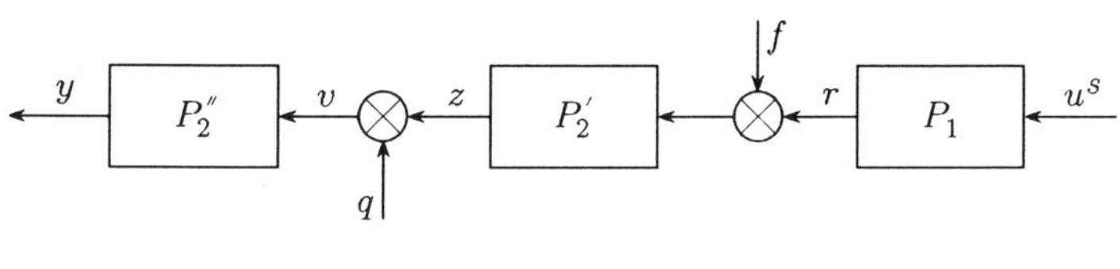

Figure 1.23.

Let us now justify it formally. When the output r of the subsystem P_1 is known, i.e., $r = P_1 u^s$, the required function $q(f)$ that compensates for the effect produced by f on v can be constructed as follows. We begin with measuring the signal z, then transform it by the operator R_1, and then subtract the obtained signal from the programmed control u^s. As a result, we obtain a block diagram of the control system in which $z = P_2' f + P_2' r$, $r = -P_1 \, R_1 z + P_1 u^s$

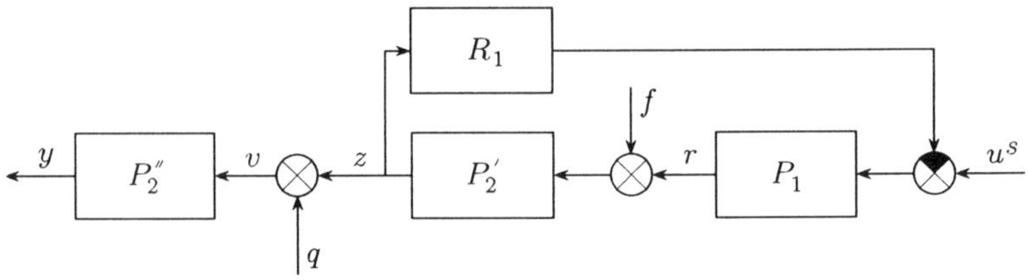

Figure 1.24.

(Fig. 1.24). Substituting the first expression into the second, we get a relation $r = -P_1\,R_1\,(P_2'f + P_2'r) + P_1u^s$. We solve this relation for r and find that the signal

$$r = -\frac{P_1\,R_1\,P_2'}{1 + P_1\,R_1\,P_2'}\,f + \frac{P_1}{1 + P_1\,R_1\,P_2'}\,u^s \tag{1.18}$$

depends on the external disturbance f, which we did not have before. Therefore, following the recommendations given in the preceding sections and choosing an appropriate operator R_2 in the system shown in Fig 1.25, we can expect to find the required dependence $q(f)$, i.e., the function that would compensate for the effect produced by f on v. In this case, a second, additional, channel of the extension of disturbance f is formed (it is shown by dash line in the figure), and this explains the name of the invariance principle being considered.

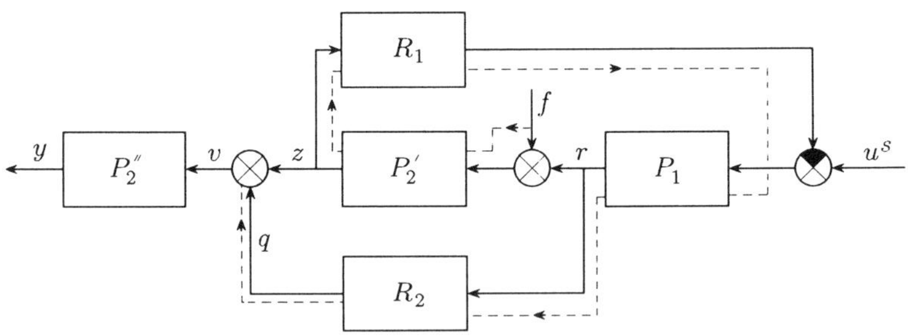

Figure 1.25.

Let us analyze the obtained control system. Since the signal z is given in the block diagram shown in Fig. 1.25 by the expression

$$z = P_2'(f + r),$$

we can substitute r from (1.18) into this relation and obtain the relationship between the signals z, f, and u^s in the form

$$z = \frac{P_2'}{1 + P_1\,R_1\,P_2'}\,f + \frac{P_2'\,P_1}{1 + P_1\,R_1\,P_2'}\,u^s.$$

Knowing now the dependence of the signal $v = z + q$ on f, we can easily choose a function $q(f)$ required for ensuring the invariance relative to f. To this end, we must choose the operator R_2 in the block diagram shown in Fig. 1.25 in the form

$$R_2 = \frac{1}{R_1\, P_1}. \tag{1.19}$$

Indeed, with such a choice of the operator R_2, with due account of relation (1.18), we can easily determine that

$$q = R_2 r = -\frac{P_2'}{1 + P_1\, R_1\, P_2'}\, f + \frac{1}{R_1(1 + P_1 R_1 P_2')}\, u^s.$$

Consequently, the signal v is expressed by the relation

$$v = z + q = \frac{1}{R_1}\, u^s$$

and does not depend on the uncontrolled disturbance f, and this is what we had to prove.

The block diagram of the synthesized invariant stabilization system is shown in Fig. 1.26.

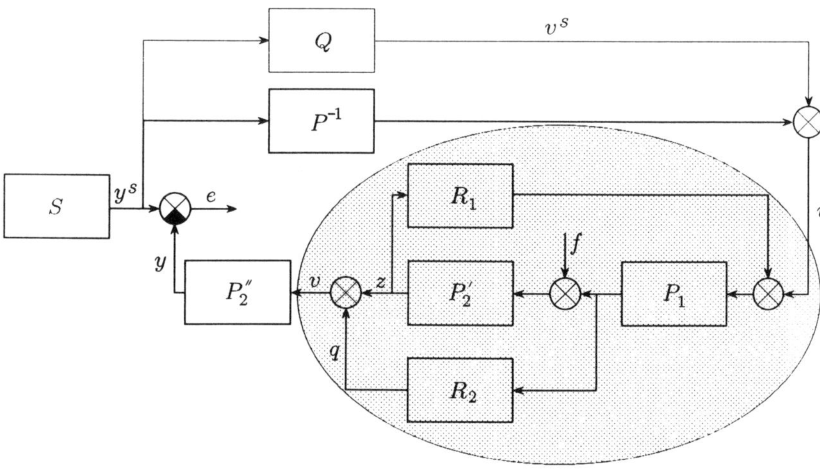

Figure 1.26.

Thus, the independence of the output y of the object of the disturbance f is ensured but, generally speaking, at the cost of the loss of the equality $y = y^s$ required in the stabilization problem since we now have

$$y = P_2''\, v = \frac{P_2''}{R_1}\, u^s = \frac{P_2''\, P^{-1}}{R_1}\, y^s = \frac{R_2}{P_2'}\, y^s,$$

but the validity of the equality

$$R_2 = P_2' \tag{1.20}$$

was never presupposed. Consequently, in order to find the exact solution of the stabilization problem by the method indicated above, we must either satisfy condition (1.20) or introduce a reference action operator y^s, $v^s = Q\, y^s$ which would correct the programmed control u^s (Q is a feedforward correcting operator which is hatched in Fig. 1.26).

In the latter case, we have obvious relations

$$y = \frac{P_2''}{R_1}\,(u^s + v^s) = \left(\frac{R_2}{P_2'} + R_2\, P_1\, P_2''\, Q\right) y^s = \frac{R_2}{P_2'}\,(1 + P\, Q)\, y^s$$

in whose definition the compensation conditions (1.19) are taken into account as well as the notation

$$P = P_1\, P_2, \quad P_2 = P_2'\, P_2''$$

introduced earlier.

Under a correcting operator, in the choice of the operators R_1 and R_2 there appears a degree of freedom needed for the weakening (but not for the removal) of the requirements to the physical realizability, restricted only by the compensation condition

$$P_1\, R_1\, R_2 = 1$$

and by the condition of unbiasedness of the solution of the stabilization problem

$$R_2\,(1 + P\, Q) = P_2'.$$

Therefore, among the principles of direct compensation for external disturbance, the double-channel principle obviously has the largest sphere of applicability. However, it is not free from serious drawbacks either.

- In the first place, when we use this principle, or rather, other principles of compensation, we have to confine ourselves to stable objects.
- Second, a local feedback loop with parameter $P_1\, P_2'$ in the direct channel and operator R_1 in the feedback channel appears in the block diagram of the corresponding stabilization system (Fig. 1.25). The stability of this loop is not at all inevitable, and since the choice of the operator R_1 is restricted by the compensation condition, most likely we will have to take special measures for the stabilization of motions in this internal loop.
- Third and fourth, we can see from the compensation condition $R_1\, R_2\, P_1 = 1$ that in a nontrivial case (when P_1 is a physically realizable dynamical link and, consequently, the degree of the polynomial appearing in the numerator of the corresponding transfer function is lower than the degree of the polynomial appearing in its denominator) the exact fulfilment of the compensation condition in the class of physically realizable operators

R_1, R_2 is impossible. Therefore we can only speak about an approximate compensation for the disturbance. We come to the same conclusion when we take into account that the condition of complete compensation is expressed by an exact equality, and in order to take this circumstance into account, we need information concerning the exact values of the parameters of the object, which is usually absent in practical applications. In other words, the use of this principle of control does not, in general, lead to stable control systems.

- Finally, the conditions of applicability of the double-channel principle are very specific and presuppose not only the availability of information concerning the internal coordinates of the object, but also the possibility of an active, as a matter of fact, controlling action on the internal coordinates. It stands to reason that these possibilities and, to be more precise, their combination, do not often exist in practice.

This is why the control theory turned to a more active use of feedback. We will discuss this fact in greater detail in the next section and will now consider a simple example illustrating the peculiarities of the application of the double-channel principle.

Example 4 (Stabilization with the use of the double-channel principle). Consider a third-order control object decomposed in accordance with the general theory of the double-channel method into three subsystems (Fig. 1.27). The transfer functions of the operators P_1, P_2', and P_2'' have the form $W_1(s) = 1/s$, $W_2'(s) = 1/(s+\lambda_1)$, $W_2''(s) = 1/(s+\lambda_2)$, where λ_1 and λ_2 are known constants, and let the coordinates z and r be measurable. We have to solve the stabilization problem without measuring the disturbance f, i.e., achieve an exact satisfaction of the relation $y = y^s$. Here y^s is a given time function.

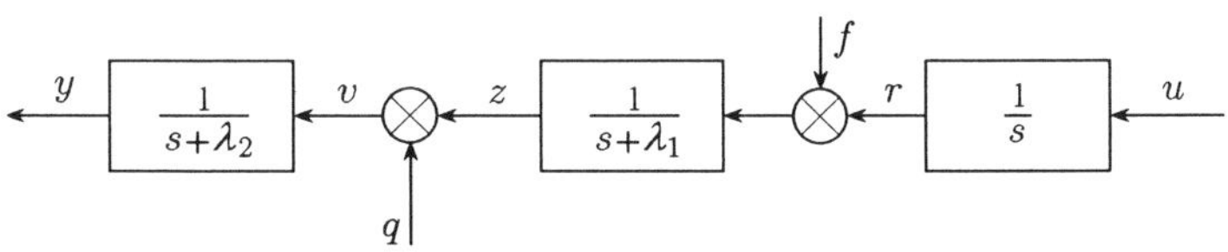

Figure 1.27.

Note that this control object can be described by the differential equations

$$\dot{y} = -\lambda_2 y + z, \quad \dot{z} = -\lambda_1 z + r + f, \quad \dot{r} = u.$$

In accordance with the theory of the double-channel principle exposed above, we supplement Fig. 1.27 by a local feedback with operator R_1 and a local feedforward with an operator R_2 (Fig. 1.28), where $W_1(s)$ and $W_2(s)$ are the required transfer functions corresponding to R_1 and R_2. We exclude the variables q and z from the equations

$$y = \frac{1}{s+\lambda_2} v, \quad v = z + q, \quad q = W_2(s), \quad z = \frac{1}{s+\lambda_1}(f + r), \quad sr = u^s - W_1(s) z$$

that describe this diagram and find the relation

$$r = \frac{s+\lambda_1}{s(s+\lambda_1) + W_1(s)} u^s - \frac{W_1(s)}{s(s+\lambda_1) + W_1(s)} f$$

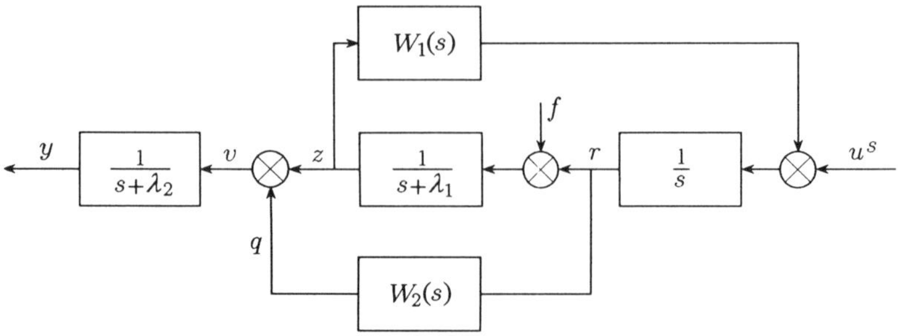

Figure 1.28.

and then the relationship between the variable $v = z + q$, disturbance f, and the input u^s in the form

$$v = \frac{1 + W_2(s)\,(s + \lambda_1)}{s(s + \lambda_1) + W_1(s)}\, u^s + \frac{s - W_1(s)\, W_2(s)}{s(s + \lambda_1) + W_1(s)}\, f.$$

When the condition of full compensation (1.19) is fulfilled, the transfer functions $W_1(s)$, $W_2(s)$ satisfy the relation

$$W_1(s)\, W_2(s) = s. \tag{1.21}$$

It is clear that this relation cannot hold for the physically realizable transfer functions. For instance, if $W_2(s) = 1$, then $W_1(s) = s$, and it is physically unrealizable.

It stands to reason that the choice of transfer functions $W_1(s)$ and $W_2(s)$ is still more restricted when we have to ensure the exact equality $y = y^s$. Indeed, when the compensation condition (1.21) is fulfilled, the relations

$$y = \frac{1}{s + \lambda_2}\, v = \frac{1 + W_2(s)\,(s + \lambda_1)}{s(s + \lambda_1) + W_1(s)}\, \frac{u^s}{s + \lambda_2}\bigg|_{W_1 W_2 = s} = \frac{u^s}{W_1(s)\,(s + \lambda_2)}$$

hold true. If we now use the standard approach to the formation of programmed control when $u^s = P^{-1} y^s$ and P is the operator of the object with the transfer function

$$W(s) = \frac{1}{s(s + \lambda_1)(s + \lambda_2)},$$

then the output of the system is transformed in accordance with the expression

$$y = \frac{s(s + \lambda_1)}{W_1(s)}\, y^s,$$

and the relation

$$W_1(s) = s(s + \lambda_1)$$

must be satisfied for ensuring the required condition $y = y^s$. From Eq. (1.21) we find the transfer function of the operator P_2:

$$W_2(s) = \frac{1}{s + \lambda_1}.$$

If R_2 is a Volterra operator (i.e., satisfies the principle of causality), then R_1 is a non-Volterra operator since it contains a double differentiation. Consequently, such a control system is physically unrealizable.

This means that

- in the framework of the double-channel principle it is impossible to achieve an exact compensation for disturbance and, at the same time, ensure an exact solution of a stabilization problem using only Volterra operators.

Therefore a problem arises of an approximate realization of the compensation condition or of the use of load correcting operator y^s for achieving the exact equality $y = y^s$. Since the second possibility was discussed in detail in the exposition of the theory of this compensation method, we shall investigate in detail only those circumstances which arise in the approximate realization of the compensation condition by Volterra operators.

In (1.21) we first choose operators with transfer functions of the form

$$W_1(s) = s, \quad W_2(s) = 1$$

and then replace the non-Volterra operation of exact differentiation by a Volterra operation of approximate (real) differentiation and, instead of $W_1(s)$, use

$$\widetilde{W}_1(s) = \frac{s}{\tau s + 1},$$

where τ is a small time constant. Under this assumption, the compensation condition is, of course, not fulfilled since now

$$\widetilde{W}_1(s)\, W_2(s) = \frac{s}{\tau s + 1},$$

and, consequently, for the variable v we have an expression

$$v = \frac{1 + W_2(s)\,(s + \lambda_1)}{s(s + \lambda_1) + \widetilde{W}_1(s)}\, u^s + \frac{\tau s}{(\tau s + 1)(s + \lambda_1)}\, f,$$

and, hence, the output $y = v/(s + \lambda_2)$ of the object depends on the disturbance f. However, this dependence diminishes together with the time constant τ which can be taken sufficiently small. To satisfy the condition of unbiasedness of the solution of the stabilization problem, we must use the correcting operator indicated above. The final form of the block diagram of the stabilization system

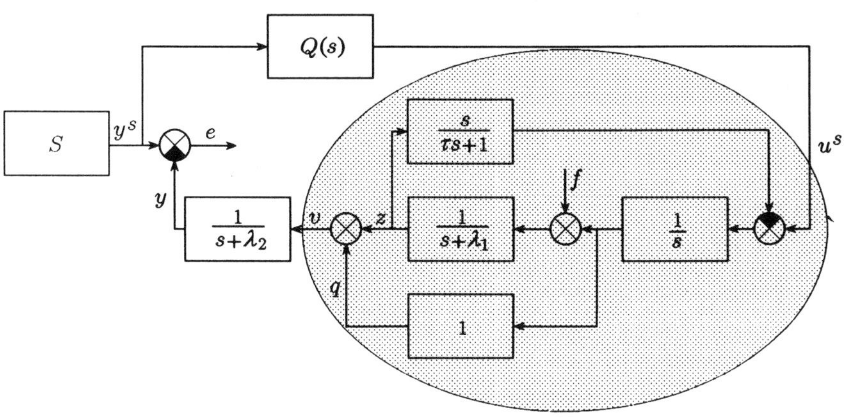

Figure 1.29.

that we obtain is shown in Fig. 1.29. In this figure, the transfer function $Q(s)$ is chosen from the condition

$$\frac{1+(s+\lambda_1)}{s(s+\lambda_1)+s/(\tau s+1)}\,\frac{Q(s)}{s+\lambda_2}=1$$

which ensures a programmed control u^s such that the transfer function from y^s to y is equal to unity, i.e.,

$$y=y^s+\frac{\tau s}{(\tau s+1)(s+\lambda_1)}f. \qquad (1.22)$$

This diagram can define an efficient stabilization problem only if λ_1 and $\lambda_2<0$, i.e., when the proper motions of the operator

$$P_2=\frac{1}{(s+\lambda_1)(s+\lambda_2)}$$

are exponentially stable.

Analyzing relation (1.22), we can easily see that the synthesized control system exactly solves the stabilization problem if the unknown disturbance f is constant. Indeed, in this case, the second term in the sum (1.22) vanishes. This, seemingly particular, observation constitutes the main content of the principle of internal model or the K-image principle which we shall discuss in the next section. We shall begin with investigating the problem of stability of the inner loop of the transfer function $W_1(s)=s/(\tau s+1)$ in feedback (Fig. 1.29). Acting in a standard way, we find the polynomial, which is responsible for the stability of this loop, in the form

$$\varphi(s)=\left[\,\tau s^2+(\lambda\tau+1)s+(\lambda+1)\,\right]s.$$

The existence of a zero root means that the loop is on the boundary of stability, and, consequently, the hidden parameters may lead to the instability of proper motions of the inner loop, and this, in turn, will lead to a degraded performance of the whole system. This fact testifies to the nonrobustness of the synthesized control system.

We shall conclude this subsection by showing the peculiarities of the application of the double-channel principle in the state space. The initial equations for the object in question were given above. Under the assumptions accepted above (the signals z and r are measured and the input of the operator P_2'' with the transfer function $W_2''(s)=1/(s+\lambda_2)$ can be acted upon), we can rewrite these equations as

$$\dot{y}=-\lambda_2 y+z+q, \quad \dot{z}=-\lambda_1 z+r+f, \quad \dot{r}=u. \qquad (1.23)$$

It is obvious that the coordinate y does not depend on the disturbance f if the sum $v=z+q$ does not depend on this disturbance. However, this means that the rate of variation of the coordinate v which is described by the equation

$$\dot{v}=\dot{z}+\dot{q}=-\lambda_1 z+r+f+\dot{q}$$

does not depend on f either. We set $q = r$. Then, by virtue of (1.23), the preceding equation yields an expression

$$\dot{v} = -\lambda_1 z + r + f + u, \tag{1.24}$$

in which the control u must be chosen such that its right-hand side would depend only on the signal v and program u^s. In this case, the control can depend only on the signals z, r, and u^s.

We find the expression for the function f on the right-hand side of the second equation in (1.23) and substitute the result into (1.24). As a result, we obtain $\dot{v} = \dot{z} + u$. It is clear now that if

$$u = -\dot{z} + u^s + kv, \tag{1.25}$$

where k is a constant coefficient, then the signal v satisfies the equation

$$\dot{v} = kv + u^s$$

and does not depend on the disturbance f, and, hence, the output of the object does not depend on this disturbance either, and this is what we had to prove. Note that this result coincides with the result described above which was obtained by the structural method for $k = 0$.

1.6. The K-Image Method or the Method of an Internal Model

The method of compensation for the disturbance $f(t)$ based on the exact knowledge of the model that generates this disturbance is very popular in the classical and contemporary control theory. In the linear theory, the model of disturbance is defined either by the vector differential equation

$$\dot{x} = Hx, \quad f = hx, \tag{1.26}$$

where H, h are a constant matrix and a constant vector of appropriate dimensions, or by the scalar differential operator $K(s)$ that annihilates the disturbance $f(t)$, i.e.,

$$K(s)\, f = 0. \tag{1.27}$$

Here $K(s) = s^n + a_n s^{n-1} + \cdots + a_1$ and the parameters a_i are constant for all $i = 1, \ldots, n$. In the first case, the disturbance can be uniquely defined by the initial phase vector x^0 since

$$f = h(sE - H)^{-1} x^0,$$

and in the second case it is defined by the values $f(0)$, $f^{(1)}(0)$, ..., $f^{(n-1)}(0)$. The definition of the disturbance as a wave model (1.26) or in the form of an annihilating operator (1.27) is the question of convenience. Earlier we used only

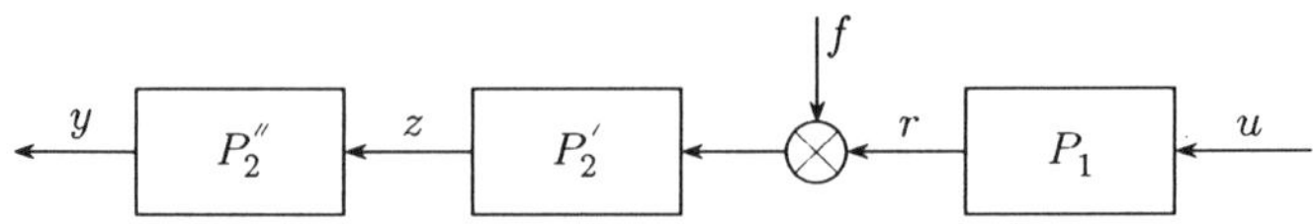

Figure 1.30.

the form (1.27), and precisely for this form did Academician V.S. Kulebakin develop the corresponding section of the invariance theory. We are reminded of this by the letter K which is traditionally used in the notation of the annihilating polynomial.

Thus, let us consider an object (Fig. 1.30) subjected to the action of disturbance $f(t)$ and control u. The annihilating operator $K(s)$ and the internal variable $z(t)$ are assumed to be known. We have to choose the control $u(t)$ such that the validity of the exact equality $y(t) = y^s(t)$ would be ensured for any $f(t)$ satisfying (1.27).

We can easily see from Fig. 1.30 and from the equations for the control object

$$y = P_2'' z, \quad z = P_2'(f + r), \quad r = P_1 u$$

that y does not depend on the interference f if the coordinate z does not depend on f. However, in the original structure the last fact is observed only if $P_2' f \equiv 0$, i.e., if P_2' is an annihilating operator for f. Since there is no hope to satisfy this identity, we shall try to change the operator acting from f to z with the aid of the local feedback with operator R (Fig. 1.31). In the system modified in this way, we use the relations

$$z = P_2' \,(f + r), \quad r = P_1 \,(u - Rz)$$

to find the dependence of z on f.

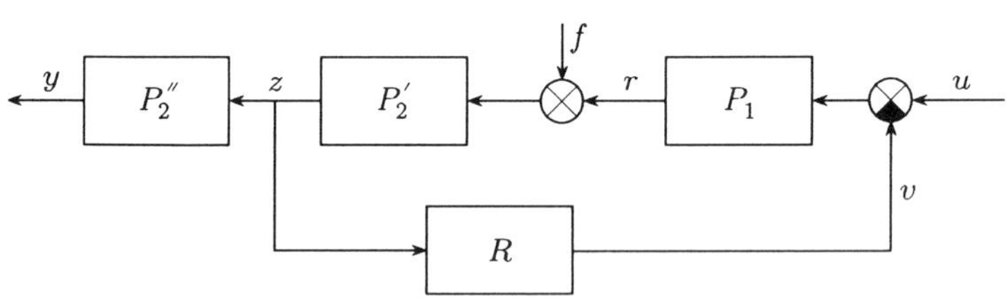

Figure 1.31.

After substituting the second relation into the first and collecting terms, we find the required dependence in the form

$$z = \frac{P_2'}{1 + P_2' \, P_1 \, R} f + \frac{P_2' P_1}{1 + P_2' \, P_1 \, R} u. \tag{1.28}$$

The coordinate z does not depend on the interference f if there exists an operator R such that for a certain operator L there holds a relation

$$\frac{P_2'}{1 + P_2' P_1 R} = LK, \tag{1.29}$$

where K is a polynomial from (1.27). Solving Eq. (1.29) for R, we find the required operator R:

$$R = \left(\frac{P_2'}{LK} - 1\right)\frac{1}{P_2' P_1} = \frac{N}{K}, \tag{1.30}$$

where $N = (P_2' - LK)/(LP_2' P_1)$.

Relation (1.30) shows that the local feedback that ensures the invariance condition must contain the inverse of the operator that annihilates the disturbance. The second name of the method that we describe, namely, the method of internal model, is connected precisely with this circumstance.

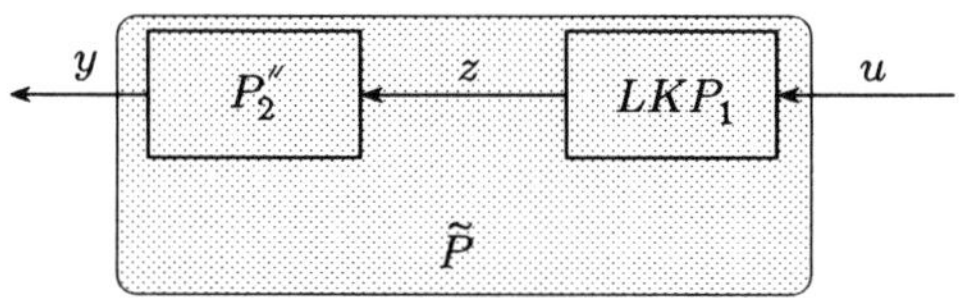

Figure 1.32.

When the condition of invariance (1.29) is fulfilled, the equalities

$$z = \frac{P_2'}{1 + P_2' P_1 R} P_1\, u = LKP_1\, u$$

hold true, and, to complete the solution of the stabilization problem ($y = y^s$), we must choose a load correcting feedback for the new object with operator $\widetilde{P} = P_2'' LKP_1$ (Fig. 1.32). If $\widetilde{P}$ is a minimally phase object, i.e., the polynomials in its numerator and denominator are stable, then we can use the standard load feedforward (Fig. 1.33) which solves the problem, in other words, keeps the relation $y = y^s$ in the equilibrium state.

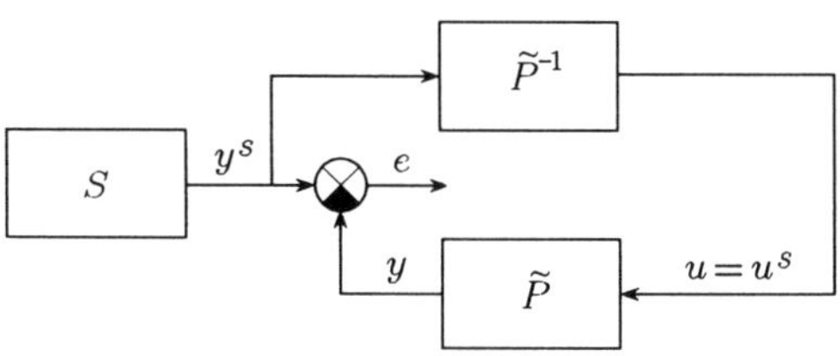

Figure 1.33.

Thus, the minimal phase property of the object $\widetilde{P}$ is a necessary but not the only condition for the application of the K-image method. Another constraint is connected with the physical realizability of the inverse operator $\widetilde{P}^{-1}$.

Indeed, the local feedback operator R must be physically realizable, but then the operator $\widetilde{P}$ must also be physically realizable and, consequently, the operator $\widetilde{P}^{-1}$ cannot be realized. We can only speak about a physical realizability of the approximation $\widetilde{\widetilde{P}}^{-1}$ of the operator $\widetilde{P}^{-1}$, and the stabilizaton problem can, evidently, be solved only approximately.

To weaken the requirement that $\widetilde{P}^{-1}$ should possess the minimal phase property and the physical realizability, along with the load dependence we can use a stabilization error feedback $e(t)$ represented in Fig. 1.34 in which Q and M are operators of the corresponding de‹ endences.

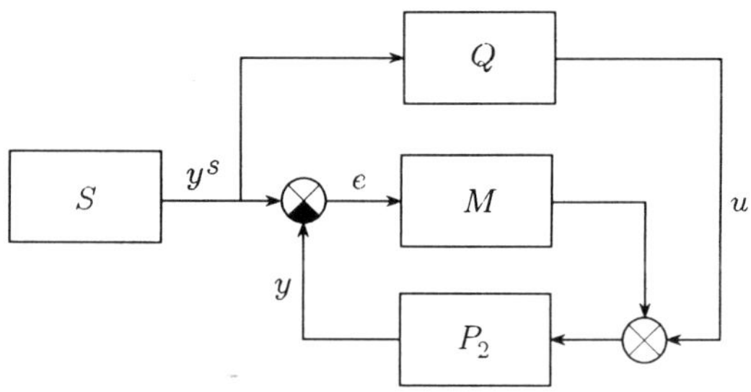

Figure 1.34.

The equations

$$y = \widetilde{P}\,(u + Me), \quad e = y^s - y$$

give the relationship between y and y^s in the form

$$y = \frac{\widetilde{P}(Q+M)}{1+\widetilde{P}M}\, y^s = W(s)\, y^s . \tag{1.31}$$

Thus, the stabilization problem cannot be solved exactly when the feedback operators M are restricted since we find from the relation

$$\frac{\widetilde{P}(Q+M)}{1+\widetilde{P}M} = 1 \tag{1.32}$$

that $\widetilde{P}Q = 1$ and, hence, Q is physically unrealizable. However, by an appropriate choice of the operator M we can influence the stability of the denominator of the transfer function $W(s)$ and achieve a better approximation of relation (1.32) by an appropriate choice of the operator Q. The requirement of the stability of zeros of the operator $\widetilde{P}$ is preserved since this modification of the stabilization system does not affect the location of these zeros.

Note that if we discard the load feedforward, i.e., accept $Q \equiv 0$ in the block diagram in Fig. 1.34, and lay the solution of the problem on the feedback, then we get the following relationship between y and y^s:

$$y = \frac{\widetilde{P}M}{1 + \widetilde{P}M} y^s .$$

It is clear that only under the condition $|\widetilde{P}M| \gg 1$ can we hope to get a satisfactory solution of the stabilization problem.

However, this is a different theme, the theme of high gain feedback which we shall consider in the next section, and here we shall consider an example referring to the K-image method.

Example 5 (Stabilization by the K-image method). In order to estimate the possibilities and drawbacks of the K-image method when we operate with transfer functions and in the state space, we shall consider the cases of a constant and an exponential disturbance.

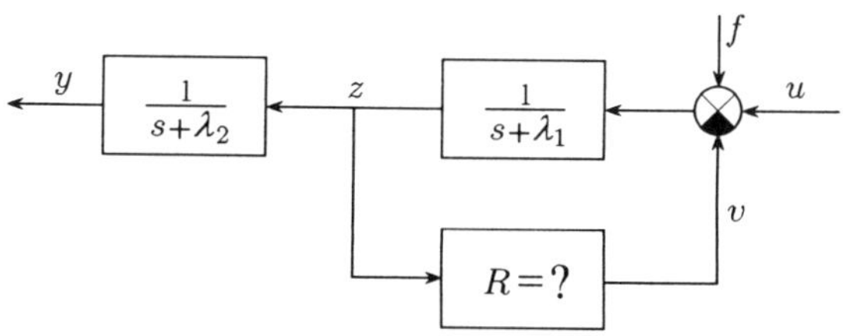

Figure 1.35.

Case A. Suppose that the disturbance f = const acts on an object in accordance with the block diagram given in Fig. 1.35. We have to find an operator R under the action of which the coordinate z does not depend on f and, in addition, the relation $y = y^s$ holds true. Since f = const, the simplest operator annihilating the constant is a differentiation operator, i.e., $K = s$. Consequently, according to the theory exposed above, we take $R = N(s)/s$. We verify that under this operator R the coordinate z does not depend on f.

Indeed,

$$z = \frac{f - Rz + u}{s + \lambda_1} = \frac{sf - N(s)z + su}{s(s + \lambda_1)}.$$

However, $sf \equiv 0$, and, hence,

$$z = \frac{-N(s)z + su}{s(s + \lambda_1)}.$$

The operator of transfer from u to z is defined by the expression

$$z = \frac{s}{s(s + \lambda_1) + N(s)} u,$$

and, consequently,

$$y = \frac{s}{(s + \lambda_2)\left[s(s + \lambda_1) + N(s)\right]} u.$$

If we try to ensure the equality of the operators $\widetilde{P} = P$, i.e.,

$$\frac{s}{(s + \lambda_2)[s(s + \lambda_1) + N(s)]} = \frac{1}{(s + \lambda_2)(s + \lambda_1)},$$

then we necessarily obtain $N(s) = 0$. Therefore we shall choose an operator $N(s)$ that will ensure the achievement of stability of the "renovated" object $\widetilde{P}$ or, what is the same, the achievement of the stability of the polynomial

$$\varphi(s) = s(s + \lambda_1) + N(s)$$

under the stability of the object P.

The stability of the polynomial can be attained, for instance, for $N = k = \text{const} > 0$ since, as a result, we get $\varphi(s) = s^2 + s\lambda_1 + k$. For $\lambda_1 > 0$ (this condition is fulfilled according to the assumption) and $k > 0$, this is a Hurwitz polynomial.

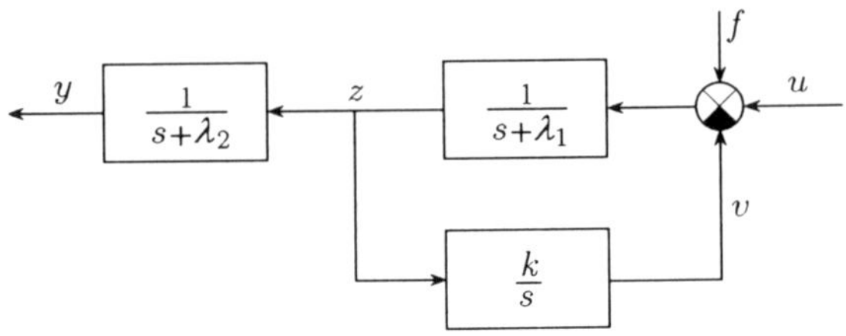

Figure 1.36.

As a result, the feedback compensating for the disturbance turns out to be integral (Fig. 1.36). It is easy to calculate that the transfer function of such an object is associated with the expression

$$\widetilde{W}(s) = \frac{s}{(s+\lambda_2)\left[s(s+\lambda_1)+k\right]}.$$

Since the transfer function $\widetilde{W}(s)$ has stable poles, in order to solve the stabilization problem

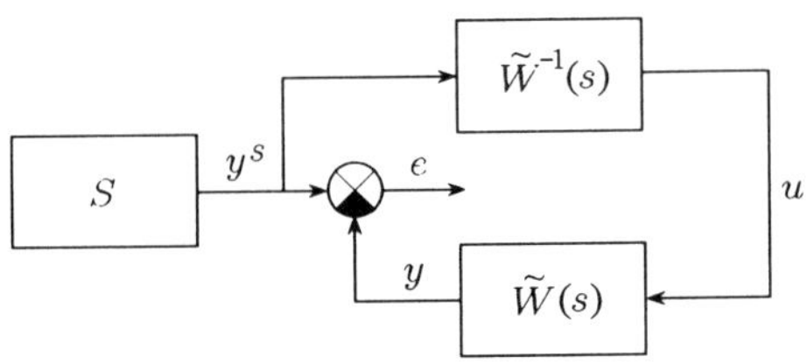

Figure 1.37.

we can use the load control principle (Fig. 1.37) with a transfer function inverse to $\widetilde{W}(s)$:

$$\widetilde{W}^{-1}(s) = (s+\lambda_2)(s+\lambda_1) + k\frac{s+\lambda_2}{s}.$$

We can see that $\widetilde{W}^{-1}(s)$ is physically unrealizable, and, consequently, we can only speak about its approximate realization.

Case B. The problem becomes unrealizable by this method if an exponential disturbance $f = e^{\alpha t}$, $\alpha = \text{const}$ acts at the input of the object.

The operator $K = s - \alpha$ is annihilating for $f(t)$. Indeed,

$$Kf = \left(\frac{d}{dt}f - \alpha f\right)\bigg|_{f=e^{\alpha t}} = \alpha f - \alpha f = 0.$$

Following the recommendations of the theory, we obtain

$$R = N(s)/s - \alpha,$$

and the relationship between z and u is defined by the expression

$$z = \frac{(s-\alpha)}{(s-\alpha)(s+\lambda_1)+N(s)}u.$$

We choose the operator $N(s)$ from the considerations of stability of the polynomial

$$\varphi(s) = (s-\alpha)(s+\lambda_1) + N(s).$$

It is clear that we can set $N(s) = k_2 s + k_1$. Then, if the inequalities $k_1 > \lambda_1\alpha$, $k_2 > \alpha - \lambda_1$ are satisfied, $\varphi(s)$ is a Hurwitz polynomial.

As a result, the transfer function of the "renovated" object is defined by the expression

$$\widetilde{W}(s) = \frac{s-\alpha}{(s+\lambda_2)\varphi(s)},$$

and the inverse transfer function

$$\widetilde{W}^{-1}(s) = \frac{(s+\lambda_2)\varphi(s)}{s-\alpha}$$

has an unstable pole, and, hence, we cannot apply the load control principle in this case. The error feedback $e = y^s - y$ does not change anything in this diagram either since this feedback does not eliminate the right-hand zeros of the transfer functions.

Case C. We shall again consider the object represented by the block diagram in Fig. 1.35 with disturbance $f = \text{const}$, but we shall now use differential equations to describe the K-image method. This object is described by the differential equations

$$\begin{aligned} \dot{y} + \lambda_2 y &= z, \\ \dot{z} + \lambda_1 z &= f + u - v. \end{aligned} \tag{1.33}$$

After differentiating Eq. (1.33) with respect to t, with due account of the identity $\dot{f} \equiv 0$, we have

$$\ddot{z} + \lambda_1 \dot{z} = \dot{u} - \dot{v},$$

and, consequently, z does not depend on f. We set

$$\dot{v} = kz. \tag{1.34}$$

Then $\ddot{z} + \lambda_1 \dot{z} + kz = \dot{u}$, and if $k > 0$, then this equation is stable and, by an appropriate choice of $\dot{u}(y^s)$, we can obtain $y \to y^s$, and this solves the problem. Equation (1.34) describes the integral feedback since

$$v = k\int z\,dt.$$

1.7. High Gain Factor

Let us consider one of the most powerful methods of solving stabilization problems under the conditions of uncertainty, namely, the high gain feedback method which, for a linear case, reduces to the use of a high gain factor in the controller.

1.7.1. Statement of the problem, its peculiarities and the idea of its solution

Suppose that a disturbance f acts on an object with operator $P = P_1 P_2$ (Fig. 1.38). We have to ensure the independence of the controlled coordinate y of the disturbance f by the choice of the control u and, in addition, provide for the achievement and preservation of the equality

$$y = y^s.$$

Note some peculiarities of the statement of the stabilization problem. First, we do not suppose, as before, that the object P is necessarily stable. Second, the disturbance f is not necessarily small or vanishing with time. Finally, we do not presuppose that the disturbance f can be directly or indirectly measured.

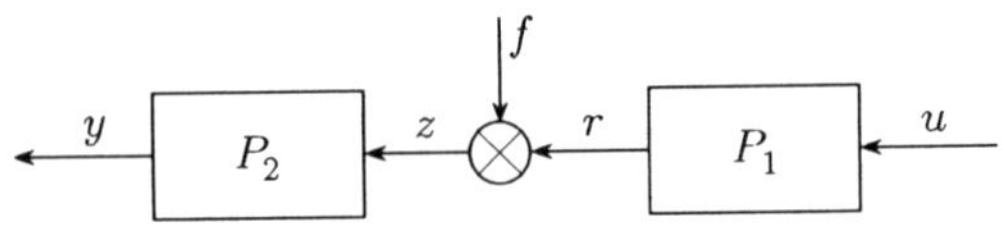

Figure 1.38.

Under these conditions, it is impossible to apply the methods described above, and therefore it is necessary to use either the feedback or the error control method in the framework of the block diagram shown in Fig. 1.39, in which we must choose a feedback operator R', or a combination of the load control principle and feedback in the framework of the block diagram shown in Fig. 1.40

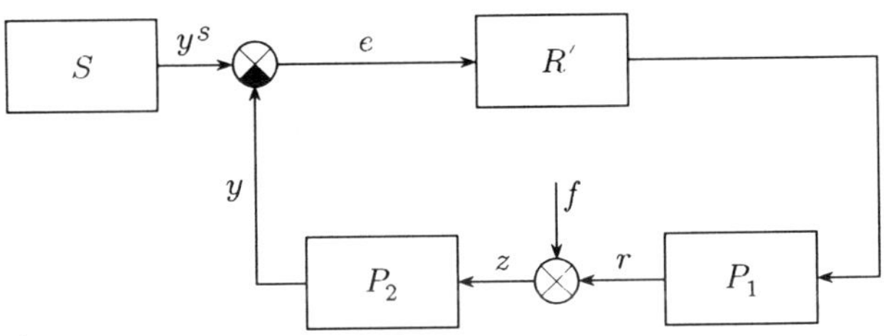

Figure 1.39.

in which, in addition, we must construct a load feedback operator Q. The *idea of solving the problem* is based on the transition from the diagram shown in Fig. 1.39 to the diagram presented in Fig. 1.41. This transition is realized by

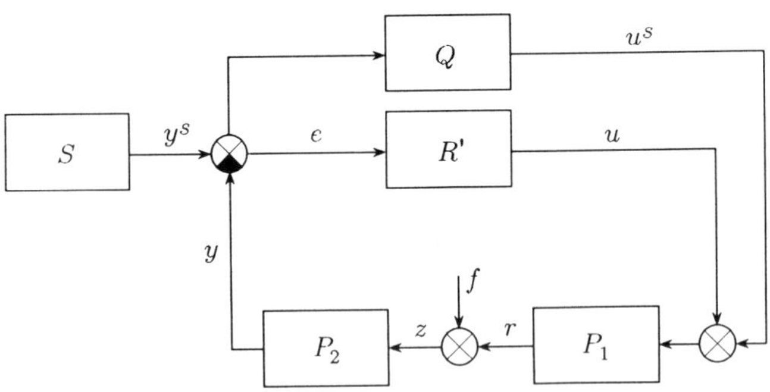

Figure 1.40.

means of the substitution $R' = kR$, where k is a scalar gain factor and R is the required feedback operator. The equations that describe the obtained closed control system have the form

$$y = P_2(f + r), \quad r = P_1 u, \quad u = kR(y^s - y).$$

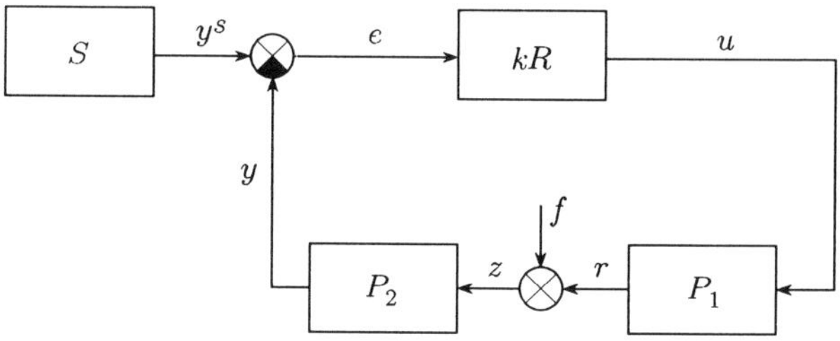

Figure 1.41.

After eliminating the variables u and r, we find the relationship between the main variables y, y^s, and f first in the form

$$(1 + kP\,R)y = P_2 f + kP\,R\,y^s \tag{1.35}$$

and, after the division by $1 + kPR$, in the final version

$$y = \frac{P_2}{1 + kP\,R} f + \frac{kP\,R}{1 + kP\,R} y^s . \tag{1.36}$$

We can see that as $k \to \infty$, the first term in expression (1.36) diminishes to zero and the second term tends to the reference action y^s, and this solves the stabilization problem in question if, of course, the feedback system is asymptotically stable as every value of $k \to \infty$.

In addition, the use of high gain factors leads to an increase in the values of the variables of the system, and since constraints imposed on amplitudes always exist in real systems, it is necessary to study the effect they produce on the properties of the system. Finally, we must find out how sensitive is the solution given by high gain feedback to the variations of the conditions of the problem, i.e., to regular and singular perturbations. The properties of a system which continuously depend on these perturbations are said to be robust.

1.7.2. Problems and limitations of the high gain feedback method

In this section and in the subsequent sections we shall consider some of the problems mentioned above. Here we shall study the conditions of stability of the limit and prelimit systems and the effect produced by amplitude constraints on the control.

From Eq. (1.36) of a feedback system we obtain

$$(1 + kP\,R)\,y = P_2\,f + kP\,R\,y^s ,$$

and, consequently, as $f \to \infty$, we must study the stability of free motion or, what is the same, the stability of the zero solution of the equation

$$(1 + kPR)\,y = 0. \tag{1.37}$$

We denote the transfer function of the object P by

$$W_P(s) = \frac{b_{m+1}s^m + b_m s^{m-1} + \cdots + b_1}{s^n + a_n s^{n-1} + \cdots + a_1} = \frac{b(s)}{a(s)},$$

and the transfer function of the controller R by

$$W_R(s) = \frac{p_{n+1}s^n + p_n s^{n-1} + \cdots + p_1}{q_{n+1}s^n + q_n s^{n-1} + \cdots + q_1} = \frac{p(s)}{q(s)}.$$

Substituting these expressions into Eq. (1.37), we get

$$\left[1 + k\frac{b(s)\,p(s)}{a(s)\,q(s)}\right] y = \frac{a(s)\,q(s) + kb(s)\,p(s)}{a(s)\,q(s)} y,$$

and the *stability problem* reduces to finding out whether, as $k \to \infty$, the parametric family of polynomials

$$\varphi_k(s) = a(s)\,q(s) + kb(s)\,p(s) \stackrel{\text{def}}{=} \varphi_{a,q}(s) + \varphi_{b,p}(s) \tag{1.38}$$

is Hurwitzian.

It follows from (1.38) that as $k \to \infty$, some of the zeros of the polynomial under investigation coincide with the zeros of the polynomial $\varphi_{b,p}(s)$ which, of course, must have negative real parts, and this is the *first limitation* of the method. Furthermore, by the condition of physical realizability, the degree of the polynomial of the numerator is lower than the degree of the polynomial of the denominator, i.e.,

$$\deg \varphi_k(s) < \deg \varphi_{a,q}(s), \tag{1.39}$$

and, consequently, in addition to the zeros of the polynomial $\varphi_{b,p}(s)$ the polynomial $\varphi_\infty(s)$ has other zeros which must also belong to the open left half-plane of the variable s. We can see from the analysis of the prelimit polynomial (1.38) that the absolute values of its zeros $\lambda_i(k)$, which do not tend to the zeros of the polynomial $\varphi_{b,p}(s)$, as $k \to \infty$, increase to infinity since the relation

$$\varphi_k\big(\lambda_i(k)\big) = a\big(\lambda_i(k)\big)q\big(\lambda_i(k)\big) + k\,b\big(\lambda_i(k)\big)\,p\big(\lambda_i(k)\big) = 0$$

must be satisfied for any k. It is clear that from the consideration of stability these zeros must vanish, as $k \to \infty$, in the left open half-plane of the complex variable s, and this is the *second limitation* of this method.

In order to express conditions (1.39) in terms of the parameters of the system, we use Euclid's algorithm to divide the polynomial $\varphi_{a,q}(s) = a(s)\,q(s)$ by the polynomial $\varphi_{b,p}(s)$. As a result, we get a relation

$$a(s)\,b(s) = \Theta(s)\,b(s)\,p(s) + r(s), \tag{1.40}$$

where $\Theta(s)$ is the quotient and $r(s)$ is the remainder of the division, and the degree of the remainder is lower than that of the divisor, i.e.,

$$\deg r(s) < \deg\big(b(s)\,p(s)\big). \tag{1.41}$$

Using relation (1.40), we represent the polynomial $\varphi_k(s)$ as

$$\varphi_k(s) = b(s)\,p(s)\left[\Theta(s) + \frac{r(s)}{b(s)\,p(s)} + k\right].$$

It is clear now that the infinitely increasing, as $k \to \infty$, zeros of the polynomial $\varphi_k(s)$ tend to the zeros of the polynomial

$$\vartheta(s) = \Theta(s) + k$$

since, by what was proved above and by virtue of inequality (1.41), we have

$$\lim_{k\to\infty} \frac{r(\lambda_i(k))}{b(\lambda_i(k))\,p(\lambda_i(k))} = 0.$$

Thus, only the following two possibilities lead to the stability of the polynomial $\varphi_k(s)$ when $k \to \infty$ and when the polynomial $\varphi_{b,p}(s)$ is stable:

if $\deg \Theta(s) = 1$, i.e., $\Theta(s) = \Theta_2 s + \Theta_1$, then $\Theta_2 > 0$,

if $\deg \Theta(s) = 2$, i.e., $\Theta(s) = \Theta_3 s^2 + \Theta_2 s + \Theta_1$, then $\Theta_2 > 0$, $\Theta_3 > 0$.

All other situations, the inequality $\deg \Theta(s) > 2$ inclusive, lead to instability.

Example 6 (Stability of systems with feedback). We shall illustrate the theory exposed above by a simple example, for which purpose we shall consider the object shown in Fig. 1.42. The parameters λ_1 and λ_2 of the object are not necessarily positive numbers, i.e., the object is not necessarily stable.

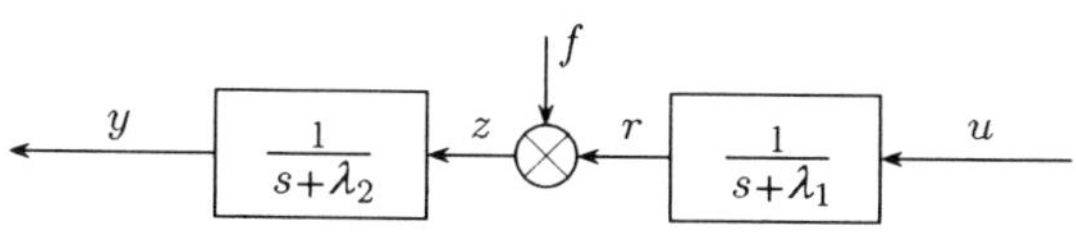

Figure 1.42.

We apply the error feedback $e = y^s - y$ with a transfer function

$$W_R(s) = k\frac{p(s)}{q(s)}, \quad k = \text{const}, \quad \deg q(s) \geq \deg p(s).$$

As a result, we obtain a feedback control system whose block diagram is shown in Fig. 1.43. It is easy to find out that the relationship between the variables y, y^s, and f is defined by the expression

$$\left[1 + \frac{kp(s)}{(s+\lambda_2)(s+\lambda_1)\,q(s)}\right] y = \frac{f}{s+\lambda_2} + \frac{kp(s)}{(s+\lambda_2)(s+\lambda_1)\,q(s)} y^s,$$

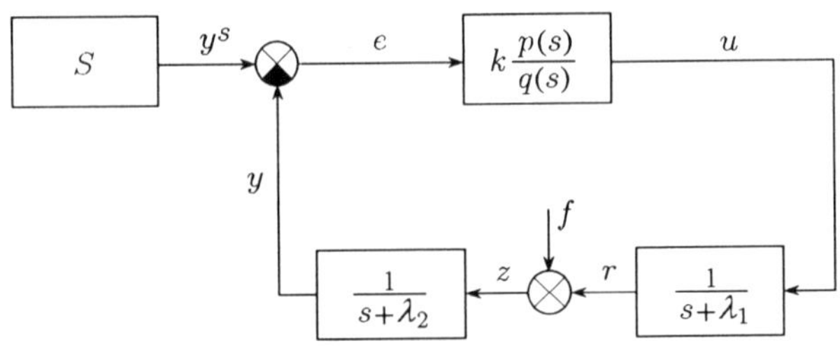

Figure 1.43.

which, after the reduction to a common denominator, assumes the form

$$\left[(s+\lambda_2)(s+\lambda_1)\,q(s)+kp(s)\right]y(s)=q(s)\,(s+\lambda_1)\,f+k\,p(s)\,y^s.$$

If the feedback parameter $k\to\infty$ in the last relation, then, in the limit, we have the required relation

$$y=y^s.$$

It also follows from the theory that the limiting process, as $k\to\infty$, is justified if the following two polynomials are stable:

$$\varphi_k(s)=(s+\lambda_2)(s+\lambda_1)\,q(s)+kp(s),\quad p(s).$$

If, for instance, $p(s)=1$, then necessarily $q(s)=1$, and the stability of the limit system is defined by the stability of the polynomial

$$\varphi_k(s)=s^2+(\lambda_1+\lambda_2)s+\lambda_1\lambda_2+k$$

which occurs only when the inequality

$$\lambda_1+\lambda_2>0$$

is satisfied.

It is clear that the last inequality may not be satisfied. Now if $p(s)=s+c$, where $c=\text{const}>0$, then, for $q(s)=1$, the stability of the polynomial

$$\varphi_k(s)=s^2+\left[(\lambda_1+\lambda_2)+k\right]s+\lambda_1\lambda_2+kc$$

occurs for any parameters λ_1, λ_2 and for a sufficiently large value of the feedback factor k. We should bear in mind that the transfer function of the controller

$$W_R(s)=k\frac{p(s)}{q(s)}=k(s+c)$$

is physically unrealizable since $\deg q(s)<\deg p(s)$ and, consequently, in the commutations presented above we must set at least

$$q(s)=s+q,\quad q=\text{const}>0.$$

Then we can establish that the polynomial responsible for the stability of the feedback system has the form

$$\varphi_k(s)=s^3+(\lambda_1+\lambda_2+q)s^2+\left[\lambda_1\lambda_2+(\lambda_1+\lambda_2)q+k\right]s+\lambda_1\lambda_2 q+kc,$$

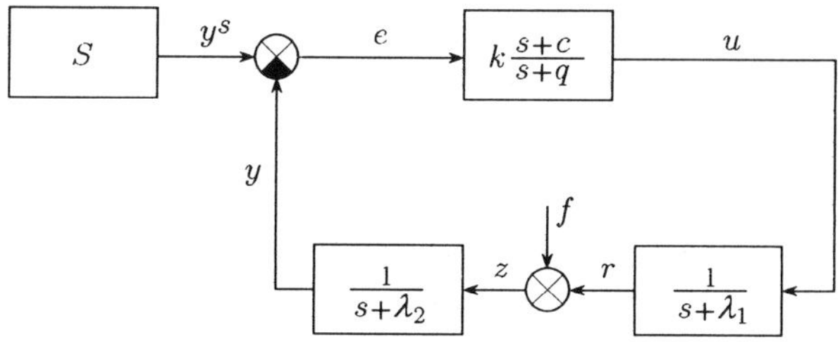

Figure 1.44.

and the stability occurs for sufficiently large values of the parameters k and q. In this way we can obtain the structure of the feedback control system (Fig. 1.44) which is stable under an infinite increase of the feedback factor ($k \to \infty$) and which solves the stated stabilization problem.

1.7.3. On the robustness of systems with a high gain factor

Let us consider the effect produced by two types of variations of the operator of the object:

- regular variations or variations of the parameters,
- irregular variations or singular perturbations which change the order of the object.

To analyze the consequences which follow regular perturbations, it suffices to consider the polynomial

$$\varphi_k(s) = a(s)\,q(s) + kb(s)\,p(s)$$

which is responsible for the stability of the feedback system. It is obvious that if $\varphi_{b,p}(s) = b(s)\,p(s)$ and $\varphi_k(s)$ are Hurwitz polynomials (the latter for a sufficiently large value of k), then, by virtue of the continuous dependence of the spectrum on the parameters, small variations of the parameters of the polynomials $a(s)$ and $b(s)$ (i.e., parameters of the object) do not qualitatively change the situation.

In other words,

- systems with a high gain factor are robust to regular perturbations.

Under a singular perturbation, the order of the object changes, say, instead of the transfer function $W(s) = b(s)/a(s)$ we deal with a transfer function of the form

$$W_\tau(s) = \frac{b(s)}{a(s)\,\tau(s)},$$

where $\tau(s)$ is a stable but unknown polynomial of a degree not lower than the first, i.e., $\deg \tau(s) \geq 1$. Then a polynomial of the form

$$\varphi_k^\tau(s) = \tau(s)\, a(s)\, q(s) + kb(s)\, p(s)$$

is responsible for the stability of the feedback system, and it is clear that if the inequality $\deg \tau(s) \geq 2$ is valid, then the polynomial $\varphi_k^\tau(s)$ is always unstable as $k \to \infty$ for the physically realizable feedback. The hope for the stability remains only when $\deg \tau(s) = 1$. However, this hope is illusive since the order of a real object is always higher than that of its mathematical model and, consequently,

- the practical application of feedback with high gain factor always leads to instability.

Consequently, the value of the gain factor, which is maximally admissible from the considerations of stability, is bounded by a certain critical value $k_{\rm cr}$, i.e., $0 \leq k \leq k_{\rm cr}$. It stands to reason that this fact does not allow us to obviate the influence of the disturbance f on the controlled coordinate y and, as a consequence, obtain the required equality $y = y^s$, i.e., does not allow us to get an exact solution of the stabilization problem. In other words,

- systems with a high gain factor are nonrobust relative to singular perturbations.

Example 7 (Instability of systems with feedback relative to singular perturbations). Suppose that in the system with feedback given in Example 6 there exists a singular perturbation, acting in the way described above, with an operator

$$\tau(s) = \tau s + 1, \qquad \tau = \text{const} > 0.$$

Then the polynomial of the feedback control system is described by the expression

$$\varphi_k^\tau(s) = (\tau s + 1)(s + \lambda_2)(s + \lambda_1)(s + q) + k(s + c)$$

and it is clear that there is no positive value of τ for which it would be Hurwitzian as $k \to \infty$. Thus, the stabilization system constructed in Example 6 is inefficient as $k \to \infty$. Note that for $k < k_{\rm cr}$ the polynomial $\varphi_k^\tau(s)$ is stable, but the required relation $y = y^s$ is not satisfied, and, moreover, the stabilization error depends on the disturbance $f(t)$.

Let us find out whether the additional load feedforward brings in anything principally new in accordance with the block diagram shown in Fig. 1.45. The equations corresponding to this structure have the form

$$y = P_2(f + r), \quad r = P_1(u + u^s), \quad u^s = Qy^s, \quad u = kR(y^s - y).$$

After the elimination of the variables u, u^s, and r we find the required relationship between the variables y, y^s, and f in the form

$$(1 + kPR)y = P_2 f + (kPR + PQ)y^s.$$

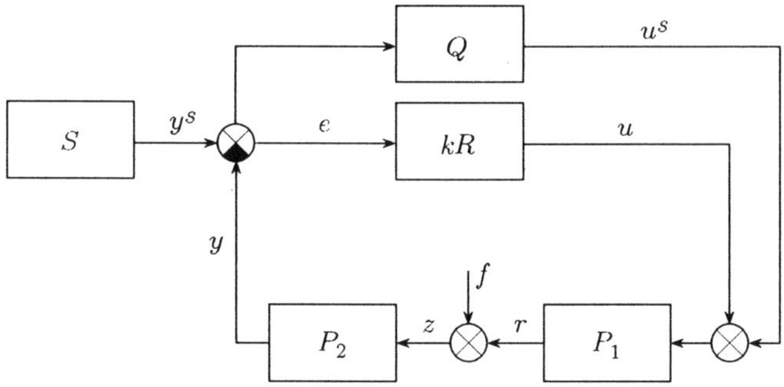

Figure 1.45.

It follows from this expression that the load feedforward under consideration does not affect the stability of the system since the polynomial $1 + kPR$ responsible for stability does not depend on the operator Q. The precision of the compensation for the disturbance f does not depend on the operator Q either. The advantage of feedforward is that for a finite gain factor ($k \ll k_{\text{cr}}$) we can use the operator Q in order to increase the accuracy of the preservation of the required relation $y = y^s$.

1.7.4. The state space method in the analysis of systems with a high gain factor

Let us solve the problem of stabilization of the object from Example 6 (Fig. 1.42) by the state space method. The differential equations which describe the object in the state space have the form

$$\begin{aligned} \dot{y} + \lambda_2 y &= f + r, \\ \dot{r} + \lambda_1 r &= u. \end{aligned} \tag{1.42}$$

We apply the operator $d/dt + \lambda_1$ to Eq. (1.42) assuming, of course, that the function f is differentiable. Then, using the standard notation

$$a_2 = \lambda_1 + \lambda_2, \quad a_1 = \lambda_1 \lambda_2,$$

we can write the equation of the object as

$$\ddot{y} + a_2 \dot{y} + a_1 y = F + u, \tag{1.43}$$

where $F = \dot{f} + \lambda_1 f$ is disturbance reduced to the input of the object, i.e., to the control input. Let

$$\ddot{y}^s + a_2 \dot{y}^s + a_1 y^s = Y^s.$$

We subtract Eq. (1.43) of the object from this expression and, since $e = y^s - y$, get an equation of the object in deviations:

$$\ddot{e} + a_2\dot{e} + a_1 e = Y^s - F - u.$$

We define the Cartesian coordinates $x_1 = e$, $x_2 = \dot{e}$ and write the equation for the control object in the phase space (x_1, x_2). We have

$$\dot{x}_1 = x_2, \quad \dot{x}_2 = -a_1 x_1 - a_2 x_2 - u + Y^s - F.$$

Under the feedback

$$u = k\sigma, \quad \sigma = x_2 + cx_1, \quad c = \text{const} > 0,$$

we obtain a feedback control system of the form

$$\dot{x}_1 = x_2, \quad \dot{x}_2 = -a_1 x_1 - a_2 x_2 - k(x_2 + cx_1) + Y^s - F. \tag{1.44}$$

As $k \to \infty$, the last differential equation degenerates into an algebraic equation $x_2 + cx_1 = 0$. After the substitution $x_2 = -cx_1$ performed in the differential equation (1.44), we get an equation of the limit motion in the form

$$\dot{x}_1 = -cx_1.$$

Since this equation is exponentially stable and

$$y^s - y = e = x_1,$$

it follows that $y^s - y \to 0$, and this completes the solution of the stabilization problem.

1.7.5. Geometrical interpretation of systems with a high gain factor

If we set

$$a_1 = \text{const}, \quad a_2 = \text{const}, \quad Y^s - F = A = \text{const}$$

in (1.44), then the equations of a feedback system assume the form

$$\dot{x}_1 = x_2, \quad \dot{x}_2 = -a_1 x_1 - a_2 x_2 - k(x_2 + cx_1) + A, \tag{1.45}$$

and we can give a geometrical interpretation of the obtained analytical results (Fig. 1.46). For a finite value of the parameter k, the qualitative behavior of phase trajectories of system (1.45) is illustrated by Fig. 1.46*a* in which $\lambda_1(k)$

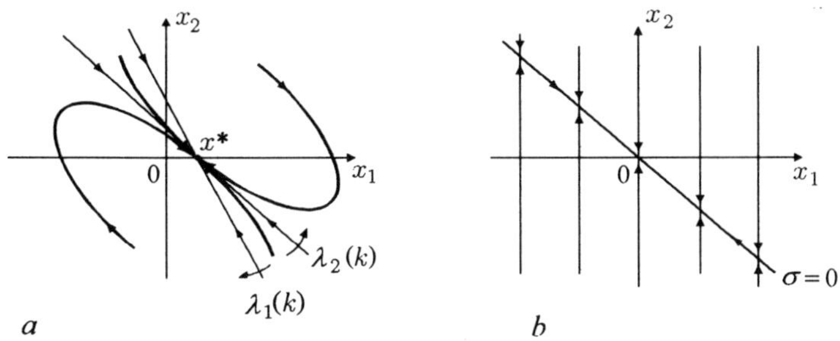

Figure 1.46.

and $\lambda_2(k)$ are zeros of the characteristic polynomial $\varphi_k(s)$ which define partial motions and $x^* = A/(kc + a_1)$ is a static error. As $k \to \infty$, the straight line $\lambda_1(k)$ tends to the straight line $\sigma = 0$, the straight line $\lambda_2(k)$ assumes a vertical position, and the point x^* tends to zero. As a result, we obtain a phase portrait of the limit system shown in Fig. 1.46*b*.

1.7.6. The effect produced by an amplitude constraint on systems with a high gain factor

Suppose that there is a constraint on the amplitude of the signal in the control channel (Fig. 1.47). Then, in contrast to the preceding case, a qualitatively new situation occurs when feedback with high gain is used. We shall explain this. Let sat$(\cdot)$ describe a saturation function or a linear zone with restrictions (Fig. 1.48*a*). If $u = k\sigma$, then the function $v(\sigma)$ is similar to sat(u) but has a changed zone of linearity (Fig. 1.48*b*). It is clear that the function sat$(k\sigma)$ becomes an ideal relay function as $k \to \infty$ (Fig. 1.48*c*). As a result, instead of the smooth control $u = k\sigma$ we get a discontinuous control $u = \operatorname{sgn} \sigma$ and instead of a linear system we get a relay system whose fragment is shown in Fig. 1.49.

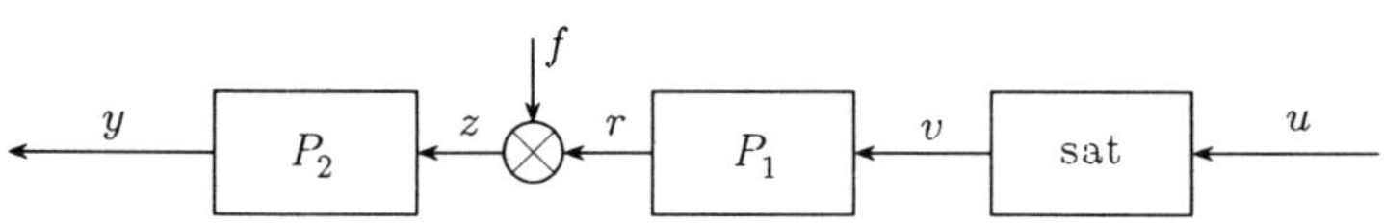

Figure 1.47.

The behavior of relay systems differs essentially from the behavior of linear systems. We consider some important peculiarities of the behavior of relay systems in the next chapter.

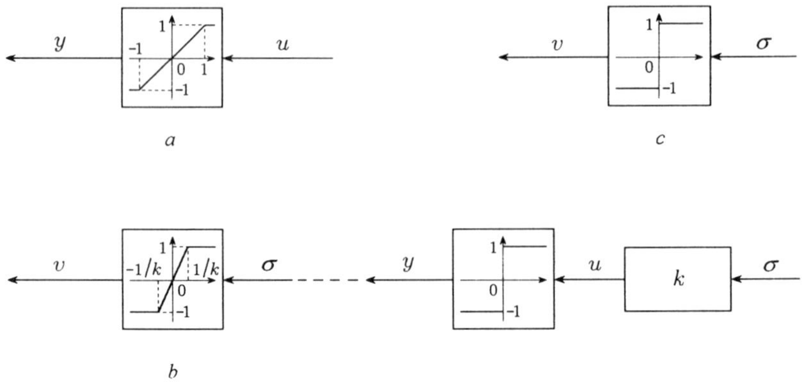

Figure 1.48.

To sum up the discussion of this theme, we can make the follosing conclusion:

- there is no linear control which would robustly solve the stabilization problem when there exists an unknown external disturbance.

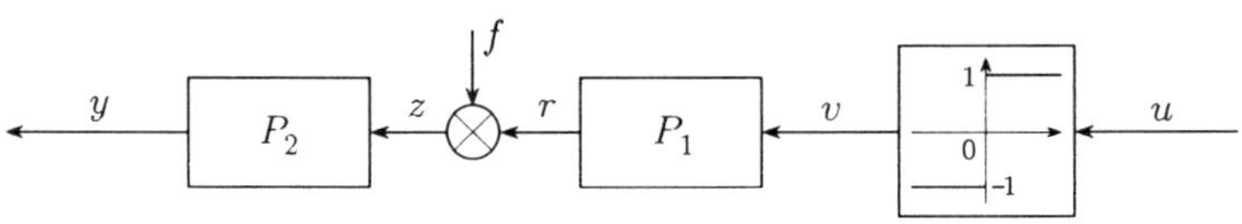

Figure 1.49.

1.8. Bibliographical Comments

The papers [13, 56, 58, 77, 78, 95–97] laid the theoretical groundwork for the classical automatic control theory. Indeed, the English scientist J. Maxwell [95] was the first to carry out a mathematical investigation of the problem of stability of the "steam engine-controller" couple and to formulate the criterion of asymptotic stability of a third-order linear equaton.

In his article [13] the Russian scientist I.A. Vyshnegradskii refined Maxwell's model and carried out an exhaustive qualitative analysis of an arbitrary three-dimensional system. He was the first to pose the problem concerning the quality of the transient process in an automatic control system and describe in terms of mathematics the boundaries of domains in the space of parameters of the equation of a system with different types of qualitative behavior of its solutions. He also gave practical recommendations for the adjustment of J. Watt's controller.

The Slovac engineer A. Stodola formulated the necessary conditions for the stability of an arbitrary polynomial and drew the attention of the German mathematician A. Hurwitz to the problem of necessary and sufficient conditions for stability. Hurwitz solved this problem in 1895 with the use of Ch. Hermite's results. However, even earlier, in 1877, the equivalent conditions of stability were found by E.J. Routh [97].

In his dissertation [56], A.M. Lyapunov gave the most general methods of investigation of stability and the principal concepts of the modern theory of the stability of motion.

The frequency methods and block diagrams became widely used in control theory after the works of H. Nyqwist [96], H. Black [77], H. Bode [78], and A.V. Mikhailov [58]. The scientists carried out an exhaustive investigation of the stability of feedback systems using the frequency characteristics of the object and the controller.

The main statements of problems and the basic principles of the automatic control theory have roots stretching far back. It is believed that the American M. Boulton in 1788 and the Frenchman J.-V. Poncelet in 1826 brought into general use the principle of feedforward control. The authorship of the feedback principle is dated 1788 and is attributed to J. Watt. The integral component and feedback were introduced by the Frenchman L. Molinie in 1837, and the derivative was introduced by Germans, brothers William & Werner Siemens in 1844. The load control was proposed, for the first time, by C.T. Porter in 1908, the relay feedback was apparently used for the first time by C.L. Schofield as early as 1836, and proportional control was used in 1870 by the American D.A. Woodbury.

The reader can find the contemporary interpretation of the direct control methods in [1, 6, 11, 12, 64, 67]. The double-channel principle is described in [61, 62, 67]. Kulebakin's method is described in detail in [49, 52]. The theory and methods of a high gain factor were apparently systematically exposed for the first time in [57].

Chapter 2

Synthesis of Nonlinear Controllers

Relay systems constitute the simplest type of principally nonlinear dynamical systems. They are described by differential equations with discontinuous characteristics and are therefore often called discontinuous systems. In the automatic control theory, the interest in the investigation of discontinuous systems is due to the use of servomechanisms of the full speed ahead–full speed astern type and also because of amplitude constraints imposed on signals.

In this chapter we consider some special problems of the theory of relay systems which are of interest in the general context of the monograph and, first of all, in connection with the analysis of the possibilities of relay stabilization in the conditions of uncertainty.

2.1. Relay Feedback

2.1.1. Basic concepts

A relay system is a system whose block diagram includes at least one relay element (RE). As a rule, in the automatic control theory, the feedback shown in Fig. 2.1 is called a relay feedback. In this figure, the sign (sgn) is a nonlinear operation which establishes a relationship between the control u and the output

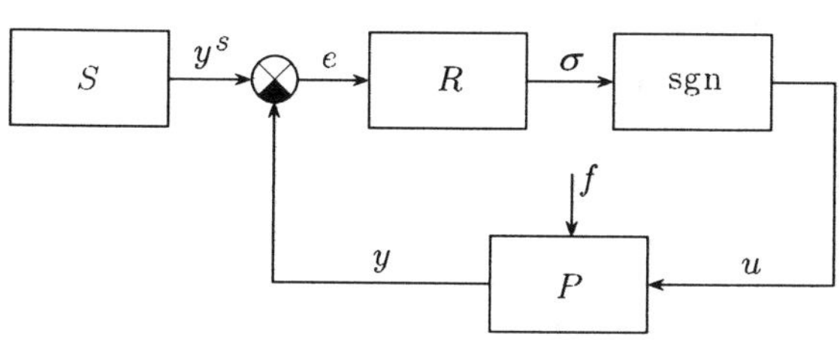

Figure 2.1.

σ of the controller (regulator R) according to the relation $u = \operatorname{sgn}\sigma$. If there are several relay elements, then the system is called a cascade relay system.

A characteristic feature of a relay is discontinuity, and this brings about an essential peculiarity in the methods of analysis and synthesis of relay systems and also affects their qualitative behavior and functional possibilities. Typical models of a relay are shown in Fig. 2.2.

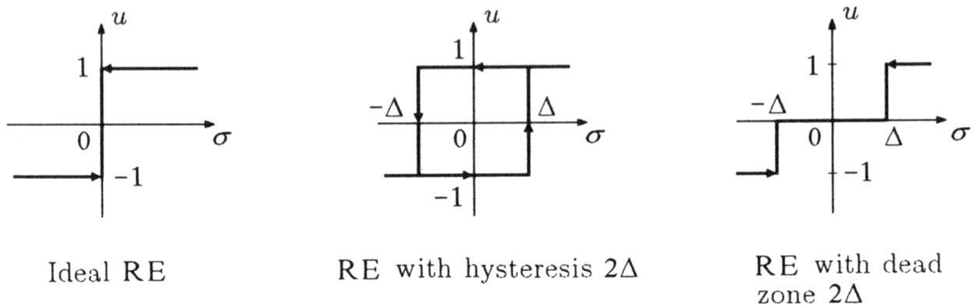

Figure 2.2.

A necessity of considering a relay system arises when an actual drive is used in the control system (a power drive of the "full speed ahead–full speed astern" type) or when a relay system serves as a convenient abstraction, say, as a limit system (Fig. 2.3) for a system with high gain feedback when there exists an amplitude constraint in the control channel.

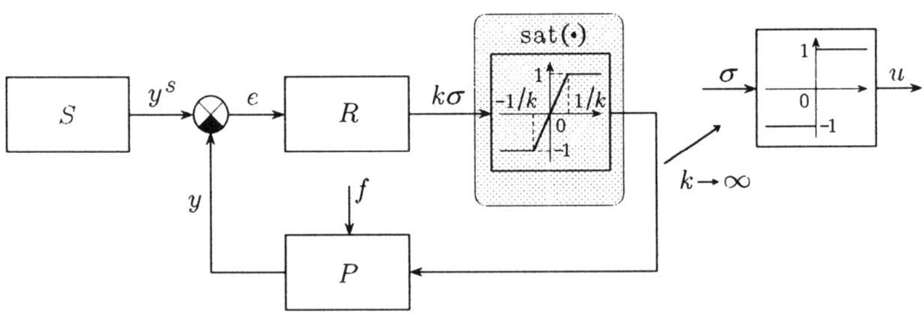

Figure 2.3.

Example 8 (The simplest relay system). Let us consider a simple first-order tracking system shown in Fig. 2.4.

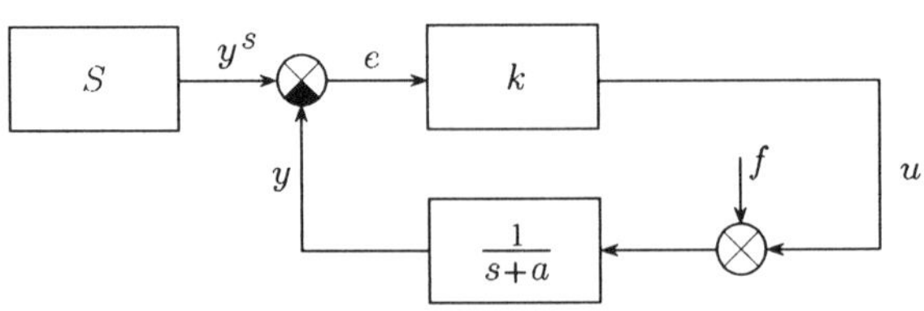

Figure 2.4.

The motion of this system can be described by equations

$$\begin{aligned} \dot{y} + ay &= u + f, && \text{an object} \\ u &= ke, && \text{a control} \\ e &= y^s - y, && \text{tracking error.} \end{aligned} \tag{2.1}$$

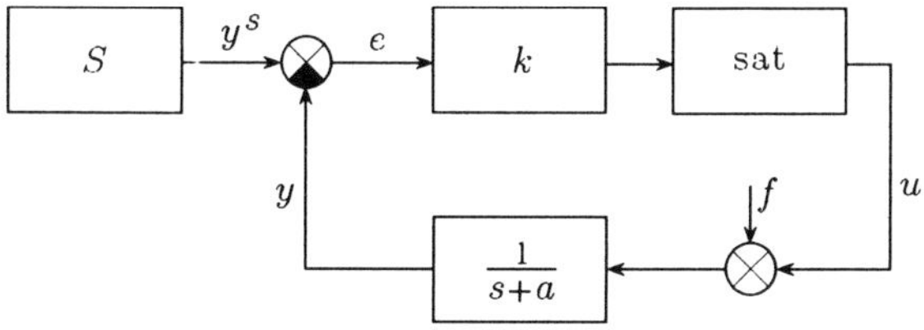

Figure 2.5.

In order to analyze the properties of system (2.1), it is convenient to rewrite the equation of its motion relative to the "error". Direct calculations yield

$$\dot{e} = \dot{y}^s - \dot{y} = \dot{y}^s - u - f - a\,y\big|_{y=y^s-e} = -(k+a)\,e + \dot{y}^s + a\,\dot{y}^s - f. \tag{2.2}$$

Let $e(t,k)$ be a solution of Eq. (2.2). It is clear that

$$\lim_{k\to\infty} e(t,k) = 0$$

for any parameter a, a "practically" arbitrary reference action $y^s(t)$, and disturbance $f(t)$. Suppose now that a constraint of the type sat$(\cdot)$, i.e., a saturation, exists in the feedback channel of the system under consideration (Fig. 2.5). Then, as $k \to \infty$, a qualitatively new situation arises, and the "equation in errors" has the form

$$\dot{e} = -a\,e - \text{sat}(ke) + F,$$

where, for convenience, we use the notation

$$F = \dot{y}^s + a\,y^s - f.$$

In the limit, as $k \to \infty$, we deal with a relay system

$$\dot{e} = -ae - \text{sgn}\,e + F. \tag{2.3}$$

System (2.3) differs significantly from system (2.2). It is clear that for an unstable object (i.e., when $a < 0$), the tracking problem can be solved by system (2.3) only if the inequality

$$|ae + F| < 1$$

is satisfied, in other words, only under definite initial conditions and only for a constrained disturbance F. Consequently,

- as compared to a linear feedback, a relay feedback stabilizes smaller classes of objects and only under disturbances with constraints imposed on their amplitude.

2.1.2. Sliding mode at a point

A sliding mode is often observed in relay systems. For clarity, let us consider the following example:

$$\dot{e} + ae = -\operatorname{sgn} e, \quad a = \text{const}. \tag{2.4}$$

The phase space of system (2.4) is one-dimensional (Fig. 2.6).

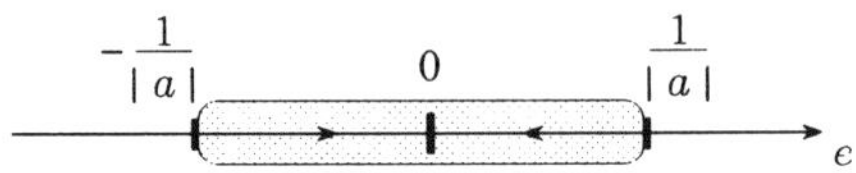

Figure 2.6.

It is easy to see from relation (2.4) that an unstable object (i.e., when $a < 0$) is stabilized at zero under any initial conditions $e(0)$ from the interval $\left(-\frac{1}{|a|}, \frac{1}{|a|}\right)$ whereas a stable object (i.e., when $a \geq 0$) is stabilized at zero throughout the straight line under any initial conditions $(-\infty, \infty)$. In the neighborhood of the point 0, the phase trajectories of system (2.4) are directed towards each other, and, consequently, the phase point cannot leave the point 0. The solution $e = 0$ is not "classical" and must be understood in some other sense, for instance, in the sense of Filippov [72]. This solution is associated with infinitely frequent switchings of the relay. These switchings are associated with a sliding mode.

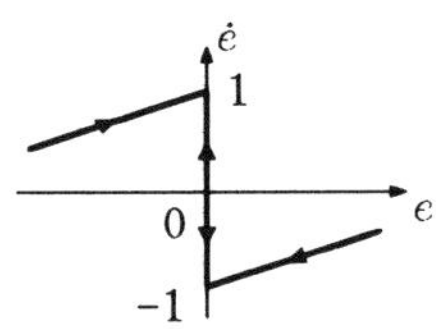

Figure 2.7.

Let us investigate the sliding mode in system (2.4) in greater detail. We shall represent it geometrically as a manifold in the plane $(e, \dot{e})$ (Fig. 2.7). On the left of zero the motion is along the straight line $\dot{e} + ae = 1$ and on the right of zero the motion is along the straight line $\dot{e} + ae = -1$. The interval $[1, -1]$ on the axis $e = 0$ is a segment of the sliding mode when the motion equation changes

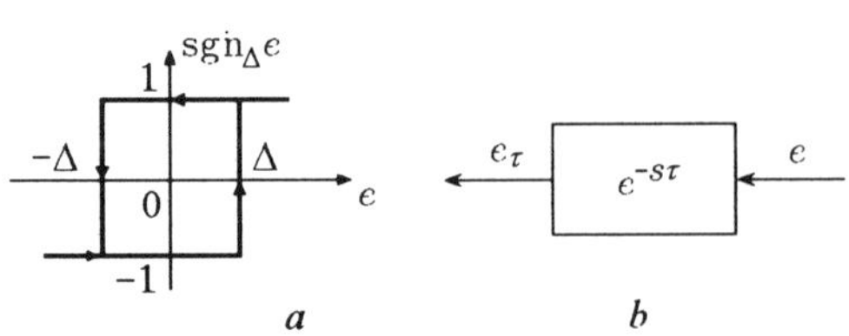

Figure 2.8.

in a jump. Let us find out what will happen to the system when a delay occurs in the switching of the relay. It may be a space or time delay. In the former case (Fig. 2.8*a*), the relay has a hysteresis loop 2Δ wide and is denoted by $\mathrm{sgn}_\Delta\, e$ and in the latter case (Fig. 2.8*b*) the switching occurs in time τ after the input signal changes sign. We denote the corresponding element by $\mathrm{sgn}_\tau\, e = \mathrm{sgn}\, e_\tau$, where τ is a time delay constant.

The motion equations for a relay system with a delay are $\dot{e} + ae = -\,\mathrm{sgn}_\Delta\, e$ and $\dot{e} + ae = -\,\mathrm{sgn}_\tau\, e$. Figures 2.9*a* and 2.9*b* correspond to these equations. Although the delay phenomena differ essentially in the two cases, qualitatively the behavior of the system in cases *a* and *b* in Fig. 2.9 is the same, namely, the sliding mode at a point is violated and a switching mode arises with a limit

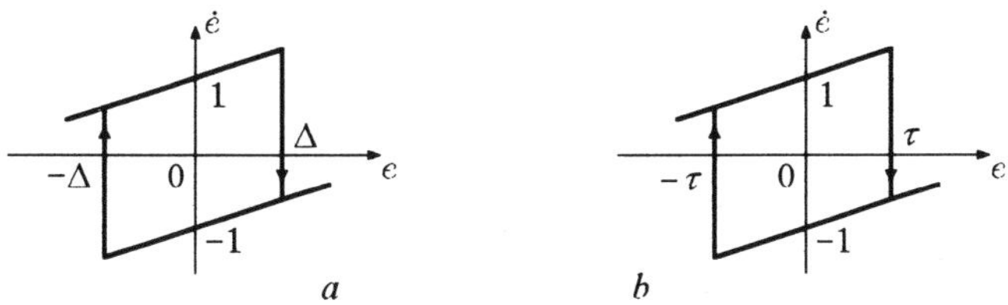

Figure 2.9.

cycle in the neighborhood of zero whose "dimension" with respect to the input variable is proportional to the constant Δ or τ which characterizes the delay. This allows us to make the following conclusion:

- the sliding mode at a point is not robust and, consequently, the relay stabilization problem does not have a robust solution.

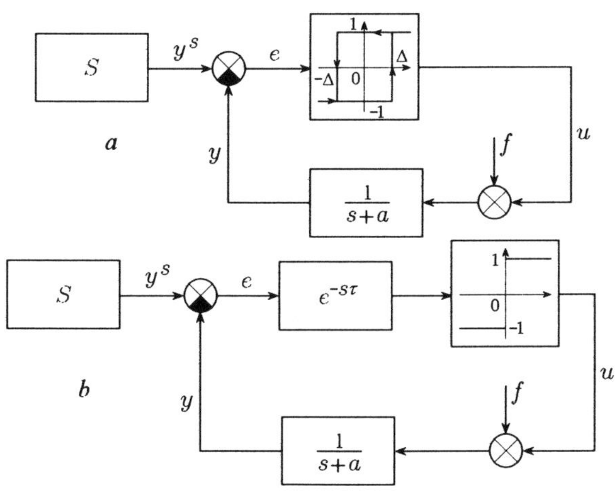

Figure 2.10.

In conclusion, we shall give the block diagrams of a relay system with a space (Fig. 2.10*a*) and time (Fig. 2.10*b*) delays in the switchings.

Example 9 (Relay stabilization of a second-order system). Let us consider a more complicated problem, namely, let us investigate the peculiarities of a relay feedback which stabilizes a second-order object (Fig. 2.11). The motion equation of such a system for $y^s \equiv 0$ obviously has the form

$$\ddot{e} = -\operatorname{sgn} e - f. \tag{2.5}$$

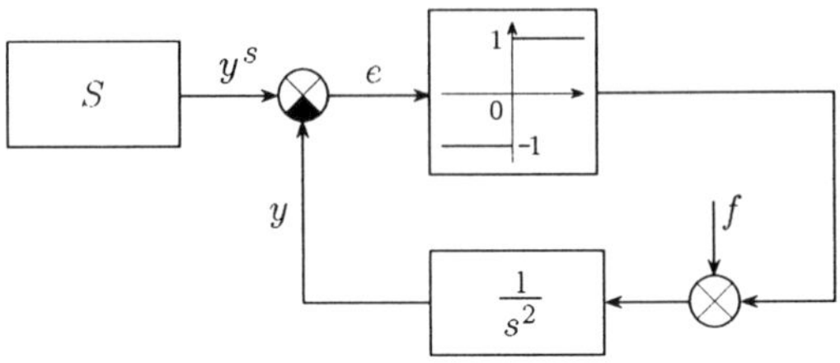

Figure 2.11.

Clearly, the necessary condition of stabilizability is the inequality $|f| < 1$ which restricts the class of admissible disturbances. If this condition is fulfilled, then, without loss of generality, we can consider the homogeneous equation

$$\ddot{e} = -\operatorname{sgn} e \tag{2.6}$$

instead of (2.5).

2.1.3. Switching mode

The phase portrait of system (2.6) is formed by "sewing" the phase trajectories from two half-planes

$$\ddot{e} = -1 \quad (I), \qquad \ddot{e} = 1 \quad (II)$$

along the axis $e = 0$. Phase trajectories are segments of parabolas with vertices on the axis $\dot{e} = 0$ (Fig. 2.12), and therefore system (2.6) is conservative.

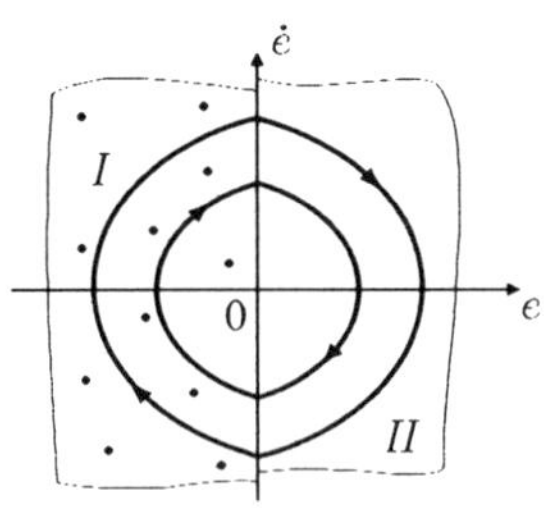

Figure 2.12.

The changes that occur in the phase portraits when there is a space or time delay are shown in Figs. 2.13a and 2.13b respectively. We can see from these figures that instability inevitably occurs in the systems under consideration. In other words, relay systems with a high relative order are nonrobust. Recall that a relative order of a linear stationary object is equal to the difference between the powers of polynomials of the denominator and numerator of a transfer function.

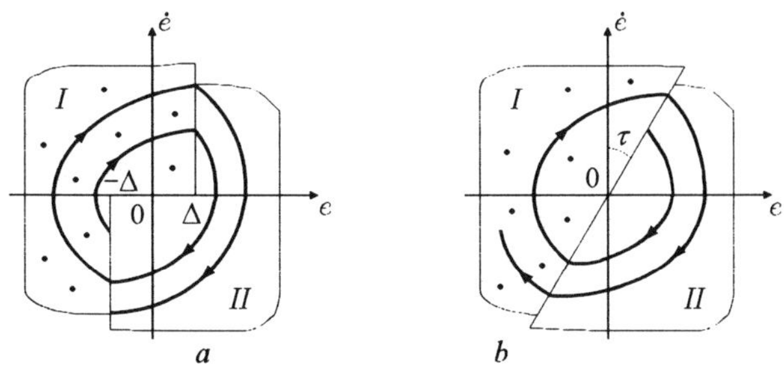

Figure 2.13.

We can make the following conclusion from Example 9:

- in order to make a relay system robust, we must lower the relative order of the transfer function from the input of the relay element to its output to unity.

Example 10 (Relay stabilization of an object with a relative order equal to two). Let us consider a relay system of stabilization of an object with a relative order equal to two (Fig. 2.14). Again, without loss of generality, we set $y^s \equiv 0$. Then the motion equation

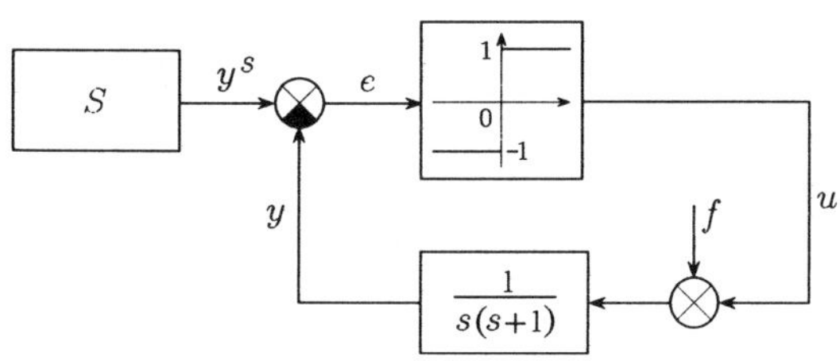

Figure 2.14.

of the system shown in Fig. 2.14 relative to the error coordinate has the form

$$\ddot{e} + \dot{e} + \operatorname{sgn} e = -f. \tag{2.7}$$

The necessary condition for the stabilizability at zero is given by the obvious inequality $|f| < 1$. Together with stability at zero of the equation

$$\ddot{e} + \dot{e} + \operatorname{sgn} e = 0, \tag{2.8}$$

this inequality guarantees the stabilizability of system (2.7) at zero.

To analyze the stability of Eq. (2.8), we shall use geometrical representations, i.e., the phase plane method. The phase trajectories of Eq. (2.8) for $e > 0$ defined by the equation

$$\ddot{e} + \dot{e} + 1 = 0$$

and for $e < 0$ when we have an equation

$$\ddot{e} + \dot{e} - 1 = 0$$

are shown in Figs. 2.15*a* and 2.15*b* respectively, The "sewing" of phase trajectories results in the phase portrait of the system shown in Fig. 2.15*c*. The analysis of this phase portrait shows

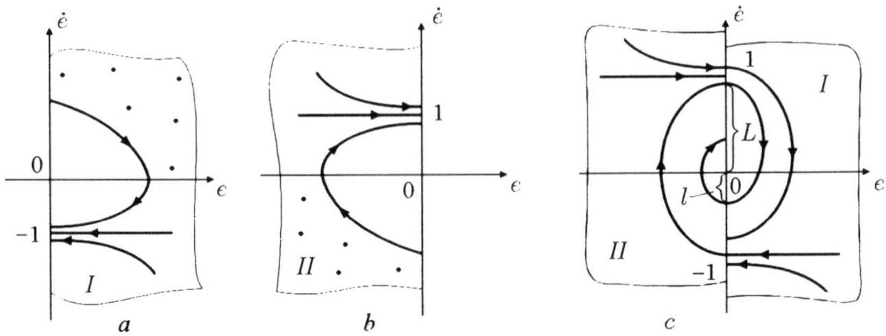

Figure 2.15.

that $l/L < 1$, and, hence, each trajectory "twists" to the origin. We can also establish this fact analytically, with the use of the Lyapunov function

$$v = |e| + \frac{\dot{e}^2}{2}.$$

By virtue of Eq. (2.8), its derivative has the form

$$\dot{v} = \dot{e}\,\mathrm{sgn}\,e + \dot{e}\ddot{e} = \dot{e}\,\mathrm{sgn}\,e + \dot{e}(-\dot{e} - \mathrm{sgn}\,e) = -\dot{e}^2.$$

Since the manifold $\{\dot{e} = 0\} \setminus \{0\}$ does not contain integer trajectories, the zero of Eq. (2.8) is asymptotically stable.

2.1.4. On the robustness of the switching mode

Let us make sure that the stability of zero, $\ddot{e} + \dot{e} + \mathrm{sgn}\,e = 0$, of Eq. (2.8) is not preserved when a space or time delay occurs in the switching. Indeed, suppose that a space delay of magnitude Δ occurs. Then we can describe the behavior of the system by the equation

$$\ddot{e} + \dot{e} + \mathrm{sgn}_\Delta\, e = 0 \tag{2.9}$$

rather than by Eq. (2.8) and illustrate it by Fig. 2.16. After "sewing" the phase

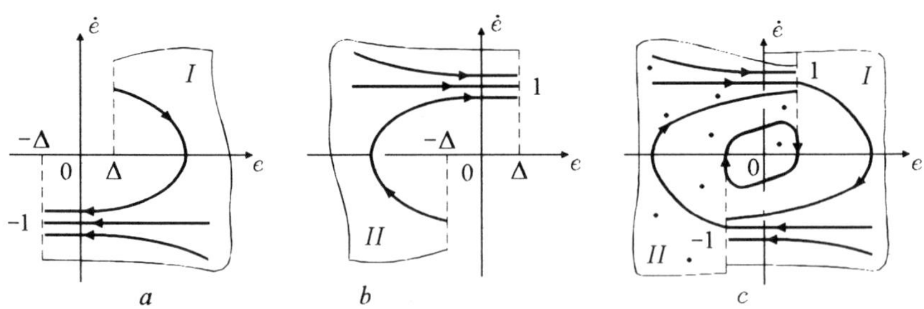

Figure 2.16.

trajectories along the lines of discontinuity, we get a summarizing phase portrait of Eq. (2.9) in which we can easily see the existence of a limit cycle (Fig. 2.16*c*). Consequently, the relay stabilization system (2.8) is nonrobust since its behavior changes qualitatively when the conditions of the problem vary.

2.1.5. Relay stabilization of an object with self-regulation

It is interesting to find out whether the conclusion made in the preceding section remains valid for simpler objects. Let us consider the capabilities of a relay feedback when an object with self-regulation is stabilized, i.e., an object which is asymptotically stable under zero control. For simplicity, we shall only consider the object shown in Fig. 2.17. For $y^s \equiv 0$ we must analyze the equation

$$\ddot{e} + \dot{e} + e + \operatorname{sgn} e = -f.$$

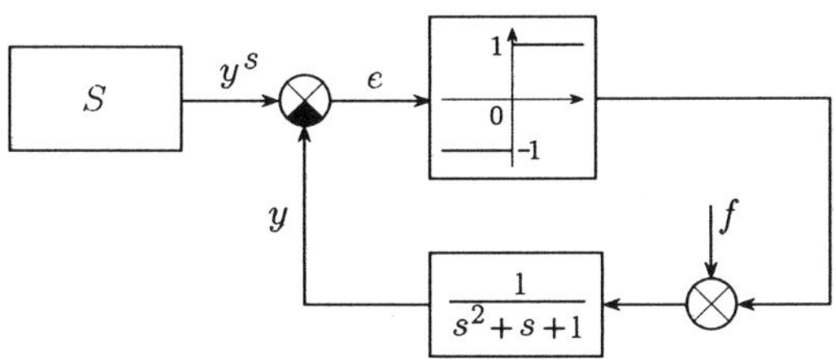

Figure 2.17.

It is clear that when the condition $|f| < 1$ is fulfilled, the investigation of the stabilizability reduces to the analysis of stability of free oscillations of the control system described by the equation

$$\ddot{e} + \dot{e} + e + \operatorname{sgn} e = 0. \tag{2.10}$$

With this aim in view, we again use geometrical representations, i.e., a phase plane. The phase trajectories of Eq. (2.10) consist of segments of phase trajectories of the equation

$$\ddot{e} + \dot{e} + e + 1 = 0$$

which acts for $e > 0$ (Fig. 2.18*a*) and phase trajectories of the equation

$$\ddot{e} + \dot{e} + e - 1 = 0$$

which acts for $e < 0$ (Fig. 2.18*b*). After "sewing" the trajectories along the lines of discontinuity $e = 0$, we obtain a summarizing phase portrait (Fig. 2.18*c*) in which we can see a tendency of the trajectories of the system to twist to the origin. This is due to the fact that $l/L < 1$. The analytical investigation of the stability of Eq. (2.10) can be carried out with the use of the Lyapunov function $v = \dot{e}^2 + e^2 + |e|$ whose derivative $\dot{v} = -\dot{e}^2$.

If a space delay Δ occurs in the switchings of Eq. (2.10), the phase portraits undergo natural changes and assume the form shown in Fig. 2.19. In Fig. 2.19*c* we can observe a limit cycle in the neighborhood of zero, which is equivalent to the nonrobustness of the relay system under investigation.

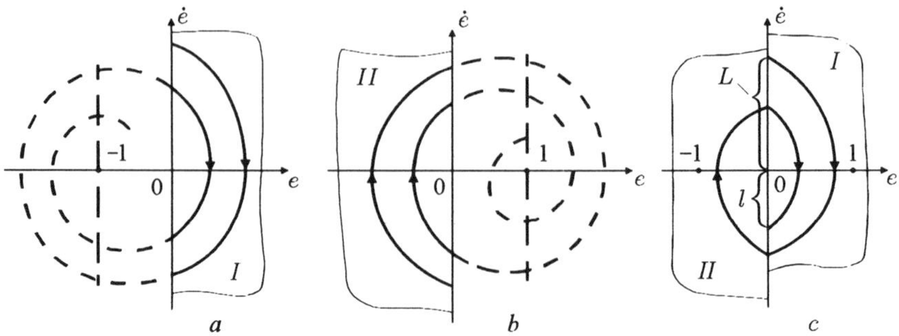

Figure 2.18.

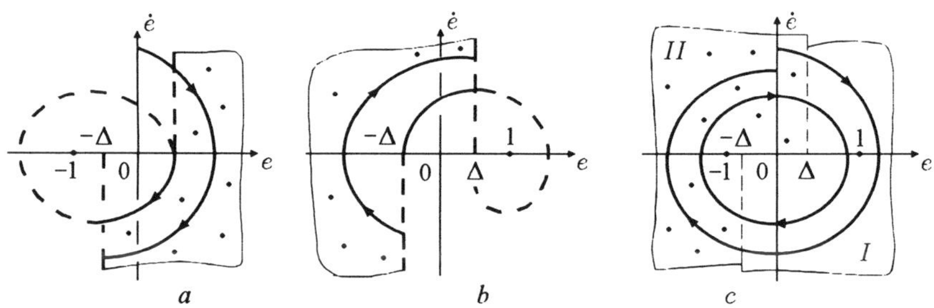

Figure 2.19.

2.1.6. Stabilization of an object with a high relative order

Consider an arbitrary linear object with a relative order $r = n - m$, where $r \geq 3$, n and m are the orders of polynomials in the denominator and numerator respectively (for instance, the object shown in Fig. 2.20). For $y^s \equiv 0$, we must investigate the stability of zero of the equation

$$e^{(n)} + a_n e^{(n-1)} + \cdots + a_1 e + \operatorname{sgn} e = -f.$$

Again, under the condition

$$|f| < 1$$

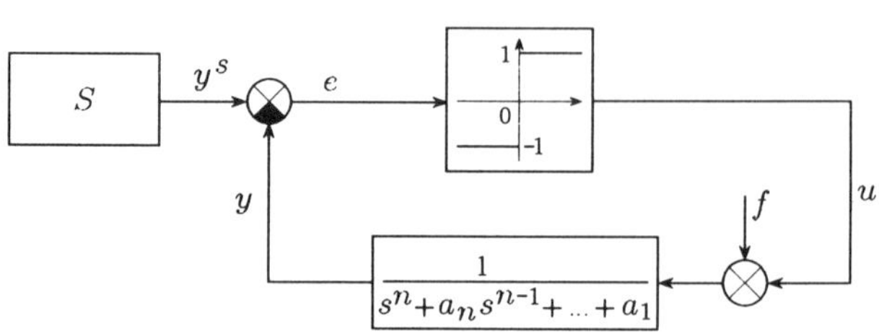

Figure 2.20.

everything reduces to the stability of free oscillations

$$e^{(n)} + a_n e^{(n-1)} + \cdots + a_1 e + \operatorname{sgn} e = 0. \tag{2.11}$$

However, the zero of the last equation is always unstable for $n \geq 3$, which follows from the instability of the characteristic equation corresponding to the situation that is observed after the replacement of the relay element $\operatorname{sgn} e$ by a linear gain factor ke for a sufficiently large value of the gain factor k. We proved above that such a replacement during the analysis of the stability of zero was valid.

2.1.7. Robust stabilization: discontinuity, continuity, and information about the state

The cases considered above allow us to trace the obvious relationship between continuity, discontinuity, and the amount of information necessary for the stabilization of linear objects. Here are typical examples. Let us stabilize the object

$$\dddot{e} + a_2\ddot{e} + a_1\dot{e} = u \tag{2.12}$$

with the aid of linear feedback $u = -ke$ when the relations $a_1a_2 > k > 0$, $a_1 > 0$ are satisfied. For none of the values of the parameters a_1, a_2 can the object (2.12) be stabilized at zero by the relay feedback $u = -\operatorname{sgn} e$. However, if we introduce a priori information about the derivatives $\dot{e}$ and $\ddot{e}$, say,

$$u = -\operatorname{sgn}(e + \dot{e}), \quad u = -\operatorname{sgn}(e + \dot{e} + \ddot{e}),$$

into the law of relay feedback, then the stability of zero for $a_1, a_2 > 0$ will be guaranteed. In other words,

- there exist situations when the transition from a discontinuous element to a continuous one, in particular, to a linear element makes it possible to solve the problem when the information that we have about its state is smaller in volume.

A similar relationship exists between the robustness of the system and the continuity of its nonlinear elements. We can elucidate this fact by considering the equation

$$\dot{e} + ae + \operatorname{sgn} e = 0, \quad a = \text{const} > 0.$$

In the plane $(e, \dot{e})$ it is associated with Fig. 2.21*a* with a sliding mode at zero. When there is a delay τ in the switching of the discontinuous element, we deal with the equation

$$\dot{e} + ae + \operatorname{sgn} e_\tau = 0, \quad e_\tau = e(t - \tau),$$

which is associated with Fig. 2.21*b*. In this figure, we can see the origination of a switching mode with a limit cycle. When the delay τ is sufficiently small, we can resort to the approximation $e_\tau \cong e - \tau\dot{e}$ and, instead of $\operatorname{sgn} e_\tau$, use a discontinuous element of the form $\operatorname{sgn}(e - \tau\dot{e})$, which does not change the qualitative pattern of

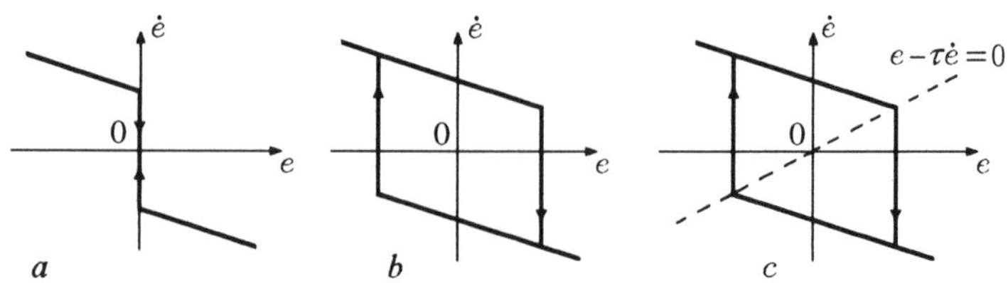

Figure 2.21.

the phenomenon (Fig. 2.21c). The situation changes qualitatively if we replace an ideal relay by a linear zone with a constraint, i.e., pass from a discontinuous element to a continuous one, say, to a saturation type element.

Then, for analyzing the stability at zero, we must use the linear equation

$$\dot{e} + a\,e + k\,e_\tau = 0, \quad k = \mathrm{const} > 0,$$

which is associated with a characteristic quasipolynomial $s + a + k\,e^{-s\tau} = 0$. For a sufficiently large value of the gain factor k, this equation always has zeros $\lambda(k)$ in the right complex half-plane which tend to the zeros of the equation $e^{-s\tau} = 0$ as $k \to \infty$. These zeros obviously have positive real parts. On the other hand, if k is finite and τ is small, then the analysis of stability can be reduced to the investigation of the first-order polynomial $(1 - k\tau)s + a + k = 0$ by means of the approximation $e^{-s\tau} \cong 1 - s\tau$. It is obvious that if the inequalities $1 - k\tau > 0$, $a + k > 0$ are satisfied, stability occurs, i.e., the limit cycle disappears.

Summing up the investigation, we can infer that the relay stabilization is suitable only for objects with a relative order $r \leq 2$. Since we considered stabilization for $r = n - m = 2$ above, we can now investigate the peculiarities of relay stabilization for $r = n - m = 1$.

2.1.8. Robust stabilization of an object of the first relative order

Consider a relay control system whose block diagram is shown in Fig. 2.22. In this case, the relative order of the object is $r = m - n = 2 - 1 = 1$. In order to introduce state variables, it is convenient to represent the control object as a

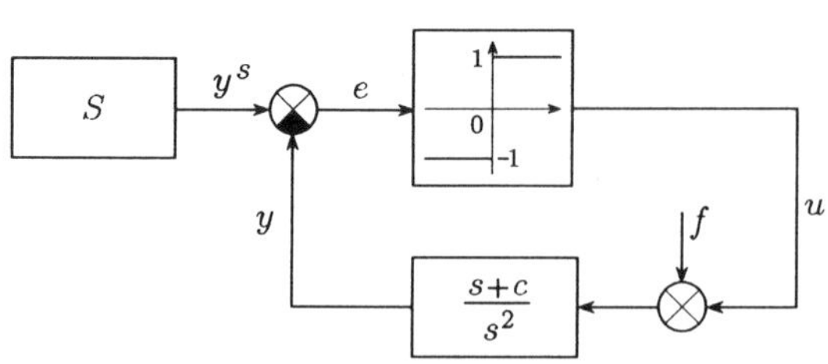

Figure 2.22.

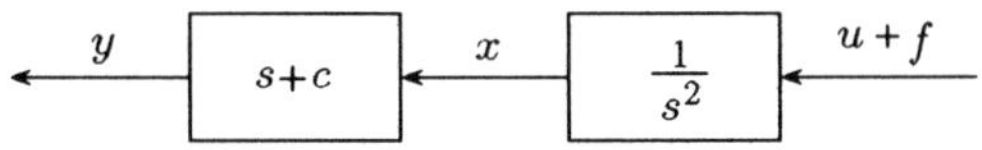

Figure 2.23.

block diagram shown in Fig. 2.23. Then the behavior of the control object can be described by the equations

$$\ddot{x} = u + f, \quad y = \dot{x} + cx, \quad c = \text{const}.$$

After introducing the state variables defined by the relations $x_1 = x$, $x_2 = \dot{x}$, under the weak condition $y^s \equiv 0$, we obtain the motion equation of the system in question in the standard form

$$\begin{aligned} \dot{x}_1 &= x_2, \\ \dot{x}_2 &= -\operatorname{sgn}(x_2 + cx_1) + f. \end{aligned} \tag{2.13}$$

If the condition $|f| < 1$ is fulfilled, the investigation of stabilizability of system (2.13) reduces to the analysis of the stability of zero of the equations

$$\begin{aligned} \dot{x}_1 &= x_2, \\ \dot{x}_2 &= -\operatorname{sgn}(x_2 + cx_1). \end{aligned} \tag{2.14}$$

In order to investigate this system, we construct its phase portrait which results from the "sewing" along the line $\sigma = x_2 + cx_1 = 0$ of phase trajectories of two systems

$$\Sigma_1 =: \begin{cases} \dot{x}_1 = x_2, \\ \dot{x}_2 = 1, \end{cases} \qquad \Sigma_2 =: \begin{cases} \dot{x}_1 = x_2, \\ \dot{x}_2 = -1 \end{cases}$$

acting in the half-planes $\sigma < 0$ and $\sigma > 0$ respectively.

2.1.9. Sliding mode on an interval

The trajectories of the systems Σ_1, Σ_2 are parabolas which, for $c > 0$, are located as shown in Fig. 2.24. Two points A and B marked in the figure are the points

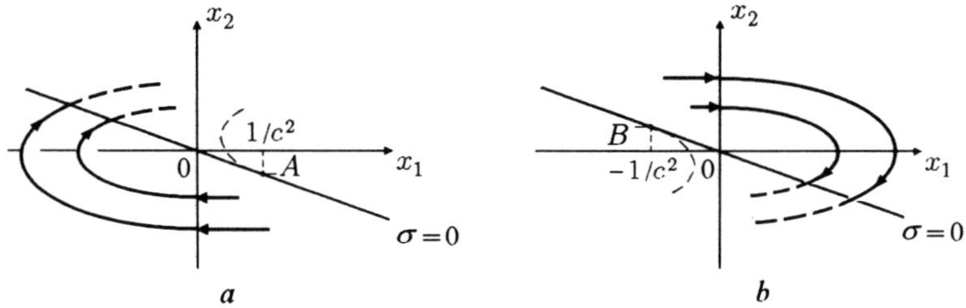

Figure 2.24.

where the phase trajectories of the systems Σ_1 and Σ_2 touch the straight line $\sigma = 0$. The coordinates of these points can be found from the conditions of tangency ($\sigma = 0$, $\dot{\sigma} = 0$): $A = \left(\frac{1}{c^2}, -\frac{1}{c}\right)$, $B = \left(-\frac{1}{c^2}, \frac{1}{c}\right)$. After "sewing" the phase trajectories, we obtain the phase portrait shown in Fig. 2.25.

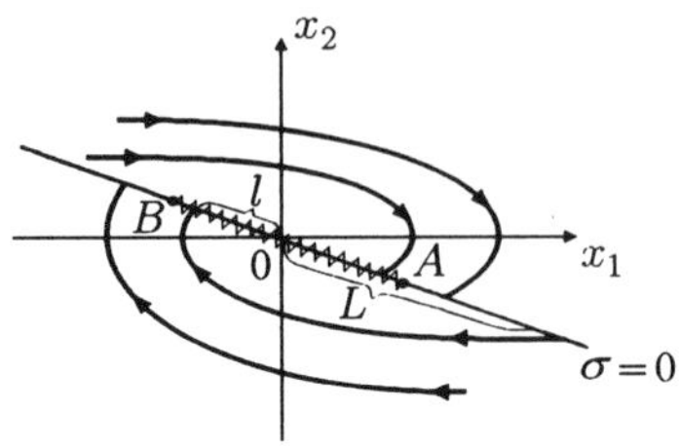

Figure 2.25.

We can infer from Fig. 2.25 that

- the ratio $l/L < 1$, and, consequently, the representative point gets into the interval AB in a finite number of switchings of the relay;
- on the interval AB an "opposite" motion of representative points occurs along the phase trajectories of systems Σ_1 and Σ_2, and, consequently, the representative point cannot leave the interval AB, i.e., a sliding mode originates.

In order to find the equation of motion in a sliding mode, we shall carry out an heuristic reasoning. In a sliding mode

$$\sigma = x_2 + cx_1 = 0,$$

but, by definition,

$$x_1 = x, \quad x_2 = \dot{x}_1 = \dot{x}.$$

Hence, we can take an equation of the form

$$\dot{x} + cx = 0$$

as an equation for sliding. For $c > 0$, it is exponentially stable, and this entails the stability of the relay system under consideration.

2.1.10. Real sliding mode on an interval

Suppose that there is a space delay Δ in switching the relay in the relay system (2.14), i.e., we actually deal with a system

$$\Sigma^\Delta =: \begin{cases} \dot{x}_1 = x_2, \\ \dot{x}_2 = -\operatorname{sgn}_\Delta(x_2 + cx_1). \end{cases} \tag{2.15}$$

In order to obtain its phase portrait, we must superimpose the phase portraits of systems Σ_1^Δ (Fig. 2.26*a*) and Σ_2^Δ (Fig. 2.26*b*). As a result, we shall have the phase portrait of system (2.15) which is given in Fig. 2.27.

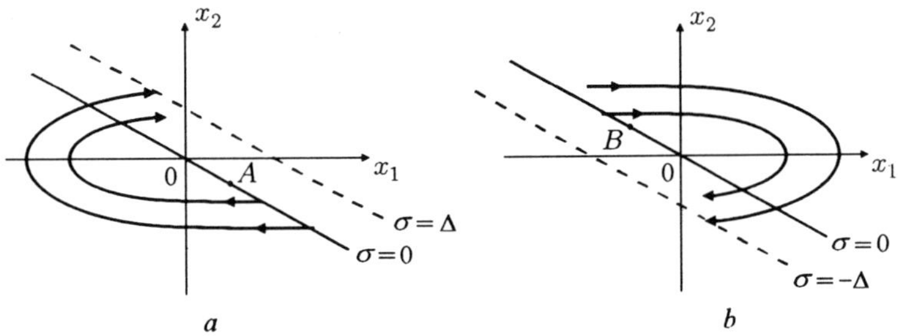

Figure 2.26.

Analyzing Fig. 2.27, we see that

- a limit cycle appears in the neighborhood of the origin,
- the sliding mode turns into a switching mode called a real sliding mode; it is characterized by a finite frequency of switchings of the relay and a finite "amplitude" of deviation of the point that represents it from the line $\sigma = 0$.

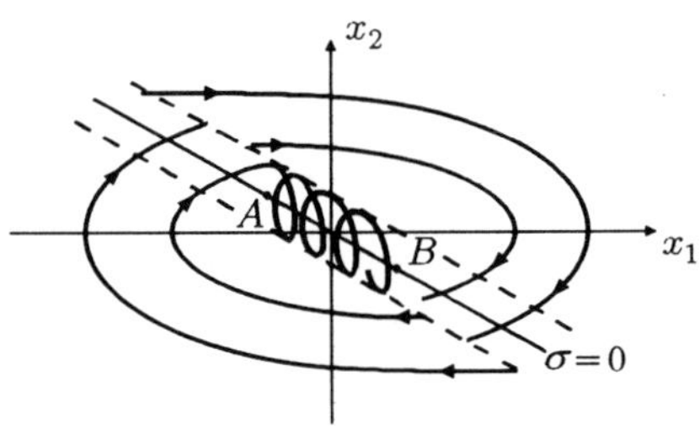

Figure 2.27.

Consequently, just as all the systems considered before, the relay control system in question is not robust, i.e., when the switchings are nonideal, the asymptotically stable system turns into a dissipative system.

2.1.11. Relay stabilization of a generalized object

Let us consider an arbitrary object with a relative order equal to unity and a relay system shown in Fig. 2.28 that controls this object. The transfer function $W(s) = b(s)/a(s)$ shown by the block diagram in Fig. 2.28 is given by the ratio of two polynomials

$$\begin{aligned} b(s) &= b_n s^{n-1} + b_{n-1} s^{n-2} + \cdots + b_1, \\ a(s) &= s^n + a_n s^{n-1} + \cdots + a_1, \end{aligned}$$

whence it follows that under the condition $|f| < 1$ the stability of the system at zero is possible only when the polynomial $b(s)$ is Hurwitz and the polynomial $a(s)$ has no zeros in the right half-plane of the variable s and no more than two of its zeros lie on the imaginary axis in the plane of the variable s. Objects of this kind are known as objects with minimal phase properties.

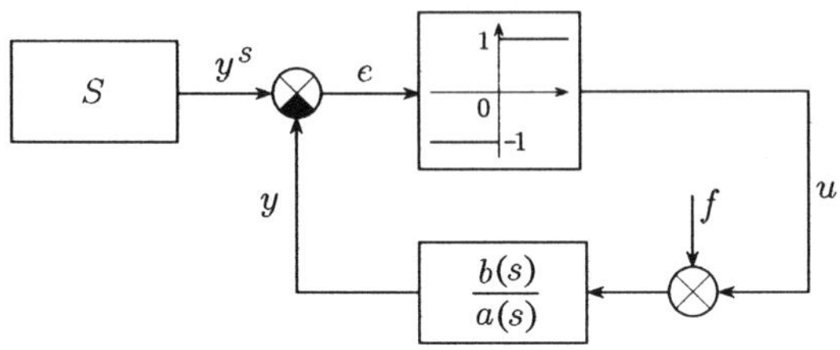

Figure 2.28.

We can infer from what was said above that a relay feedback is stabilizing if

- the object has minimal phase properties,
- the object has a relative order $r \leq 2$,
- the reference action signal and the disturbance are uniformly bounded.

Moreover, relay systems are not robust and their solutions and, hence, their properties depend on disturbances.

2.2. Stabilization of an Object with an Uncertain Operator

A qualitatively new situation in a control object is created by an uncertainty, in particular, a parametric uncertainty. In order to solve stabilization problems, we must employ new methods of synthesis and control algorithms. This section is devoted to the analysis of some of the already known approaches to this problem.

2.2.1. Generalities

When feedback is designed, a mathematical model of an object with operator P is usually used (Fig. 2.29a). However, as a rule, the operator of the control object is known with a certain error ΔP, and therefore an object with an uncertain operator $P + \Delta P$ is actually stabilized (Fig. 2.29b). If the error ΔP is small in a certain sense and the controller R calculated with the use of the exact model P makes the feedback system robust (Fig. 2.30a), then this controller is suitable for the stabilization of the uncertain object (Fig. 2.30b). The mathematical expression of robustness of a feedback control system is the preservation of the qualitative behavior of a dynamical system under different conditions of the problem, in particular, under small regular and singular perturbations.

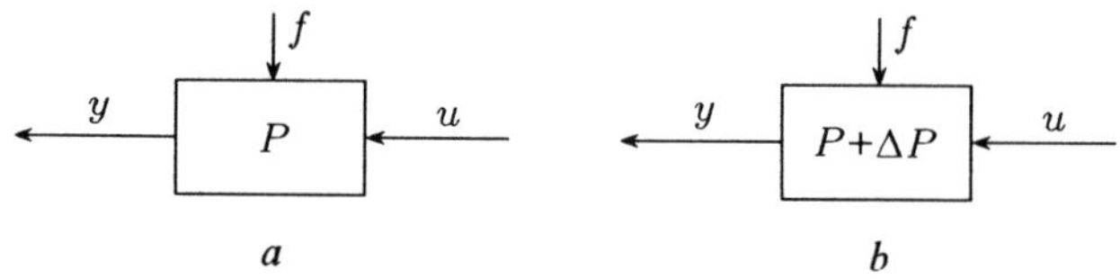

Figure 2.29.

The situation changes qualitatively when the uncertainty ΔP is not small in the description of the object and is such that the behavior of the control systems shown in Fig. 2.30 differs essentially. For instance, the system shown in Fig. 2.30*a* is stable and that shown in Fig. 2.30*b* is unstable, and so on.

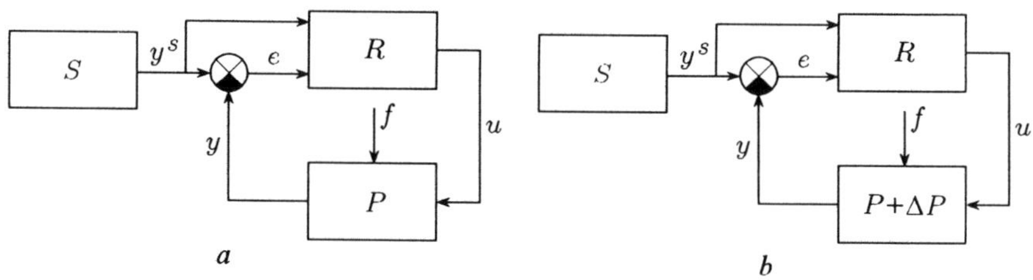

Figure 2.30.

The main difficulty encountered in situations of this kind is that the error ΔP and, hence, the information about it cannot be used in control algorithms. In other words, the feedback that stabilizes an incertain object either must give the system a very high reserve of robustness, which is hardly possible, or must change somehow itself during the control process, detecting the information about the uncertainty factors.

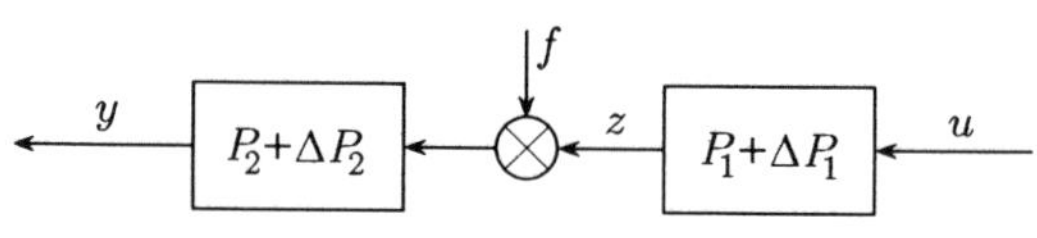

Figure 2.31.

In any case, standard methods of control synthesis based on the exact knowledge of the operator of the object are not applicable any longer. For instance, the principles of load control and disturbance control do not lead to the desired result. Indeed, a system of direct compensation (Fig. 2.31) can be described by the relation

$$\begin{aligned} y &= P\,u + \Delta P\,u + (P_2 + \Delta P_2)\,f\;\Big|_{u=P^{-1}y^s - P_1^{-1}f} \\ &= y^s + (\Delta P_2 - \Delta P P_1^{-1})\,f + \Delta P\,P^{-1}\,y, \end{aligned}$$

and it is clear that the deviation of the controlled coordinate y from the reference action y^s may be unacceptable.

What was said is also valid for the high gain method (feedback with high gain) although its dependence on the use of full information about the object and the disturbances acting on it is weaker than in all other methods.

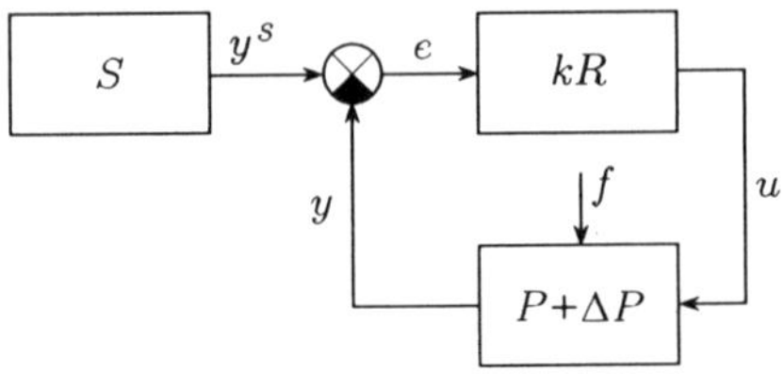

Figure 2.32.

Indeed, let us consider what happens when the gain factor k tends to infinity, see the system in Fig. 2.32 where the structure of the action of disturbance $f(t)$ is the same as that in the system shown in Fig. 2.31. The equation of this control system has the form

$$y = (P + \Delta P)kRe + (P_2 + \Delta P_2)f.$$

Since $e = y^s - y$, we have an expression $e = -k(P + \Delta P)Re - (P_2 + \Delta P_2)f$ for the stabilization error. Collecting terms and dividing the result by k, we obtain an equation

$$\left[\frac{1}{k} + (P + \Delta P)R\right] e = -\frac{1}{k}(P_2 + \Delta P_2)f.$$

In the limit, as $k \to \infty$, we have an equation $(P + \Delta P)R = e$ which is responsible for the main features of the behavior of the system with high gain. It is obvious that this behavior depends on the uncertainty factors. Consequently, feedback with high gain does not ensure, in the general case, the stabilization of the uncertain object.

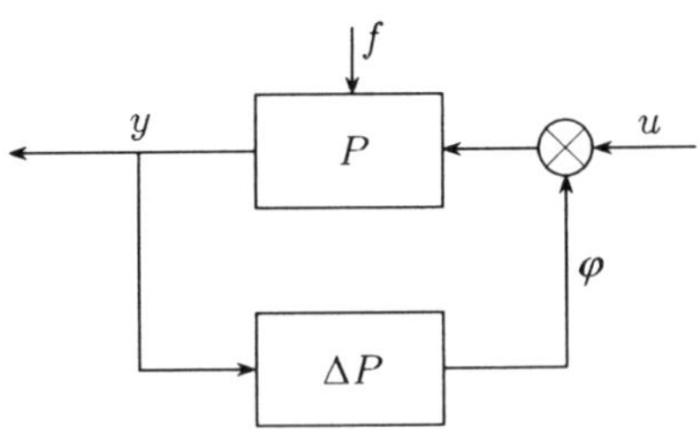

Figure 2.33.

However, if the uncertainty is structurized in some way, then the possibilities of stabilization of this uncertain object become wider. Suppose, for instance, that the uncertainty in the operator of the object ΔP can be reduced to the control input. In this case, we say that the uncertainty satisfies the matching condition. When the matching condition is fulfilled, the uncertain object can be represented by the block diagram shown in Fig. 2.33. For an object of this kind, the application of feedback with high gain may give the desired result, i.e., stabilize the uncertain object when the gain factor k of feedback tends to infinity according to the diagram shown in Fig. 2.34. Indeed, the object shown in Fig. 2.34 can be described by the equation

$$y = P\,u + P\,\Delta P\,y + P_2\,f$$

and the feedback by the equation

$$u = k\,R\,e.$$

Since $e = y^s - y$, the equation of the feedback control system for the control error has the form

$$(1 + k\,P\,R)\,e = -P\,\Delta P\,y - P_2\,f + y^s\,.$$

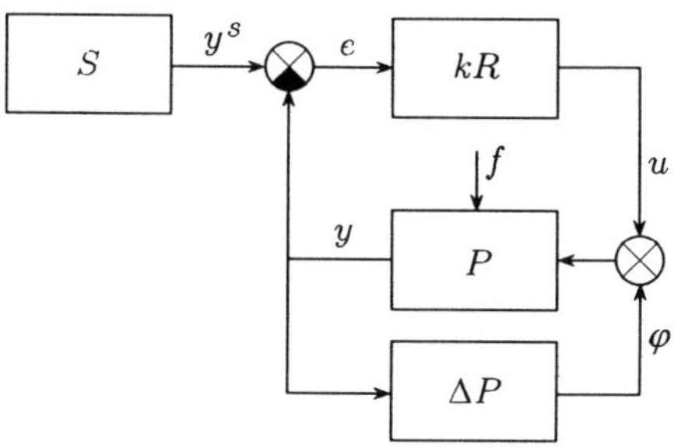

Figure 2.34.

Dividing this equation by the factor k term by term, we get a relation

$$\left(\frac{1}{k} + P\,R\right)e = -\frac{1}{k}P\,\Delta P\,y - \frac{1}{k}P_2 f + \frac{1}{k}y^s,$$

from which, as $k \to \infty$, we find the limit motion equation in the form

$$P\,R\,e = 0.$$

This equation does not depend on the uncertainty ΔP or the disturbance f and testifies to the possibility of solving the stabilization problem. It stands to reason that we must bear in mind the problems which arise when k tends to infinity, namely, nonrobustness, the strengthened influence of constraints, etc. Therefore, taking into account that the problem can, in principle, be solved, we must look for alternative methods.

2.2.2. Principle of cascade control

It has become obvious that the stabilization problem cannot, as a rule, be solved by a simple increase in the gain factor of the principal feedback (Fig. 2.34) since the limit system is unstable. Recall that we can obtain stability when introducing feedback with high gain with respect to the output only when the object has the minimal phase property and is of a relative order $r \leq 2$.

It is clear that these conditions are not always fulfilled. Therefore methods are of interest which use local feedbacks with high gain when the conditions indicated above are locally realized in some place of the block diagram of the system.

These local feedbacks with high gain, matched in an appropriate way, generate a hierarchy of gain factors and the corresponding difference of time-scale motions in different loops of feedback. At the same time, it may simplify the synthesis of control since when solving a local stabilization problem, we usually deal with a system of a lower order as compared to the order of the original system.

Frequently, the difference in time-scale motions in a system is physical in nature (for instance, in the case of electromechanical systems), and therefore this control technique does not appear to be artificial. This hierarchy of feedbacks with high gain is introduced by the principle of cascade control.

Example 11 (Hierarchy of feedbacks). Let us consider the problem of stabilization at zero of the object shown in Fig. 2.35. It is assumed that the constants λ_1, λ_2 and the external forces f_1, f_2 are unknown and that only the coordinates y and z can be used for control.

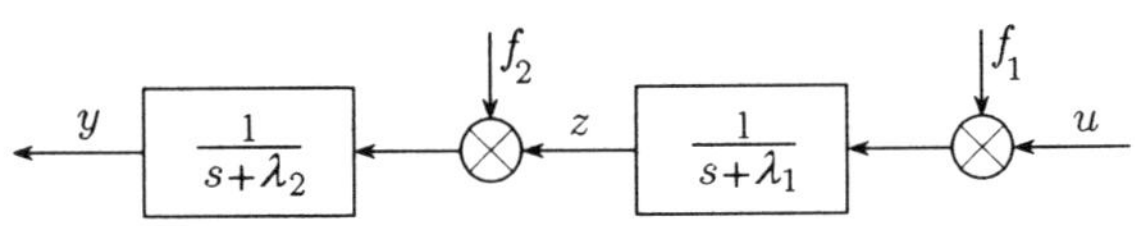

Figure 2.35.

The differential equations of this object have the form

$$\dot{y} + \lambda_2\, y = f_2 + z, \tag{2.16}$$

$$\dot{z} + \lambda_1\, z = f_1 + u. \tag{2.17}$$

It is obvious that the conditions of applicability of standard compensation methods are not fulfilled in this problem. Certain difficulties also arise when we use the ordinary feedback with high gain since the disturbance f_2 does not satisfy the matching condition. Therefore the organization of feedback with the use of the available information

$$u = -k\,(cy + z) \tag{2.18}$$

and tending k to infinity, $k \to \infty$, do not give the desired result since the limit system is described by the equation

$$\dot{y} + (c + \lambda_2)y = f_2 \tag{2.19}$$

which follows from (2.16)–(2.18). Since this equation depends on the uncertainty factors λ_2, f_2, the solution of the stabilization problem is not guaranteed (Fig. 2.36). For instance, for $c + \lambda_2 > 0$ the free motions of Eq. (2.19) are stable, but, for a finite value of the parameter c, the forced solution takes the coordinate y away from the desired zero. At the same time,

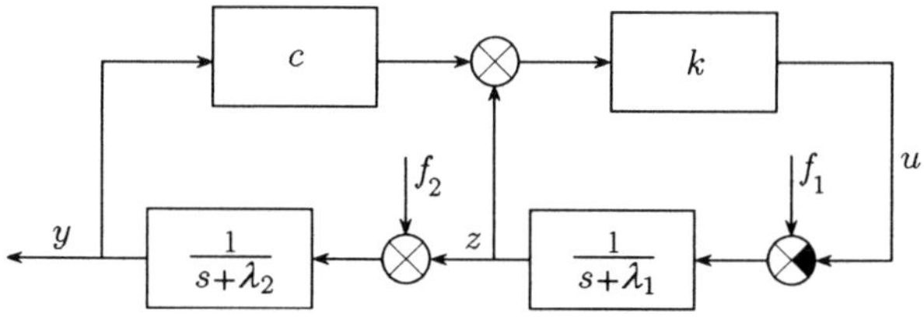

Figure 2.36.

as the parameter c tends to infinity (2.19), the stabilization problem is solved and we do not need any information concerning the parameters λ_1, λ_2 of the object and the disturbances f_1, f_2 which can be arbitrary at any stage of the solution.

Example 11 demonstrates that it is useful to tend the feedback parameters to infinity step by step. Here $k \to \infty$ and $c \to \infty$ but so that $c/k \to 0$. It is the ratio c/k that establishes the hierarchy of the gain factors that we spoke about earlier. It is precisely this idea that is exploited by the principle of cascade control.

In all fairness it should be pointed out that the use of an a priori information concerning the state of the object allows us to get by only with high gain feedback when solving the problem in question. To demonstrate this possibility, we differentiate Eq. (2.16)

$$\ddot{y} + \lambda_2 \dot{y} = \dot{f}_2 + \dot{z} \tag{2.20}$$

and multiply Eq. (2.16) termwise by λ_1,

$$\lambda_1 \dot{y} + \lambda_1 \lambda_2 y = \lambda_1 f_2 + \lambda_1 z. \tag{2.21}$$

Then we add the results (2.20) and (2.21) together and take into account that Eq. (2.17) has the form

$$\dot{z} + \lambda_1 z = f_1 + u.$$

As a result, we get the equation for the object in the form

$$\ddot{y} + (\lambda_1 + \lambda_2)\dot{y} + \lambda_1 \lambda_2 y = F + u, \tag{2.22}$$

where $F = \dot{f}_2 + \lambda_1 f_2 + f_1$.

The advantage of the resulting equation (2.22) is that now the disturbance matches the control, i.e., satisfies the matching condition, and therefore the procedure that we have used is called a reduction of disturbance to the control input. It stands to reason that, in this case, we must assume that the function $f_2(t)$ is differentiable.

For object (2.22) it is possible to use the standard feedback with high gain $u = -k(cy + \dot{y})$ according to the block diagram shown in Fig. 2.37 in which the tending of k to infinity for $c > 0$ solves the stabilization problem since the limit system is described by the first-order exponentially stable equation

$$\dot{y} + cy = 0.$$

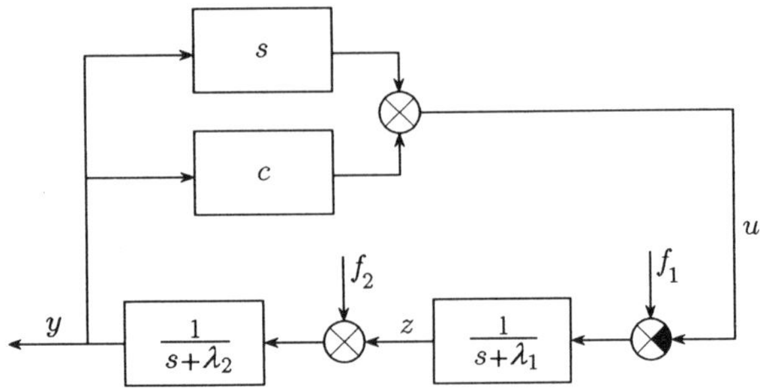

Figure 2.37.

The limitations of the method that we have considered are obvious, they are connected with

- the differentiability of the function f_2 which may not be,
- the availability of information about the derivative $\dot{y}$ which is also problematic.

Let us return to the original statement of the stabilization problem from Example 11 and solve it with the use of the available information about the object alone, following directly the main idea of the principle of cascade control. Let us consider the first equation for the object P_2:

$$\dot{y} + \lambda_2 y = f_2 + v, \tag{2.23}$$

in which we have replaced the variable z by v. We assume the signal v to be a control for the object P_2. Note that the disturbance f_2 for the object P_2 satisfies the matching condition and, consequently, it is possible to use a feedback with high gain. Let $v(y)$ be a feedback stabilizing y at zero for object (2.23). However, the signal z actually acts on the object P_2, and therefore there is an error $e = v - z$ at its input. This error can be eliminated (and this will solve the problem in question) by an appropriate choice of control u at the input of the object P_1:

$$\dot{z} + \lambda_1 z = f_1 + u. \tag{2.24}$$

To synthesize an appropriate control u, it is convenient to write the motion equation of the object P_1 (2.24) for the error $e = v - z$. Differentiating the last equation with respect to t and performing changes according to the relations

$$z = v - e, \quad \dot{z} = f_1 + u - \lambda_1 v + \lambda_1 e,$$

we find the required motion equation in the form

$$\dot{e} + \lambda_1 e = \varphi - u, \tag{2.25}$$

where $\varphi = \dot{v} + \lambda_1 v + f_1$. Object (2.25) also satisfies the matching condition, and therefore it is relevant to employ the error feedback with high gain in the standard form $u = k_1 e$. As a result, we have a feedback control system with a motion equation of the form

$$\dot{e} + (k_1 + \lambda_1)e = \varphi.$$

As $k_1 \to \infty$, object (2.25) is stabilized at zero, and this is what we had to prove. Since it is clear that as $k_2 \to \infty$, the feedback $v = -k_2 y$ stabilizes the object P_2 at zero for any λ_2 and f_2, we can now depict the block diagram of the synthesized system of cascade control (Fig. 2.38). The cascade structure of this

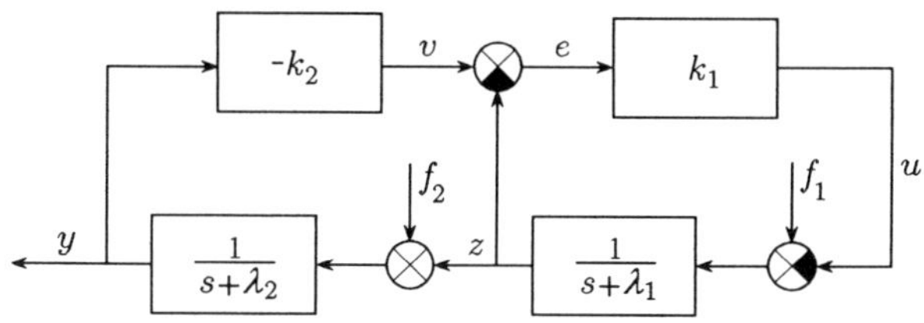

Figure 2.38.

control system follows from the fact that the feedback $u = k_1 e$ can damp the disturbance

$$\varphi = k_2\dot{y} + \lambda_1 k_2 y + f_1$$

only when the condition

$$\lim_{\substack{k_1 \to \infty \\ k_2 \to \infty}} \frac{k_2}{k_1} = 0$$

is fulfilled.

The last condition establishes the hierarchy of gain factors of the local loops of feedbacks and, at the same time, imparts the cascade structure to the system. If we represent the block diagram of the obtained stabilization system as the block diagram shown in Fig. 2.39, then the possibility arises of useful substantive

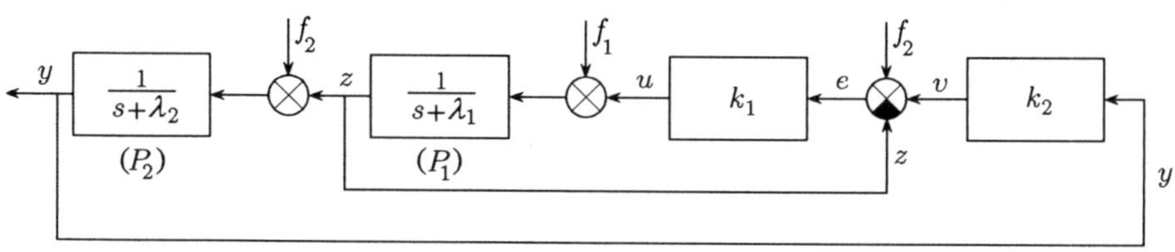

Figure 2.39.

interpretations. In this representation, we can see two loops of feedbacks, where the inner loop is "faster" ("faster" in the limit than the outer loop), and, hence, we can consider the variables of the outer loop to be as if "frozen" with respect to the variables of the inner loop. Therefore, if the inner loop is stable, then we can assume that $e = v - z = 0$ and, consequently, the inner loop as it "transfers" or induces the control v at the input of the object P_2. In this connection, the principle of cascade control is sometimes called a principle of induced feedbacks.

2.2.3. The structure of objects with cascade control

It is easy to generalize the structure of the cascade control system considered above (Fig. 2.39) to the case of an arbitrary number m of cascades (Fig. 2.40). In this case, the stabilization of the output y of the object at zero is possible when a definite hierarchy of gain factors

$$\lim_{k_i \to \infty} \frac{k_{i+1}}{k_i} = 0, \quad i = 1, \dots, m-1,$$

is established for local feedbacks with high gain.

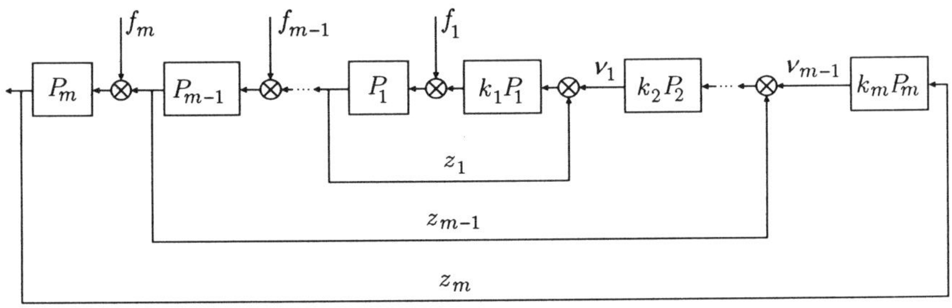

Figure 2.40.

However, far from all objects can be stabilized in this way, but only those which admit a block composition shown in Fig. 2.40. The essence of this composition is that the behavior of each loop must be independent of variables of inner (relative to it) loops, i.e., the relations

$$z_i = P_i(f_i + z_{i+1}), \quad i = 1, \dots, m-1,$$
$$z_m \stackrel{\text{def}}{=} y,$$

must be satisfied.

In the language of differential equations this means that the right-hand side of the motion equation of the object in the Cauchy form must have a "triangular structure", i.e., the derivative with the largest subscript may appear only exactly in the preceding equation and must not appear in other equations with smaller subscripts. Below we give an example of motion equations of this kind in the phase space $(z_1, z_2, \dots, z_m)$:

$$\dot z_1 = \varphi_1(z_1, t) + z_2,$$
$$\dot z_2 = \varphi_2(z_1, z_2, t) + z_3,$$
$$\dots\dots\dots\dots\dots\dots\dots\dots$$
$$\dot z_m = \varphi_m(z_1, \dots, z_m, t) + u.$$

Only in this case can we interpret every highest derivative as a control for the preceding derivatives and establish a hierarchy of feedbacks. Objects of this kind are customarily called objects with a triangular structure.

What was stated above shows that the principle of cascade control is a very efficient means of stabilization of an uncertain object since

- the synthesis of feedback is iterative and at each iteration we deal with a lower-order object,
- there is no need to have full information about the phase vector of the object,
- a nonstationary uncertainty and arbitrary external actions are admissible,
- the difference in time-scale motions of physical processes in a real system are taken into account.

The system of cascade control is not free from drawbacks, namely, the use of high gain factors reduces the robustness of the system and makes the part played by amplitude constraints more important; in addition, in this case the structure of the system must be "triangular", and this restricts the class of stabilizable uncertain objects.

2.2.4. Stabilization of interval objects

Let us consider now the problem of stabilization at zero of the linear object

$$y^{(n)} + a_n y^{(n-1)} + a_1 y = u + f, \tag{2.26}$$

which we call an object P in the sequel, whose constant parameters $a_1, \ldots, a_n$ are unknown. Only the boundaries of the closed intervals containing them are given, i.e., numbers $a_1^{\pm}, \ldots, a_n^{\pm}$ are known such that $a_i \in [a_i^-, a_i^+]$. These numbers, known with an accuracy to within an interval, are called interval numbers and are denoted by

$$[a_i], \quad i = 1, \ldots, n.$$

Correspondingly, the object P is called an interval object $[P]$ and is written as

$$y^{(n)} + [a_n] y^{(n-1)} + [a_1] y = u + f. \tag{2.27}$$

Moreover, an interval object can be defined by the state equation

$$\begin{aligned} \dot{x} &= [A]x + [b]u + [d]f, \\ y &= [c]x, \end{aligned} \tag{2.28}$$

where $[A]$ is an interval matrix and $[b]$, $[d]$, $[c]$ are interval vectors. Note that the transition from notation (2.27) to (2.28) is always possible whereas the transition from (2.28) to (2.27) may not exist. When stabilizing the interval object (2.27), we actually have to deal with a family of objects, and this can be given a corresponding geometrical interpretation.

Let $a = (a_1, \ldots, a_n)$ be a vector of parameters or a point in the vector space $\mathcal{A}$ of parameters which defines a concrete object (Fig. 2.41). Then $[a] =$

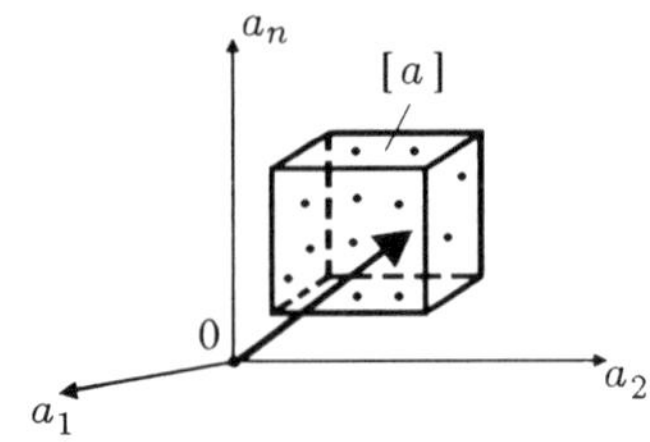

Figure 2.41.

$([a_1], \ldots, [a_n])$ is an interval vector of parameters or a set, namely, a polyhedron in the space $\mathcal{A}$ which defines the family of objects. Therefore, actually, feedback must be stabilizing for any point of the polyhedron. In this case, no problems arise when the polyhedron is "small" since there exist theorems which testify that the dependence of solutions on the parameters is continuous. But what can we do when a polyhedron cannot be considered to be "small"?

Several approaches are possible here. For instance, we can choose a median of every ith interval

$$a_i^m = (a_i^+ + a_i^-)/2$$

and introduce the half-width of this interval

$$\Delta a_i = (a_i^+ - a_i^-)/2, \quad i = 1, \ldots, n.$$

This allows us to distinguish a standard object P^m with fixed parameters (median parameters in our case)

$$y^{(n)} + a_n^m y^{(n-1)} + \cdots + a_1^m y = v \tag{2.29}$$

and define the difference between (2.29) and (2.27) as an unknown disturbance

$$\varphi(y, \ldots, y^{(n-1)}) = (a_n - a_n^m)y^{(n-1)} + \cdots + (a_1 - a_1^m)y$$

for which, however, the majorant

$$\varphi_M(y, \ldots, y^{(n-1)}) = \Delta a_n |y^{(n-1)}| + \cdots + \Delta a_1 |y|$$

is known, i.e., $|\varphi| \le \varphi_M$. Under these transformations, the problem of stabilization of the interval object P from (2.27) can be reduced to the standard problem of stabilization of the median object P^m from (2.29) when there exists an interference φ on the right-hand side, i.e.,

$$y^{(n)} + a_n^m y^{(n-1)} + \cdots + a_1^m y = u + f - \varphi. \tag{2.30}$$

If we have some information concerning the phase vector $(y, y^{(1)}, \ldots, y^{(n-1)})$, then the stabilization problem being considered can be solved by means of linear feedback with high gain

$$u = -k\left(c_1 y + c_2 y^{(1)} + \cdots + c_{n-1} y^{(n-2)} + y^{(n-1)}\right), \tag{2.31}$$

only if the polynomial

$$\varphi(s) = s^{(n-1)} + c_{n-1}s^{(n-2)} + \cdots + c_1,$$

corresponding to feedback (2.31) as $k \to \infty$ is a Hurwitz polynomial. Indeed, when the interval object $[P]$ has feedback (2.31), it is described by the equation

$$y^{(n)} + [a_n^m + k]y^{(n-1)} + \cdots + [a_1^m + kc_1]y = f - \varphi.$$

After termwise division by k we have

$$\frac{1}{k}y^{(n)} + \left[\frac{a_n^m}{k} + 1\right] y^{(n-1)} + \cdots + \left[\frac{a_1^m}{k} + c_1\right] y = \frac{1}{k}(f - \varphi)$$

and after tending the feedback factor k to infinity we get a limit motion equation

$$y^{(n-1)} + c_{n-2}y^{(n-2)} + \cdots + c_1 y = 0,$$

which, by assumption, is asymptotically stable.

It stands to reason that the possibilities of this approach are limited by standard (for systems with an infinitely large gain factor) requirements to robustness, etc., which can be somewhat weakened by the application of cascade control. However, on the whole, the problem of stabilization of the general interval object with a restricted gain in feedback remains urgent.

2.2.5. Interval stability

The method based on the theory of interval stability is an alternative, to a certain degree, of feedback with high gain.

This theory is based on different criteria of localization of zeros, in particular, on the criteria of stability of polytopes which are polynomials with interval coefficients. For instance, according to Kharitonov's criterion, for the polytope

$$[\varphi] = s^n + [a_n]s^{(n-1)} + \cdots + [a_1]$$

to be stable, it is necessary and sufficient that the following four polynomials be stable:

$$\begin{aligned}
\varphi_{--}^{++} &= a_1^+ + a_2^+ s + a_3^- s^2 + a_4^- s^3 + \ldots \\
\varphi_{++}^{--} &= a_1^- + a_2^- s + a_3^+ s^2 + a_4^+ s^3 + \ldots \\
\varphi_{-+}^{+-} &= a_1^+ + a_2^- s + a_3^- s^2 + a_4^+ s^3 + \ldots \\
\varphi_{+-}^{-+} &= a_1^- + a_2^+ s + a_3^+ s^2 + a_4^- s^3 + \ldots
\end{aligned}$$

Generally speaking, Kharitonov's conditions are redundant, i.e., in some cases it is sufficient to verify the stability of a smaller number of polynomials. For instance, for the second-order polytope

$$[\varphi] = s^2 + [a_2]s + [a_1]$$

it suffices to verify the stability of only one polynomial

$$\varphi = s^2 + a_2^- s + a_1^- ,$$

which fact follows directly from Fig. 2.42. Since the necessary and sufficient conditions of stability of polytope $[\varphi]$ consist in the positiveness of its coefficients, it suffices to verify only the point $(=)$. However, if we want to localize the zeros

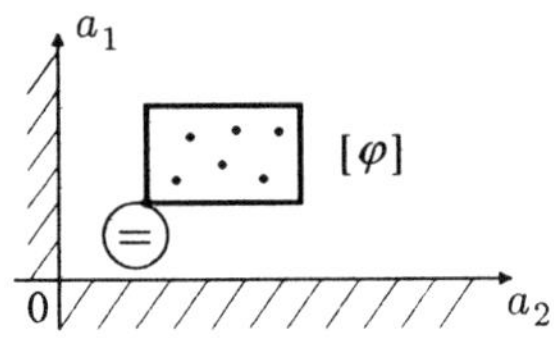

Figure 2.42.

of the polytope more exactly, then we need additional conditions. For instance, the right-hand branch of the parabola

$$4a_1 = a_2^2 \tag{2.32}$$

defines the boundary of aperiodic stability in the plane (a_1, a_2), and it is clear that for purely aperiodic processes the rectangle $[a] = \big([a_1],[a_1]\big)$ must lie entirely below the parabola (2.32) and for oscillatory processes it must lie above it (Fig. 2.43). Therefore it is necessary to impose constraints on the points $(=)$, $(\mp)$ in the first case and on the points $(=)$, $(\pm)$ in the second case.

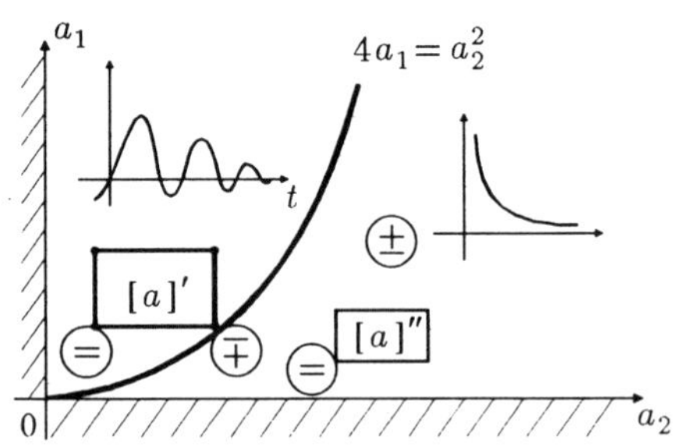

Figure 2.43.

If we require, in addition, that the degree of stability of the polytope $[\varphi] = s^2 + [a_2]s + [a_1]$ be $\eta > 0$, then the conditions of verification of this property become even more complicated. This can be seen from Fig. 2.44 where the dash line shows the boundary of the domain G with a degree of stability not less than η. Polytopes corresponding to the rectangle $[a]'$ always have a complex conjugate pair of zeros with the real part not larger than $-\eta$ whereas the polytope corresponding to the rectangle $[a]''$ always has only real zeros not exceeding the number $-\eta$.

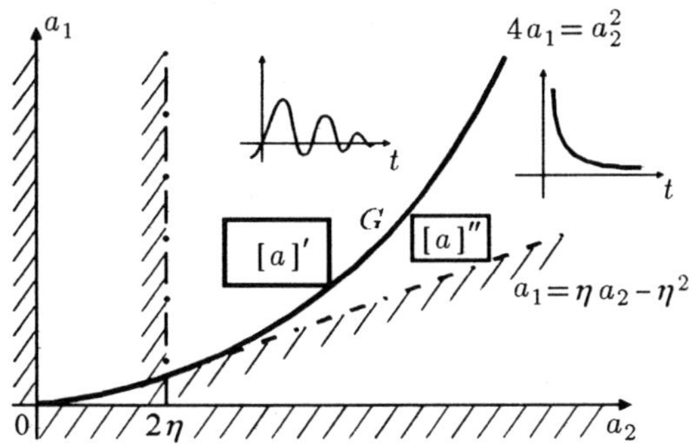

Figure 2.44.

Note that today the theory of interval stability is a well developed division of the theory of stability of motion. Here we restrict ourselves to the remarks made above and pass to the problems of using the criteria of interval stability in stabilization problems, in particular, consider the method of inclusion.

Example 12 (Inclusion method). Consider the interval object

$$\ddot{y} + [a_2]\dot{y} + [a_1]y = [b]u, \tag{2.33}$$

where $[b]$ is an interval gain factor, $b \in [b^-, b^+]$ with $b^- > 0$.

It is obvious that whatever feedback with finite gain factors we take, the feedback system will be uncertain, and for a linear feedback it will be an interval system. Therefore it is natural to try and assign the properties of the feedback system with the aid of the desired interval polynomial

$$[\varphi]_{\text{des}} = s^2 + [a_2]_{\text{des}} s + [a_1]_{\text{des}}.$$

Then the task of control will be achieved for the linear feedback

$$u = -k_1 y - k_2 \dot{y} \tag{2.34}$$

if its parameters k_1, k_2 satisfy the inclusions

$$[a_1 + bk_1] \subseteq [a_1]_{\text{des}}, \quad [a_2 + bk_2] \subseteq [a_2]_{\text{des}}.$$

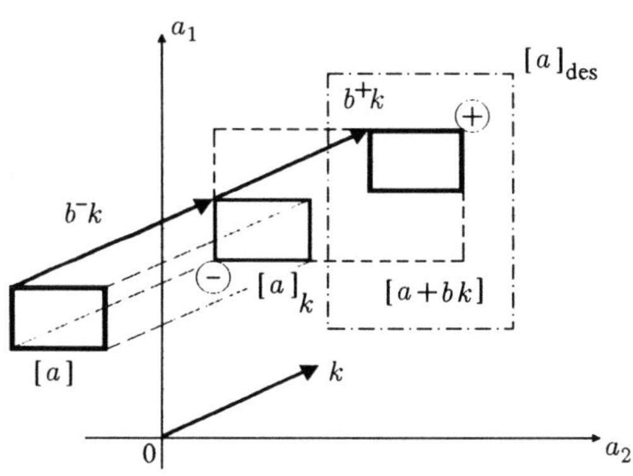

Figure 2.45.

The difficulties encountered when we use this method can be seen from Fig. 2.45. Indeed, in order to achieve an asymptotic stability, we must use the vector $k = (k_2, k_1)$ to "transfer" the initial triangle $[a]$ to the open positive unit vector of the plane (a_2, a_1). Then the boundaries of the intersected rectangle $[a]_k$ are extended since the coefficient b is unknown and varies from b^- to b^+.

Furthermore, in order to achieve the desired quality of control, the "fuzzy" rectangle $[a]_k$ must get inside the rectangle $[a]_{des}$, and this is not an easy task. For this purpose, we must verify whether two vertices of the diagonals of the rectangle $[a]_k$, say, $\oplus$ and $\ominus$, belong to the rectangle $[a]_{des}$.

It stands to reason that if we weaken the requirements to the quality of transition processes and only try to achieve, say, the stabilizability, i.e., asymptotic stability of the feedback system, then the procedure of synthesizing the feedback becomes essentially simpler. For instance, in order to stabilize the interval object (2.33) by means of the linear feedback (2.34), it suffices to choose parameters k_1, k_2 that would satisfy the interval inequalities

$$[a_2 + bk_2] > 0, \quad [a_1 + bk_1] > 0,$$

which are equivalent to the numerical inequalities

$$a_2^- + b^- k_2 > 0, \quad a_1^- + b^- k_2 > 0.$$

The validity of this statement is illustrated by Fig. 2.46.

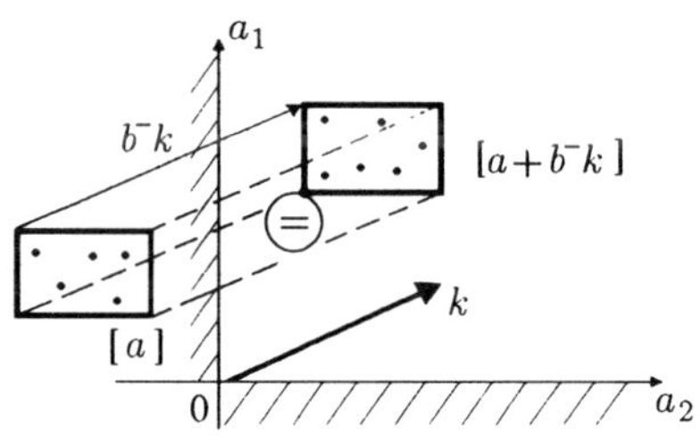

Figure 2.46.

The block diagram of a stabilization system of an interval object at zero by means of a linear feedback is shown in Fig. 2.47 and, of course, does not depend on the technique of calculation of its parameters k_1 and k_2, i.e., on whether we use the inclusion method or the method of interval stability.

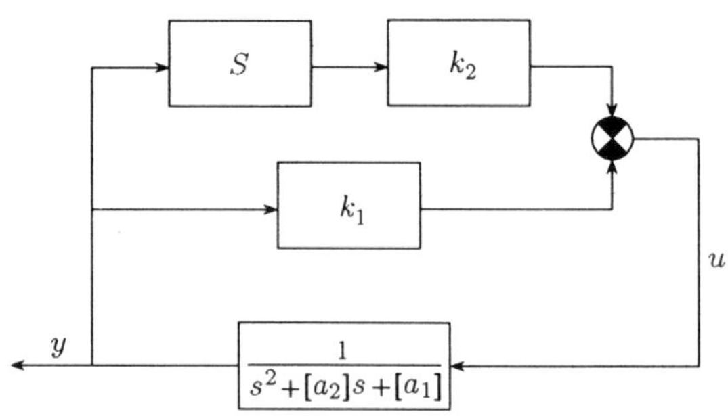

Figure 2.47.

It should be pointed out in conclusion that the stabilization methods which we have considered cannot be used if the right-hand side includes a nonvanishing, even though known, disturbance $f(t)$, i.e., when the object being stabilized has the form

$$\ddot{y} + [a_2]\dot{y} + [a_1]y = [b]u + f.$$

The absence of information about the coefficient b does not allow us to use even the method of direct compensation of disturbance, and since bounded coefficients are used in the feedback

$$u = -k_1 y - k_2 \dot{y}$$

and the interference $f(t)$ does not tend to zero as $t \to \infty$, the feedback system will be, at best, dissipative. Consequently,

- stabilization methods based on interval stability have their own field of application but do not exhaust the problem as a whole.

In particular, they cannot be used when the parameters of the object vary.

2.2.6. General features of the adaptive stabilization theory

Let us consider the problem of stabilization of an uncertain object with operator $P + \Delta P$ and assume, for simplicity, that

$$y^s \equiv 0, \quad f \equiv 0,$$

i.e., solve the problem of stabilization of free oscillations of the object.

Let us also make a standard assumption that the uncertainty ΔP satisfies the matching condition. Then the problem reduces to the choice of controller $\bar{R}$ which stabilizes the output of the object $P + \Delta P$ at zero (Fig. 2.48). Somewhat earlier we proved that such a problem could be solved by the method of a large coefficient, but this leads, as was repeatedly emphasized, to nonrobust control systems.

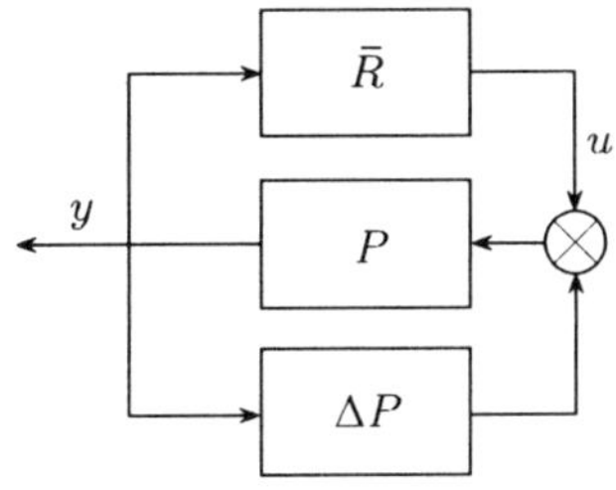

Figure 2.48.

When finite coefficients of transfer are used in feedback (and this is precisely the criterion of robustness of a system) in the theory of adaptive control, the controller $\bar{R}$ is constructed in the form of two components

$$\bar{R} = R + \Delta R,$$

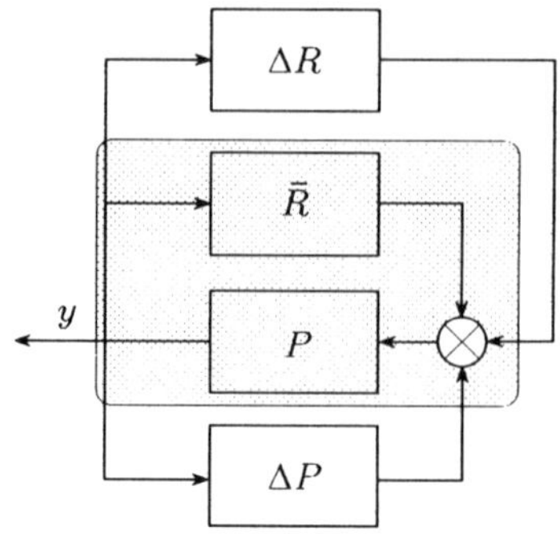

Figure 2.49.

where the choice of the first component R is aimed at the stabilization of a definite object P whereas the second component ΔR must compensate for (eliminate) the effect produced by the uncertainty ΔP on the behavior of the system.

This is the reason why we use a system whose block diagram is shown in Fig. 2.49, where the part of the system outlined by a contour is usually called a generalized object. By the choice of the controller R the operator

$$P_c = \frac{P}{1 + PR}$$

of this generalized object is endowed with all properties required in the statement of the stabilization problem, the asymptotic stability inclusive. Therefore the stabilization problem actually reduces to a mutual compensation, at the input of the object, of the signals from the feedback loops with operators ΔR and ΔP, respectively (Fig. 2.50). The operator ΔR based on the identity $\Delta R + \Delta P \equiv 0$

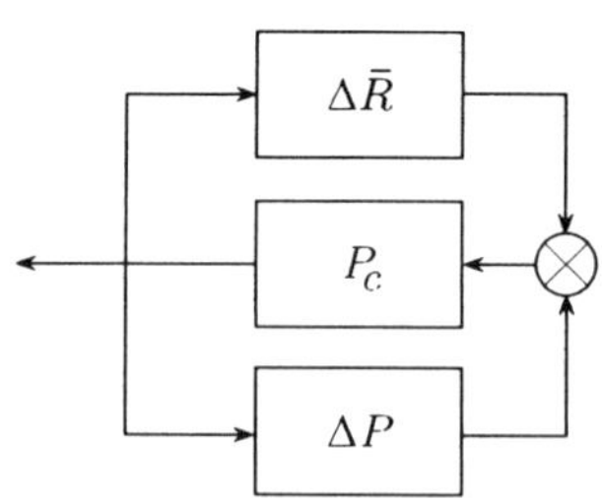

Figure 2.50.

would be an ideal solution of the problem, but the operator ΔP is unknown, and therefore we must find its asymptotic estimate $\Delta\hat{P}$ such that

$$\Delta\hat{P} \to \Delta P, \quad t \to \infty.$$

Then, setting, or, to be more precise, adjusting the feedback operator ΔR according to the law

$$\Delta R = -\Delta\hat{P}, \tag{2.35}$$

we can hope to obtain an asymptotic solution of the stabilization problem since the asymptotics

$$\Delta R + \Delta P = -\Delta\hat{P} + \Delta P \to 0, \quad t \to \infty,$$

is valid in this case.

In the adaptive control theory a special measuring element with operator D is responsible for the calculation of the estimate $\Delta\hat{P}$ and the "adapter" A is responsible for the realization of "adjustment" (2.35). As a result, we get a block diagram shown in Fig. 2.51. We can give an informative interpretation to this adaptive stabilization system in terms of the main control principles.

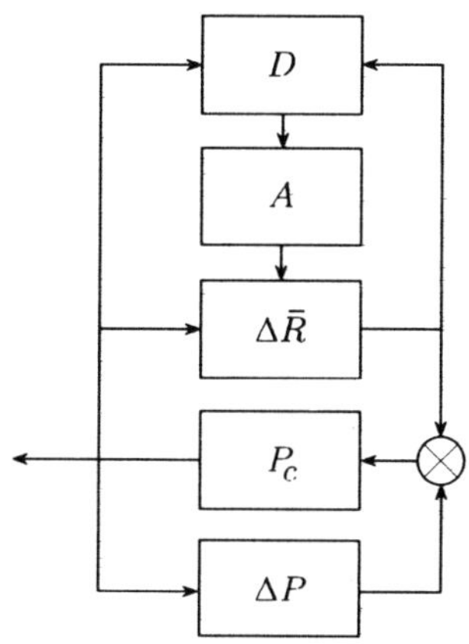

Figure 2.51.

The technique of compensation for the operator disturbance ΔP described above resembles the technique of indirect compensation for the coordinate disturbance $f(t)$ with the aid of the disturbance control principle. Indeed using available information about the parameters of the object P and about its inputs and outputs, we find the estimate of the disturbance $\hat{P}$ which is then used for the compensation $\Delta P = -\Delta\hat{P}$. The technique is like that used for the compensation for the directly nonmeasurable disturbance $f(t)$. This observation allows us to infer that the adaptive stabilization is based on the combination of two basic control principles, namely,

- feedback principle employed in the control of free motion,
- load control principle employed for compensating the operator disturbance ΔP. In Fig. 2.52 double lines depict the load control loop.

This inference is fundamental for any adaptive control schemes and can be regarded as an inalienable feature of adaptation, although certain variations in the realization of the idea of adaptation are also possible. Here are some refining remarks.

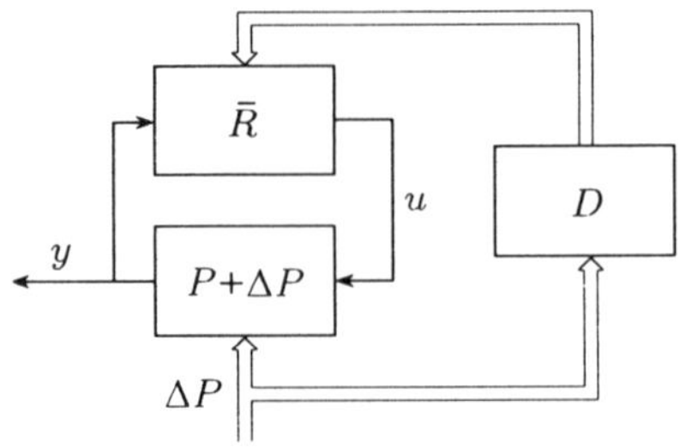

Figure 2.52.

Remark 1. There exist variations of the adaptation scheme described above, for instance, for the compensation condition

$$\Delta R + \Delta P = 0$$

to be fulfilled, not only (and, maybe, not so much as) the control operator ΔR can be changed but also the object operator ΔP. It stands to reason that, in this case, we must be able to represent ΔP as the sum

$$\Delta P = \Delta P' + \Delta P'',$$

where $\Delta P'$ is the unknown part and $\Delta P''$ is the part being adjusted. In contrast to the preceding case where we dealt with the feedback being adjusted, here we speak about the method of the object being adjusted.

Remark 2. Since the estimate $\Delta\hat{P}$ is calculated from the measurements of the coordinates of the object p or system P_c as well as from their inputs, it is obvious that it is most expedient to search for this estimate when the variations of ΔP are "slower" than the variations of these variables. In this sense we speak about a quasistationary variation of the operator ΔP and, in fact, when we calculate the feedback, we assume that the operator ΔP is stationary. If the adaptive system constructed in this way is robust, then it preserves its capacity for work when the variations are "slow" since robust systems are stable relative to constantly acting disturbances.

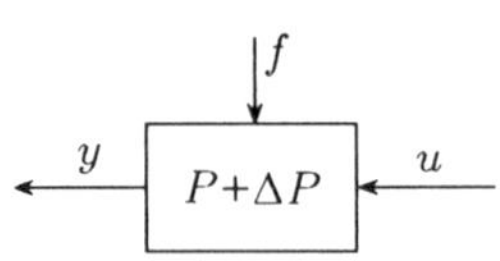

Figure 2.53.

Remark 3. The existence of external disturbance f of an uncertain control object (Fig. 2.53) makes the situation even more complicated since, in this case, an additional unremovable error is introduced into the estimate $\Delta\hat{P}$ if $\Delta f \not\to 0$ as $t \to \infty$. Therefore, instead of the relation

$$\Delta R + \Delta P = 0,$$

which is required by the compensation condition, we have a relation

$$\Delta R + \Delta P = \Delta Q$$

biased by a certain value ΔQ, and, consequently, the problem is how much the operator ΔQ changes the behavior of the object P_c (Fig. 2.54). If the disturbance is vanishing, i.e., $f \to 0$, and, hence, $\Delta Q \to 0$ as $t \to \infty$, then no problem arises in the steady-state regime. Now if the disturbance is nonvanishing, i.e., $\Delta f \not\to 0$ as $t \to \infty$, but, say, restricted, i.e., $|f| \leq f_0 = \text{const}$, then, because of the use of bounded transfer factors in the control, we can only expect, generally speaking, that the feedback control system will be dissipative with the dissipativity radius of the order f_0.

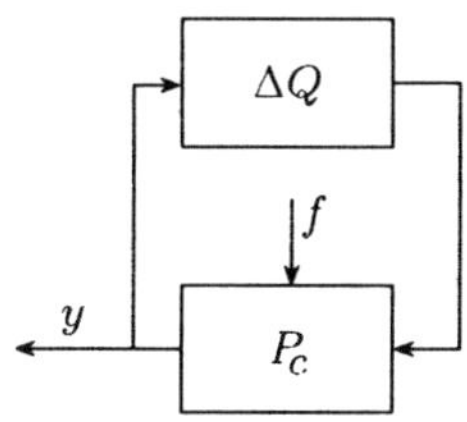

Figure 2.54.

Remark 4. Under certain conditions, the adaptive approach generates stabilizers of universal action, i.e., controllers whose application does not require the exact knowledge not only of the parameters of the object but also of its order. These controllers are said to be universal. For their application to be successful, the following conditions must be fulfilled:

- the matching conditions,
- the condition of the minimal phase property of the object P.

Indeed, in this case we can use feedback with high gain of the form kR with a gain factor k and operator R (Fig. 2.55). As $k \to \infty$, this feedback ensures stabilization if the limit motion equation

$$PRy = 0 \tag{2.36}$$

is asymptotically stable. As $k \to \infty$, this equation follows from the motion equation of the feedback system

$$(1 + kPR)y = (P_2 + \Delta P_2)f + P\Delta P y$$

which corresponds to Fig. 2.55.

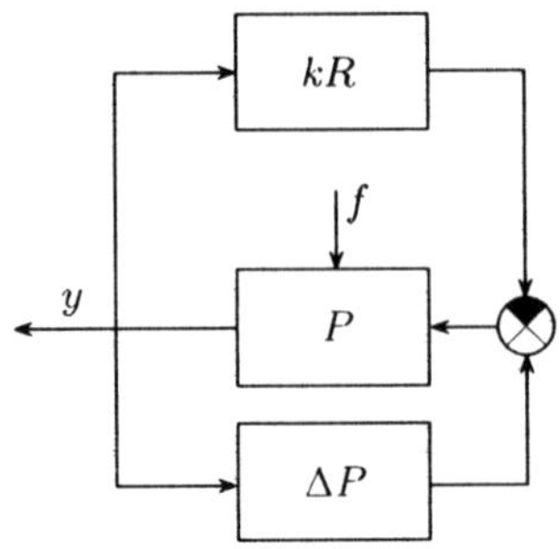

Figure 2.55.

Let us consider the condition of asymptotic stability for Eq. (2.36) in greater detail. Let the object P be stationary. Then it can be represented by the transfer function

$$W_P(s) = \frac{\beta_m(s)}{\alpha_n(s)}.$$

Similarly, we associate the operator of controller R with the transfer function

$$W_R(s) = \frac{\delta_l(s)}{\gamma_r(s)},$$

where β, α, δ, γ are polynomials of the complex variable s and the subscripts m, n, l, and r denote the degrees of the corresponding polynomials. Then the limit motion equation

$$PRy = 0$$

is equivalent to the equation

$$\beta_m(s)\,\delta_l(s)\,Y(s) = 0, \tag{2.37}$$

where $Y(s) = \mathcal{L}[y]$ is a unilateral Laplace transformation of the function $y(t)$. We can see from (2.37) that $\beta_m(s)$ must be a Hurwitz polynomial, and this means that the object P has a minimal phase property.

The nonrobustness of the feedback system with high gain in a universal stabilizer can be overcome since the feedback factor depends on the state variables, for instance, on the output of the object

$$k = k(y, \dots),$$

and increases monotonically until, under the conditions indicated above, the stability (as $f \to 0$) or dissipativity (for $|f| \le f_0$) of the equation

$$\left(\frac{1}{k(\cdot)} + PR\right) y = \frac{1}{k(\cdot)}[(P_2 + \Delta P_2)f + P\Delta P y]$$

occurs.

Example 13 (Synthesis of an adaptive control system). Using the example of stabilization at zero of the second-order object

$$\begin{aligned} \dot{x}_1 &= x_2, \\ \dot{x}_2 &= ax_1 + bu, \end{aligned} \tag{2.38}$$

with uncertain coefficients a and $b > 0$, we shall consider the most frequently employed methods of synthesis of adaptive control.

Let us stabilize object (2.38) by the feedback

$$u = -k_1 x_1 - k_2 x_2$$

and formulate the adaptation algorithms $\mathcal{A}$ of the variables ξ_1, ξ_2:

$$\dot{k}_1 = \xi_1, \quad \dot{k}_2 = \xi_2, \tag{2.39}$$

i.e., the algorithms of adjustment of the parameters k_1, k_2 such that the behavior of the feedback system Σ_k,

$$\begin{aligned} \dot{x}_1 &= x_2, \\ \dot{x}_2 &= -(bk_1 - a)x_1 - bk_2 x_2, \end{aligned}$$

would be similar (maybe even close to) the behavior of some reference system Σ_γ,

$$\begin{aligned} \dot{z}_1 &= z_2, \\ \dot{z}_2 &= -\gamma_1 z_1 - \gamma_2 z_2. \end{aligned}$$

Here γ_1 and γ_2 are assigned positive numbers; we shall also call the Σ_γ-system a model. Introducing the notation

$$b\Delta k_1 = bk_1 - a - \gamma_1, \quad b\Delta k_2 = bk_2 - \gamma_2,$$

we can represent the Σ_k-system closed by the adaptive feedback in the standard form

$$\begin{aligned} \dot{x}_1 &= x_2, \\ \dot{x}_2 &= -\gamma_1 x_1 - \gamma_2 x_2 - b\Delta k_1 x_1 - b\Delta k_2 x_2. \end{aligned}$$

We have $\Delta\dot{k}_1 = \dot{k}_1$, $\Delta\dot{k}_2 = \dot{k}_2$ by the quasistationarity hypothesis, and therefore it is formally convenient to assume that not the coefficients k_1 and k_2 themselves are adapted but their deviations from the required values, i.e., we can write the adaptation law $\mathcal{A}$ in the generalized form $\Delta\dot{k}_1 = \xi_1$, $\Delta\dot{k}_2 = \xi_2$.

In order to obtain a specific adaptation algorithm $\mathcal{A}$, we shall use the quadratic form $v(x) = \langle x, Hx\rangle$ with a positive definite (2×2)-matrix $H = [h_1, h_2]$ such that for a certain (desired) positive number $\lambda > 0$ that defines the degree of stability of the Σ_γ-system (the model of behavior of the feedback system) the derivative of this form can be written as

$$\dot{v}\big|_{\Sigma_\gamma} = -2\lambda v$$

by virtue of the Σ_γ-system. Since the choice of the parameters γ_1, γ_2 of the model is not restricted, a matrix H and a number λ of this kind exist.

Let us now introduce into consideration a quadratic form $V(x, k)$ in an extended space of variables $\{x, k\}$ by using the expression

$$V(x, k) = v(x) + \frac{b}{2}\left(\frac{\Delta k_1^2}{\beta_1} + \frac{\Delta k_2^2}{\beta_2}\right),$$

where β_1, β_2 are positive numbers. By virtue of the Σ_k-system with the adaptation algorithm $\mathcal{A}$, its derivative is given by the expression

$$\begin{aligned} \dot{V}\big|_{\Sigma_k} &= \dot{V}\big|_{\Sigma_\gamma} - \left\langle Hx, \begin{pmatrix} 0 \\ b \end{pmatrix}\right\rangle (\Delta k_1 x_1 + \Delta k_2 x_2) + b\left(\frac{\Delta k_1 \xi_1}{\beta_1} + \frac{\Delta k_2 \xi_2}{\beta_2}\right) \\ &= -2\lambda\langle x, Hx\rangle - b\left[\Delta k_1\left(x_1\sigma - \frac{\xi_1}{\beta_1}\right) + \Delta k_2\left(x_2\sigma - \frac{\xi_2}{\beta_2}\right)\right], \end{aligned} \tag{2.40}$$

where $\sigma = \left\langle Hx, \begin{pmatrix} 0 \\ 1 \end{pmatrix} \right\rangle = \langle h_2, x \rangle$. It is clear now that if we choose the right-hand sides of the adaptation algorithm in the form

$$\xi_1 = \beta_1 x_1 \sigma, \quad \xi_2 = \beta_2 x_2 \sigma,$$

then derivative (2.40) assumes the form

$$\dot V = -2\lambda \langle x, Hx \rangle$$

and takes on a negative sign in the space $\{x, k\}$, whence follows the stabilizability of the system in question.

Note that, in this case, the convergence of errors of the estimates Δk_1, Δk_2 to zero is not guaranteed since the derivative $\dot v$ must be negative definite in the extended phase space $\{x, k\}$ for this convergence to occur. It should also be noted that if there is a coordinate disturbance f, i.e., if the object P^f is stabilized in the way indicated above,

$$\dot x_1 = x_2,$$
$$\dot x_2 = -a x_1 + bu + f,$$

an additional term appears in the final expression for the derivative $\dot v$ which is proportional to the disturbance

$$\dot v = -2\lambda \langle x, Hx \rangle + b\sigma f,$$

which, for the unknown f, violates the sign negativity of $\dot v$ and, under the restriction $|f| \le f_0$, leads to the dissipativity of the system with respect to the principal variables, i.e., the stabilization problem cannot be exactly solved.

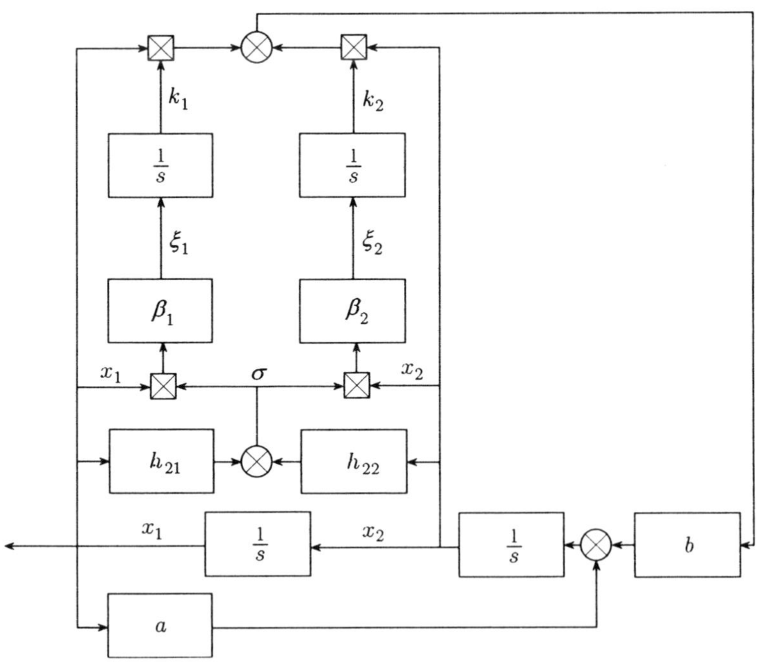

Figure 2.56.

In conclusion, we shall give a block diagram of the adaptive stabilization system synthesized in Example 13. It is shown in Fig. 2.56 in which h_{21} and h_{22} are components of the vector $h_2 = \begin{bmatrix} h_{21} \\ h_{22} \end{bmatrix}$.

In what follows, we shall need this block diagram for the sake of comparison.

2.3. Stabilization by a Variable Structure Controller

Let us now consider one more attempt to find a final solution to the problem considered in this monograph by using a controller of variable structure.

The idea of the variable structure principle consists in a jumpwise variation of the connections between the functional elements of the controller as a function of the phase state of the feedback control system which, in this case, is called a variable structure control system (VSS). If we consider a linear object and linear functional elements of the controller, then we can interpret the corresponding VSS as a collection of linear subsystems and rules of transition from one element of this collection to another when the phase point intersects the separating hyperplanes in the phase space of the system which are known as surfaces of discontinuity. Now if the object is nonlinear or the functional elements are nonlinear, then we have a collection of nonlinear subsystems and, respectively, nonlinear surfaces or manifolds of discontinuity.

In both cases the VSS, which is a nonlinear dynamical system, is described by differential equations with discontinuous right-hand sides. The synthesis of a VSS reduces to a choice of surfaces of discontinuity and an initial collection of subsystems which guarantee the solution of the posed control problem. We do not expose here the VSS theory but only want to point out that VSS serve for the robust control of an object with structural uncertainty (Fig. 2.57), where S

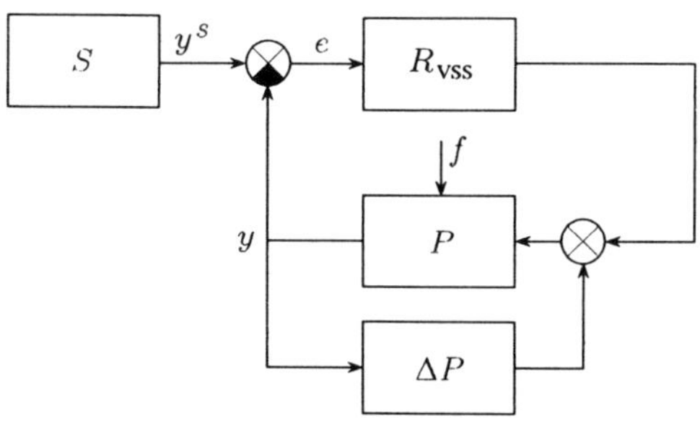

Figure 2.57.

is a reference device, P is the known operator of the object, ΔP is an unknown component of the operator of the object reduced to the control input, and R_{vss} is a variable structure controller. We shall explain the peculiarities of the synthesis of the R_{vss} by way of examples.

2.3.1. An astatic tracking system

Let us consider a simple tracking system with a rational transfer function of an object of the form $W(s) = 1/\alpha(s)$, where $\alpha(s)$ is a polynomial of degree $n = \deg \alpha(s) \geq 1$ (Fig. 2.58). If the controller R in the block diagram in Fig. 2.58 is chosen such that the tracking error $e = y^s - y$ asymptotically tends

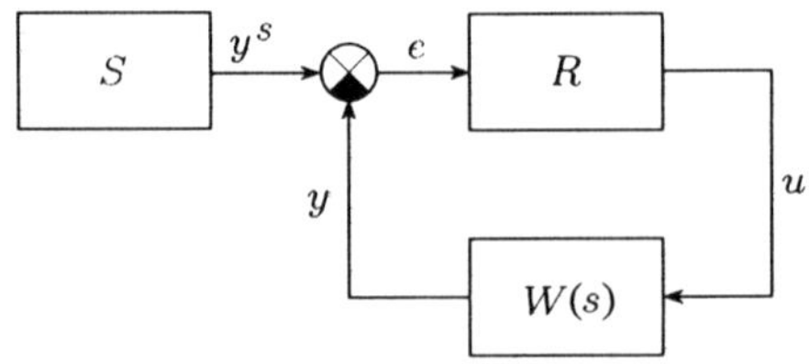

Figure 2.58.

to zero as $t \to \infty$, then such a tracking system is said to be astatic, otherwise a system is said to be static.

To discuss the main problems connected with the construction of an astatic tracking system in the framework of the linear theory, we shall write a motion equation for the system shown in Fig. 2.58 with respect to the error of tracking in operator form. We obtain equations

$$\alpha(s)e = \alpha(s)y^s - u\big|_{u=Re} = \alpha(s)y^s - Re,$$

and, after collecting terms, get an equation

$$\big[\alpha(s) + R(s)\big]e = \alpha(s)y^s. \tag{2.41}$$

It should be emphasized that $R(s)$ in (2.41) is, generally speaking, also a rational function,

$$R(s) = k\frac{\gamma(s)}{\delta(s)}, \tag{2.42}$$

with a gain factor k and polynomials $\gamma(s)$, $\delta(s)$ satisfying the condition of physical realizability

$$\deg \delta(s) \geq \deg \gamma(s).$$

We assume, for definiteness, that $\deg \delta(s) \geq 1$. Taking this assumption into account, we can reduce Eq. (2.42) to the form

$$\big[\alpha(s)\,\delta(s) + k\,\gamma(s)\big]e = \alpha(s)\,\delta(s)\,y^\delta. \tag{2.43}$$

We can see from (2.43) that, theoretically, this tracking system is astatic if and only if the polynomial $\alpha(s)\,\delta(s) + \gamma(s)$ is Hurwitz and the polynomial $\alpha(s)\,\delta(s)$ includes a factor $\mathcal{K}(s)$ which annihilates the reference action function, i.e., is such that

$$\mathcal{K}(s)y^s \equiv 0, \tag{2.44}$$

or, if the factor $\mathcal{K}(s)$ is absent, then

the polynomial $\gamma(s)$ must be Hurwitz,

in the limit, as $k \to \infty$, the behavior of the system can be described by an asymptotically stable equation

$$\gamma(s)e = 0. \tag{2.45}$$

However, since large gain factors lead to nonrobust systems, in practice, when synthesizing astatic systems, we must orient ourselves to finite transfer factors and, hence, to the stability of the polynomial $\alpha(s)\,\delta(s) + \gamma(s)$ and condition (2.44).

If the annihilating polynomial $\mathcal{K}(s)$ is Hurwitz, then the fulfilment of the conditions indicated above does not cause any difficulties. It is a different matter when the polynomial $\mathcal{K}(s)$ is unstable, say, $\mathcal{K}(s) = s^2 - a^2$, $a = \text{const} > 0$. Then we encounter serious problems. Indeed, the polynomial $\delta(s)$ cannot be unstable since, otherwise, the proper dynamics of controller (2.42) is unstable and its output $u(t)$ increases exponentially with all negative consequences, namely, the signal leaves the zone of linearity, etc. Consequently, the unstable components (factors) of the annihilating operator $\mathcal{K}(s)$ must be, at the same time, factors of the polynomial $\alpha(s)$, but this is impossible. This reasoning shows that

- it is impossible to construct an astatic system tracking the exponentially growing signal y^s in the framework of linear control theory that presupposes the use of only bounded transfer factors in the controller.

We have considered two situations, namely, exponentially stable and exponentially unstable $\mathcal{K}(s)$. Let us now analyze a situation when the reference action y^s is a polynomially growing function of time, i.e.,

$$y^s = c_{m+1}t^m + c_m t^{m-1} + \cdots + c_1,$$

where c_i are constants ($i = 1, \ldots, m+1$), and the number m is the order of the polynomial, i.e., $c_{m+1} \neq 0$. The simplest operator that annihilates polynomial (2.45) is, obviously, the operator of an $(m+1)$-fold differentiation $\mathcal{K}(s) = s^{m+1}$. It is natural to assume that it is introduced into the tracking system by the controller, i.e.,

$$\delta(s) = s^{m+1}\,\delta'(s),$$

where $\delta'(s)$ is a stable polynomial. However, in this case, by virtue of the necessary condition of stability, the inequality $\deg \gamma(s) \geq m+1$ must be satisfied for the characteristic polynomial of the feedback system

$$\alpha(s)\delta'(s)\,s^{m+1} + k\,\gamma(s) \tag{2.46}$$

to be Hurwitzian. If (2.46) is a Hurwitz polynomial, then the tracking system reproduces without error the polynomial signal (2.45), and then we say that the controller

$$R(s) = \frac{\gamma(s)}{s^{m+1}\,\delta'(s)} \tag{2.47}$$

ensures the astatism of order $m + 1$. Note that, by virtue of the well-known decomposition of the exponent

$$e^{at} = \sum_{m=0}^{\infty} \frac{(at)^m}{r!},$$

the problem of tracking an exponentially growing signal

$$y^s(t) = y^s(0)e^{at}$$

can be interpreted as a problem of constructing a system with an infinite order of astatism.

It is obvious that the technique described above cannot be used for constructing a tracking system of this kind because of the infinitely growing complexity of the controller. This observation corroborates the inference made above to the point that

- it is impossible to construct a robust astatic tracking system in the framework of a linear theory.

It may appear that this essential inference is also invalid in the framework of the principle of combined control when the controlling signal is formed as the sum of the tracking error feedback signal $u_e = Re$ and the load feedforward signal $u_y = Hy^s$ (Fig. 2.59).

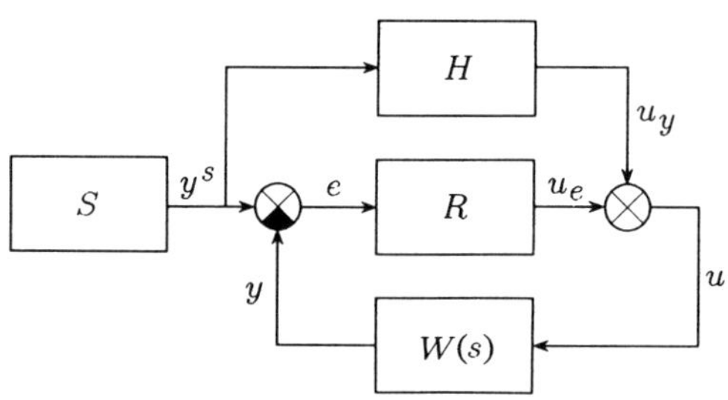

Figure 2.59.

Let us discuss this situation assuming, as before, that $W(s) = 1/\alpha(s)$, $R(s) = \gamma(s)/\delta(s)$. From Fig. 2.59 we have the following motion equation for the error:

$$\alpha(s)\, e = \alpha(s)\, y^s - u \Big|_{u=Hy^s+Re} = \left[\alpha(s) - H(s)\right] y^s - R(s)\, e.$$

After collecting terms, we have an equation

$$\left[\alpha(s)\,\delta(s) + \gamma(s)\right] e = \delta(s)\left[\alpha(s) - H(s)\right] y^s. \tag{2.48}$$

In this equation the feedforward operator $H(s)$ must be chosen such that the right-hand side becomes identically zero. There exist several possibilities of solving this problem.

Technique I. We can, for instance, set

$$H(s) = \alpha(s),$$

and then the feedforward is defined by the expression

$$u_y = \alpha(s)\, y^s$$

and presupposes an n-fold exact differentiation of the reference action signal (recall that $\deg \alpha(s) = n$). For the reason repeatedly indicated above we must reject this solution since it is nonrobust.

Technique II. The multiplicity of differentiation of the reproduced signal $y^s(t)$ can be lowered if

$$\deg \mathcal{K}(s) \leq \deg \alpha(s),$$

where $\mathcal{K}(s)$ is an operator annihilating $y^s(t)$. Indeed, we can divide the polynomial $\alpha(s)$ by $\mathcal{K}(s)$ since, generally speaking, all zeros of $\mathcal{K}(s)$ are not zeros of $\alpha(s)$. As a result of the division we have a remainder $y(s)$, i.e.,

$$\alpha(s) = \alpha'(s)\, \mathcal{K}(s) + \eta(s), \tag{2.49}$$

and the inequalities

$$\deg \eta(s) < \deg \mathcal{K}(s) < \deg \alpha(s)$$

hold true. Now, to achieve the astatism, it suffices (under the condition that $\varphi(s) = \alpha(s)\,\delta(s) + \gamma(s)$ is a Hurwitz characteristic polynomial) to set

$$H(s) = \eta(s). \tag{2.50}$$

Substituting (2.50) into (2.49) and the resulting equation into (2.48), we get an equation

$$\big[\alpha(s)\,\delta(s) + \gamma(s)\big]e = \delta(s)\big[\alpha'(s)\,\mathcal{K}(s) + \eta(s) - \eta(s)\big]y^s = 0$$

which yields what we said above.

Technique III. Now if

$$\deg \mathcal{K}(s) > \deg \alpha(s)$$

and the operator $\mathcal{K}(s)$ has stable zeros, i.e., can be represented as

$$\mathcal{K}(s) = \mathcal{K}_-(s)\, \mathcal{K}_+(s),$$

where $\mathcal{K}_-(s)$ is a Hurwitz polynomial, then these zeros can be included into the collection of zeros of the polynomial $\delta(s)$, i.e., we can take this polynomial in the form

$$\delta(s) = \delta'(s)\, \mathcal{K}_-(s).$$

If it also turns out that

$$\deg \mathcal{K}_+(s) \leq \deg \alpha(s),$$

then it is possible to synthesize the feedforward operator $H(s)$ according to the same scheme as was used in *Technique II*. If the last inequality is not satisfied, then the use of load feedforward does not cardinally affect the possibility of constructing an astatic tracking system.

It follows from the analysis carried out above that, in any case, we shall need an operation of a multiple exact differentiation, and this prevents us from saying that the problem in question is solved.

Example 14 (An exact tracking of a reference action signal). We shall illustrate the theoretical arguments given in the preceding section by a simple example and consider the following statement of the problem: we have to synthesize feedback which would ensure an exact tracking of the reference action y^s by the controlled coordinate y (Fig. 2.60). The equation of the tracking system for the error e is $\dot{e} = \dot{y}^s - u$. If $\dot{y}^s \equiv 0$, then the problem can be solved by any feedback of the form

$$u = ke, \quad k = \text{const} > 0,$$

since, in this case, the equation of the feedback system $\dot{e} = -ke$ is exponentially stable. Now if $\dot{y}^s \not\equiv 0$, say, $\dot{y}^s = \text{const}$ (i.e., the reference action grows linearly), then the static feedback $u = ke$ no longer solves the problem since the motion equation in deviations has a nonzero right-hand side in the equilibrium position $\dot{e} = -ke + \dot{y}^s$ and, consequently, a static error appears, i.e., the value of the control error $e(\infty) = \dot{y}^s/k$ which is stable as $t \to \infty$. It stands to reason that this error of tracking can be removed by increasing the gain factor k of feedback,

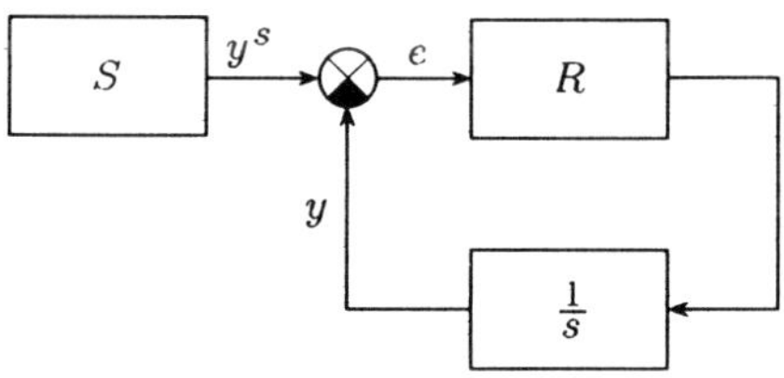

Figure 2.60.

but then problems typical of feedback systems with high gain appear about which we spoke in Chap. 1. If we additionally use a load feedforward in the system shown in Fig. 2.60, i.e., formulate the control as the sum $u = ke + \dot{y}^s$, then we obtain an astatic tracking system with a block diagram shown in Fig. 2.61 and with a motion equation of the form $\dot{e} = -ke$. This equation is asymptotically stable for $k > 0$, and this solves the problem being considered.

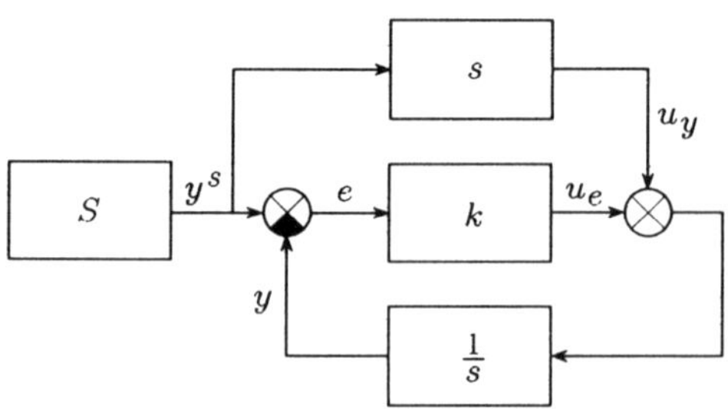

Figure 2.61.

In other words, we have constructed an astatic combined tracking system in which we must use the operation of exact differentiation for achieving astatism, and this must be done in a trivial case like this!

In this connection, it is natural to try to find out the following:

- whether the astatism (a zero stable error) can be ensured under bounded control factors and without employing the operation of exact differentiation.

2.3.2. Second-order astatism

In the preceding example we had $\dot{y}^s$ = const. We differentiate the equation for the object $\dot{e} = \dot{y}^s - u$ with respect to t and introduce a notation

$$\dot{u} = v. \tag{2.51}$$

Then we obtain an equation of the form $\ddot{e} = -v$. Consequently, we reduce the tracking problem to a problem of stabilization of free oscillations, and this problem can be solved, for instance, by means of the application of linear feedback

$$v = k_2\dot{e} + k_1 e \tag{2.52}$$

with positive factors k_1, k_2.

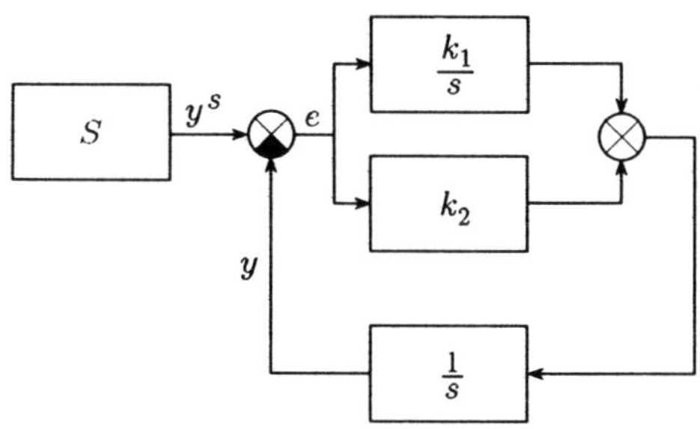

Figure 2.62.

Indeed, the corresponding feedback control system can be described by a stable equation $\ddot{e}+k_2\dot{e}+k_1 e = 0$. From (2.51) and (2.52) we obtain the required feedback in the form of a proportionally-integral law, or PI-law,

$$u = k_2 e + k_1 \int e \, dt.$$

The block diagram of the corresponding tracking system is given in Fig. 2.62.

2.3.3. Astatism of order m

It is clear by induction that if $y^s(t)$ is a polynomial with respect to t of degree m, i.e., all its derivatives beginning with the $(m+1)$th, become identically zero, then, to achieve astatism, no less than m integrations of the tracking error are needed (Fig. 2.63). As $m \to \infty$, the order of the linear control grows to infinity, and this, of course, is practically inexpedient. In particular, we cannot use this

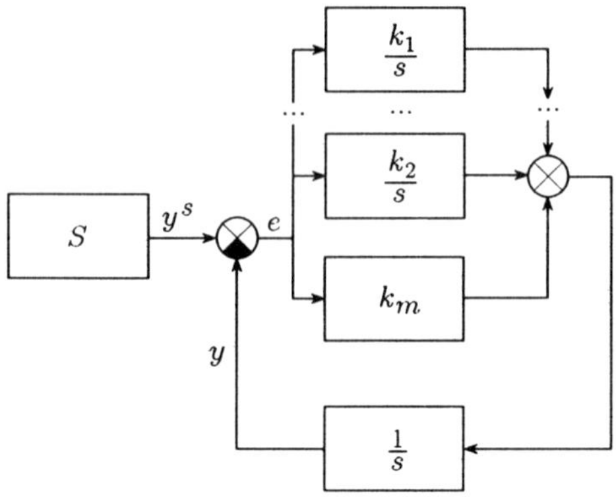

Figure 2.63.

technique to construct an astatic system for an exponentially growing reference action $y^s = e^{\alpha t}$, $\alpha = \text{const} > 0$ since

$$e^{\alpha t} = \sum_0^\infty \frac{(\alpha t)^m}{m!},$$

and an infinity-dimensional control must be used in order to employ these methods. What should be done?

2.3.4. A variable structure astatic tracking system

Let us investigate the system whose block diagram is shown in Fig. 2.64. Its peculiarity is that the sign of the feedback factor with respect to the output y is defined by the sign of the tracking error e, namely, the motion equation has the form $\dot{y} = ky \operatorname{sgn} e$, and if $k > 0$, then, for $e > 0$, the output of the object grows

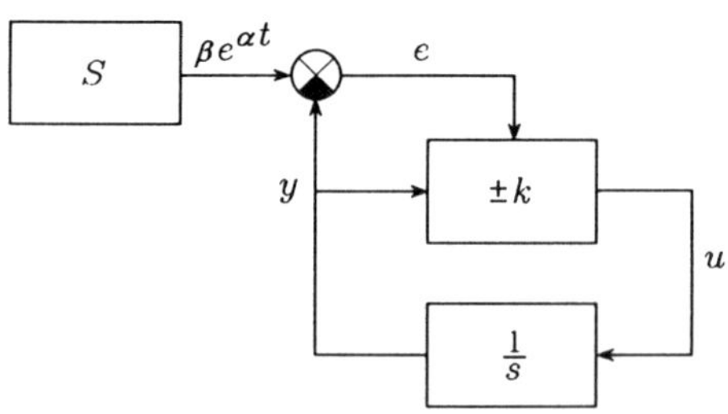

Figure 2.64.

exponentially (Fig. 2.65*a*) since $\dot{y} = ky$, $y = y(0)\,e^{kt}$ and for $e < 0$ the output of the object exponentially decreases (Fig. 2.65*b*) since $\dot{y} = -ky$, $y = y(0)\,e^{-kt}$. If we now set $k > \alpha$ and $\operatorname{sgn} y(0) = \operatorname{sgn} \beta$, then, in a finite time, the exponent $y(0)\,e^{kt}$ for $|y(0)| < |\beta|$ (and the exponent $y(0)\,e^{-kt}$ for $|y(0)| > |\beta|$) becomes

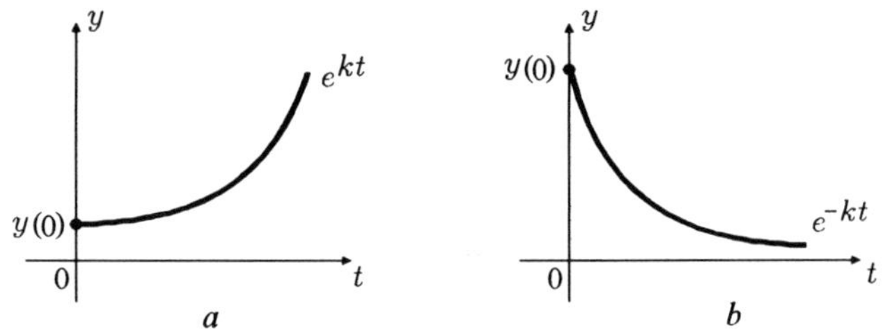

Figure 2.65.

equal to the reference exponent $\beta e^{\alpha t}$ (Fig. 2.66). We have now a sliding mode of switchings, and the output $y(t)$ of the object exactly reproduces the reference action $y^s(t)$ in the sliding mode. Consequently,

- astatism of order ∞ has been achieved with the aid of a discontinuous feedback with finite gain factors and without the use of the derivative of the reference action.

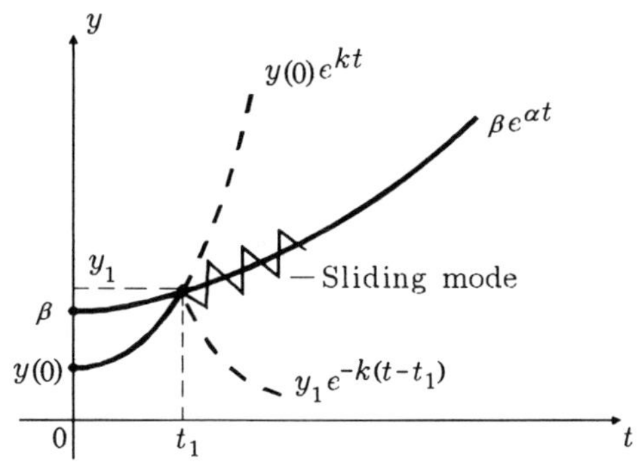

Figure 2.66.

The case that we have considered demonstrates the efficiency of using a discontinuous alternative feedback. It has turned out that this is a rule and not an exception, i.e., a regular use of unstable structures is a fundamental idea underlying the feedback theory which is an alternative to other fundamental ideas of control theory (the ideas of exact compensation and feedback with high gain) and paves the way to the construction of robust control systems, robust tracking systems inclusive. This was a key idea for the theory of variable structure control systems whose fragments are exposed in the sequel with the use of specific examples.

Example 15 (Switching mode in variable structure systems). Consider a problem of stabilization of a second-order object $\ddot{y} = u$ at zero under the condition that we have some information only about the coordinate y and about the sign of its derivative $\dot{y}$. Since the feedback

$$u = -ky \tag{2.53}$$

is not stabilizing for any fixed k, it is obvious that this problem cannot be solved by any linear means.

We shall change the feedback factor (2.53) as a function of y and $\operatorname{sgn} \dot{y}$ in the following way, for instance:

$$k(y, \dot{y}) = \begin{cases} k_1, & y\dot{y} \geq 0, \\ k_2, & y\dot{y} < 0. \end{cases}$$

Then the feedback control system has a discontinuous feedback and can be described by the equation

$$\ddot{y} = -k(y, \dot{y})\, y. \tag{2.54}$$

Let $0 < k_2 < 1 < k_1$, and then the phase trajectories of the system are ellipses extended along the $\dot{y}$-axis for $k = k_1$ (Fig. 2.67*a*) and ellipses extended along the y-axis for $k = k_2$ (Fig. 2.67*b*). The "sewing" of segments of the phase trajectories gives a resulting phase portrait of the stable system (Fig. 2.67*c*). Since $l/L = k_2/k_1$, the ordinates of points of successive switchings of the phase trajectory of the $\dot{y}$-axis form a geometric progression with ratio

$$q = \frac{k_2}{k_1} < 1,$$

and this entails an asymptotic stability of zero of system (2.54).

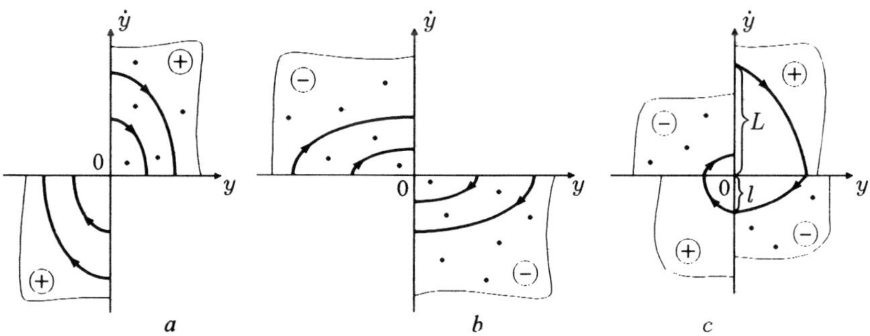

Figure 2.67.

Note that the system that we have constructed is robust, i.e., its qualitative behavior is preserved upon small delays (time or space delays) in switching. The block diagram of this system is shown in Fig. 2.68. We again have feedback whose parameters suffer discontinuities in the function of the phase state of the system.

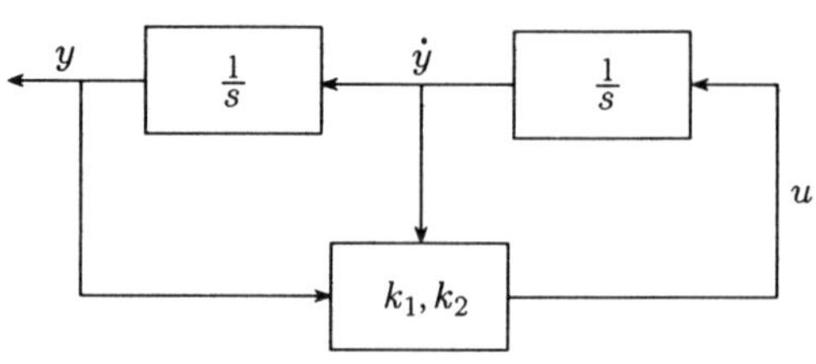

Figure 2.68.

Example 16 (Variable structure systems with motion along degenerate trajectories). Consider a problem of stabilization of the object

$$\ddot{y} = u$$

at zero under the condition that in feedback we can use information concerning the controlled coordinate y and the sign of the linear combination

$$\sigma = \dot{y} + cy, \quad c = \text{const} > 0. \tag{2.55}$$

It is clear again that the stabilization problem cannot be solved by any linear feedback of the form

$$u = -ky.$$

We make the gain factor k discontinuous, say, according to the law

$$k(y,\sigma) = \begin{cases} c^2, & y\sigma > 0, \\ -c^2, & y\sigma \le 0, \end{cases}$$

and obtain a system with discontinuous feedback described by the equation

$$\ddot{y} = -c^2 |y| \operatorname{sgn} \sigma.$$

We shall use the phase plane method to analyze this system. In the domain $y\sigma > 0$ (it is denoted by $\oplus$ in the figure) the behavior of the system is described by the equation $\ddot{y} = -c^2 y$

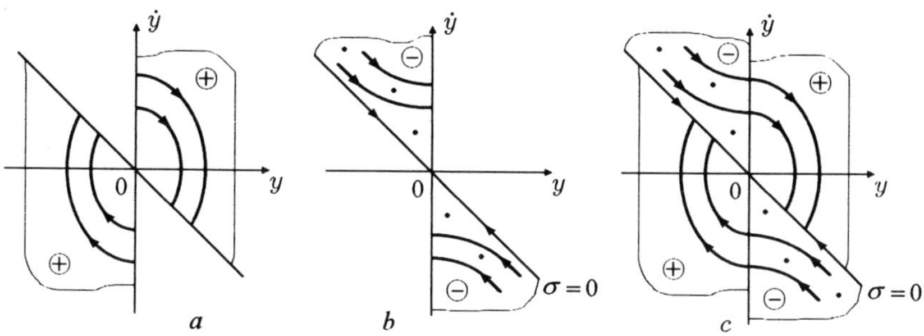

Figure 2.69.

and its phase trajectories are ellipses (Fig. 2.69*a*). For $y\sigma \le 0$ (this domain is denoted by $\ominus$ in the figure) the motion is described by the equation $\ddot{y} = c^2 y$ and its phase trajectories are hyperbolas (Fig. 2.69*b*). As a result of "sewing" phase trajectories along the lines of

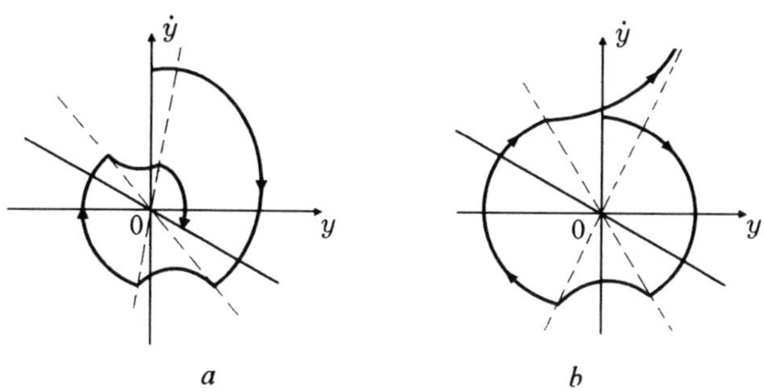

Figure 2.70.

discontinuity, we obtain an asymptotically stable system (Fig. 2.69c). In the stabilization system under consideration each trajectory attains the curve $\sigma = 0$ in a finite time interval and does not leave it. Indeed, by virtue of (2.55), the relation

$$\dot{y} + cy = 0$$

holds and $y(t) \to 0$ as $t \to \infty$, and this is what we had to prove.

The drawback of this control system is its nonrobustness. Small delays in switchings change qualitatively its behavior, namely, aperiodic transient processes may become oscillatory (Fig. 2.70a) or even unstable (Fig. 2.70b).

2.3.5. Sliding mode throughout a straight line

When investigating the modes of switching and motion along degenerate trajectories, we reveal definite merits of the variable structure principle, including the simplicity of feedback law, a decrease in the volume of information needed for stabilization of the object as compared to linear feedback. However, we can also see the drawbacks of these modes, namely, the oscillation of the transient process in the switching mode and the difficulty encountered in the organization of motion along degenerate trajectories when the parameters of the object vary and external forces act on the object. Neither do we have a simple answer to the questions connected with the robustness of control systems and with the possibility of extension of the method of synthesis to multidimensional systems. Therefore the following question is urgent:

- is it possible to remove or weaken the drawbacks indicated above and, at the same time, preserve the merits of the structure variability principle?

This question can be answered in the positive if a sliding mode on a straight line is used.

Example 17 (Sliding mode on a straight line). We shall consider a stabilization problem at zero of the object

$$\ddot{y} = u$$

under the condition that it is possible to use in feedback the information about the coordinate y and about the sign of the linear combination $\sigma = \dot{y} + cy$, $c = \text{const} > 0$. It is obvious that the linear feedback $u = -ky$ does not solve the problem for any fixed gain factor k.

We apply the variability of structure principle by setting $u = \psi(y, \dot{y})y$ and

$$\psi(y,\dot{y}) = \begin{cases} k_1, & y\sigma \geq 0, \\ k_2, & y\sigma < 0, \end{cases}$$

where the constants $k_1 < 0$, $k_2 > 0$, $c^2 > 0$. Then, in the sector $y\sigma \geq 0$ (I) in the phase plane $(y, \dot{y})$ the notion of the control system can be described by the equation

$$\ddot{y} = k_1 y, \quad k_1 < 0,$$

whose phase trajectories are ellipses (Fig. 2.71a). The equation

$$\ddot{y} = k_2 y, \quad k_2 > 0,$$

acts in the sector $y\sigma < 0$ (II) and the motion is along hyperbolic curves (Fig. 2.71b). We can see that an asymptotic stability does not occur for any positive or negative values of k. However, if we "sew" the phase trajectories along the lines of discontinuity $\sigma = 0$, $y = 0$, then we obtain an asymptotically stable system (Fig. 2.71c). The designations I and II in the figure correspond to the domains of action of structures I and II. On the straight line

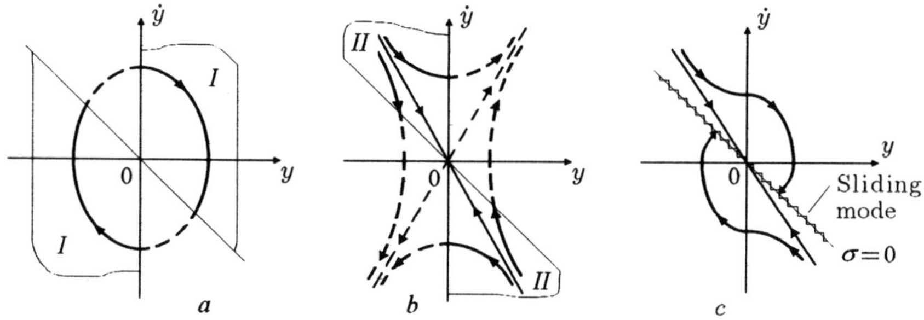

Figure 2.71.

$\sigma = 0$ the phase trajectories of equations acting in sectors I and II are directed towards each other (Fig. 2.72), where f^I, f^{II} are phase velocities. Formally, this means that $\sigma\dot{\sigma} < 0$, when $\sigma \neq 0$. Consequently, the phase point cannot leave the line of discontinuity $\sigma = 0$, the relation $\sigma \equiv 0$ holds when the motion is continued, and this motion is associated with the phase vector $f^0 = \alpha f^I + (1 - \alpha) f^{II}$, $\alpha \geq 0$, which is directed along the line of discontinuity $\sigma = 0$.

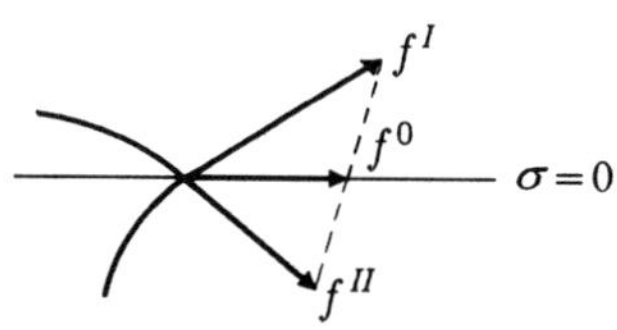

Figure 2.72.

A.F. Filippov's theory gives a strict analysis of motion of a system in a sliding mode (definition of the solution, conditions of its existence, uniqueness, continuity to the right, etc.). We give here only an heuristic reasoning. Since the relation $\sigma = \dot{y} + cy = 0$ is satisfied in a sliding mode, it follows that

$$y(t) = y(t_1)\, e^{-c(t-t_1)},$$

where t_1 is the moment of appearance of the sliding mode. Since $c > 0$, it follows that $y(t) \to 0$ as $t \to \infty$, and this is what we had to prove. This means that the stabilization problem is solved. Note that even if we had some information about the linear combination $\sigma = \dot{y} + cy$ and not about its sign as before, then, to obtain a similar result with the aid of linear feedback $u = -k\sigma$, we should require that $k \to \infty$, but here the gain factors k_1 and k_2 are finite, and this ensures the robustness of the control system. Thus,

- as a result of combining structures (I, elliptic, and II, hyperbolic), which are not asymptotically stable, a stable motion appeared which was not inherent in any one of them taken separately, i.e., a new quality appeared.

In other words, the introduction of a discontinuous static element to two inputs (ψ-cells) into the feedback (Fig. 2.73) provides the feedback control system with new qualities, namely,

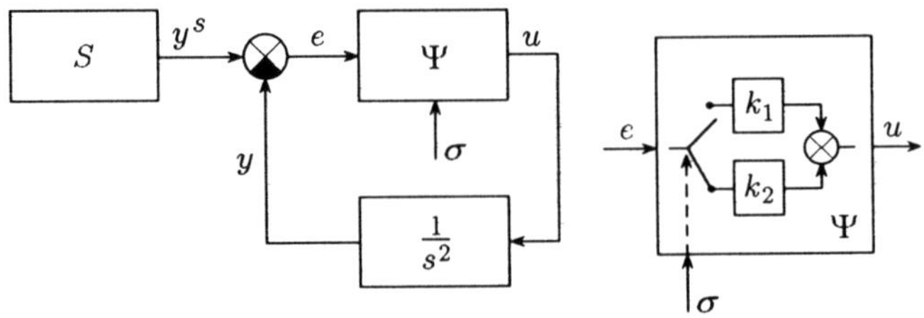

Figure 2.73.

- an asymptotic stability under weaker, as compared to a linear feedback, requirements to the volume of information,
- the lower order of the motion equation for all trajectories, except for asymptotic curves (Fig. 2.71*c*),
- nonsensitivity to the variations of the parameters of the object and to the action of the external force,
- robustness to singular perturbations.

Although the inferences made above are almost obvious, we shall give the necessary justifications.

2.3.6. Analysis of robustness of VSS relative to parametric perturbations

Let us consider the variable structure system shown in Fig. 2.74 and investigate its properties on the assumption that the parameters a_1, a_2, and b of the object are known with an accuracy to within the ranges

$$\begin{aligned} a_i^- &\le a_i \le a_i^+, \quad i = 1, 2, \\ 0 &< b \le b^-, \end{aligned}$$

i.e., we have information only about the numbers $a_i^\pm$ and b^- and do not know anything about the values of the parameters of the object and the character of their variation in time.

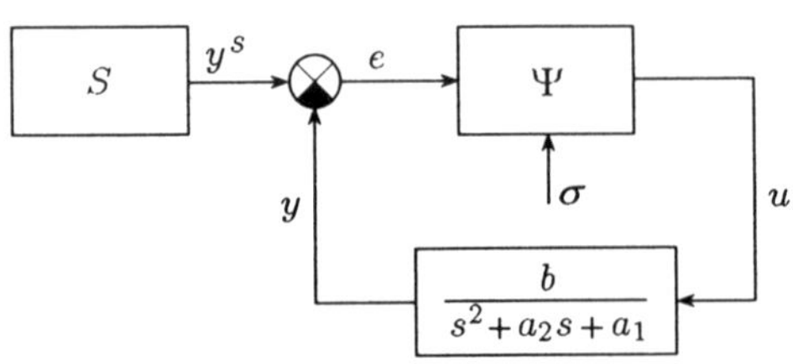

Figure 2.74.

Perturbations of an object of this kind are regular, and therefore the problem being considered is a problem of analyzing the robustness of a VSS to regular disturbances.

For convenience, we set $y^s \equiv 0$. Then we deal with a variable structure equation

$$\ddot{y} + a_2\dot{y} + a_1 y = b\psi y,$$

in which the character of switchings is defined by the action of the ψ-cell:

$$\psi = \begin{cases} k_1, & y\sigma \geq 0 \\ k_2, & y\sigma < 0, \ \sigma = \dot{y} + cy. \end{cases}$$

The equation

$$\ddot{y} + a_2\dot{y} + (a_1 - bk_1)y = 0$$

acts in the domain $y\sigma > 0$, and, when the inequality

$$a_2^2 < 4(a_1 - bk_1)$$

is satisfied, the motion is along a twisting (Fig. 2.75*a*) or untwisting (Fig. 2.75*b*) spiral depending on the sign of the parameter a_2 (all parameters are assumed to be constant). The equation $\ddot{y} + a_2\dot{y} + (a_1 - bk_2)y = 0$ acts in the domain $y\sigma < 0$, and, when the condition $a_1 - bk_2 < 0$ is fulfilled, structure II is associated with hyperbolic phase trajectories (Fig. 2.76*a*). It stands to reason that the position of asymptotic trajectories (asymptotes) varies with the variation of the parameters of the object, and if the inequality $bk_2 > c(c - a_2) + a_1$ is fulfilled for any admissible a_1, a_2, b, then the asymptote corresponding to the stable motion is disposed as shown in Fig. 2.76*a*. After "sewing" the phase trajectories of structures I and II, we get a phase portrait shown in Fig. 2.76*b* in which "skew" sectors, with origins at the points M, M' and L, L', are hatched. These sectors are covered by the phase trajectories of the VSS for certain admissible combinations of the parameters a_1, a_2, b. As before, the phase point almost always gets to the switching line $\sigma = 0$ in a finite time interval and a switching mode arises here.

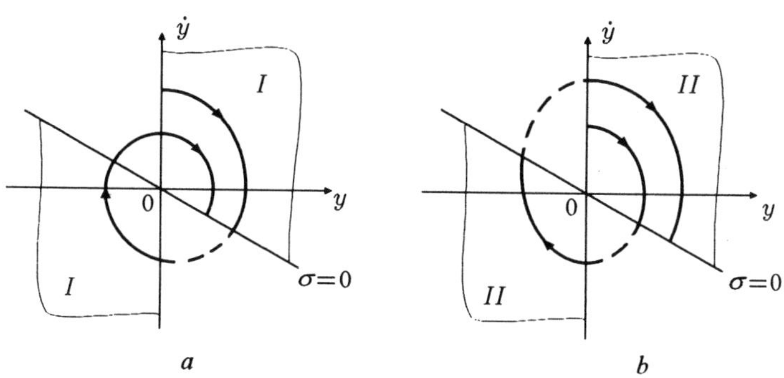

Figure 2.75.

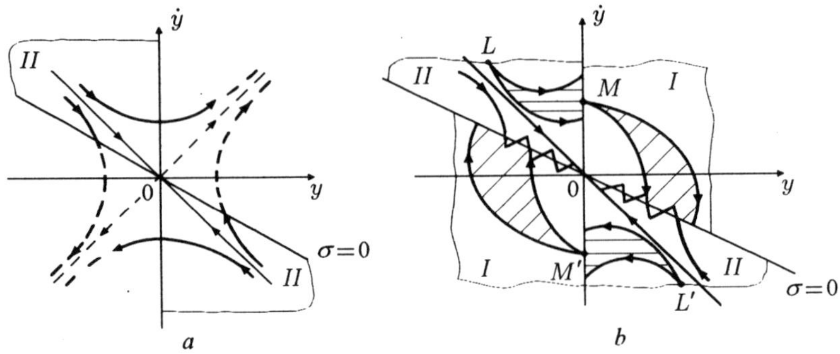

Figure 2.76.

In a sliding mode the motion is defined by the equation

$$\sigma = \dot{y} + cy = 0,$$

i.e., in this case as well, the order of the motion equation is lowered and the independence of motion of the object parameters is attained, the arbitrary variation of the parameters being admissible here.

Everything that we said above is valid under the following conditions:

- getting to the line of discontinuity $\sigma = 0$,
- existence of a sliding mode throughout the straight line $\sigma = 0$.

The necessary and sufficient condition for this hit is the absence of real positive zeros of the polynomial

$$\varphi(s) = s^2 + a_2^{\pm} s + (a_1^{\pm} - b^- k_1).$$

The sufficient condition of existence of a sliding mode has the form

$$\lim_{\sigma \to +0} \frac{d\sigma}{dt} \leq 0, \quad \lim_{\sigma \to -0} \frac{d\sigma}{dt} \geq 0 \tag{2.56}$$

and, since

$$\dot{\sigma} = (c - a_2)\sigma - [c(c - a_2) + a_1]y + b\psi y,$$

it is satisfied for

$$k_1 < \min_{a_i, b} \left[\frac{c(c - a_2) + a_1}{b} \right], \quad k_2 > \max_{a_i, b} \left[\frac{c(c - a_2) - a_1}{b} \right].$$

Note that under the conditions of stabilization by the VSS methods the constraints on the speed of variation of the object parameters are absent whereas for linear systems these constraints are present, as, for instance, in the method of frozen coefficients.

2.3.7. VSS in the presence of an external force

Consider the problem of synthesis of a variable structure controller for the system shown in Fig. 2.77.

The motion equations for a system relative to the error and its derivatives have the form

$$\ddot{e} + a_2\dot{e} + a_1 e = -bu + F, \tag{2.57}$$

where

$$F = \ddot{y}^s + a_2\dot{y}^s + a_1 y^s - bf,$$

and the parameters of disturbance F are in the ranges

$$a_i^- \leq a_i \leq a_i^+, \quad i = 1, 2; \ 0 < b^- \leq b.$$

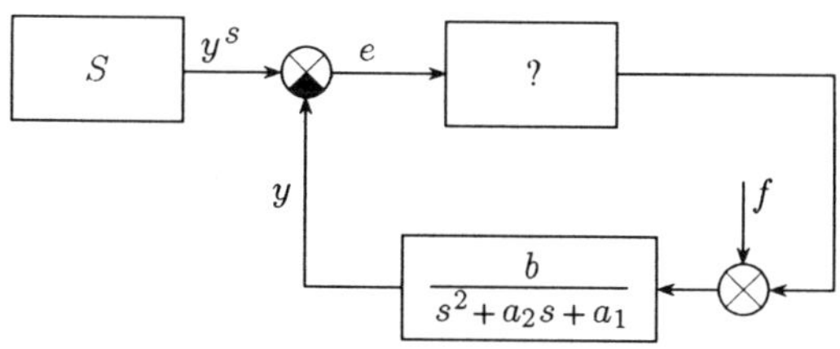

Figure 2.77.

Two versions of problem statement are possible here, namely,

1. the function $F(t)$ is known,
2. the function $F(t)$ is not measured, but its majorant $F_m(t)$ is known, i.e., the estimate $|F(t)| \leq F_m(t)$ is valid for all t.

We have to stabilize the object (2.57) at zero with the use of information concerning the error e and its derivative $\dot{e}$. Note that except for feedback with high gain, other traditional means of compensation for F are not suitable here.

We shall use the technique described in Example 17 to synthesize feedback. We take the line of switching in the plane $(e, \dot{e})$ in the form $\sigma = \dot{e} + ce$ and write the expression for its derivative with the use of the motion equation (2.57). We have

$$\dot{\sigma} = (c - a_2)\sigma - \left[c(c - a_2) + a_1\right]e - bu + F. \tag{2.58}$$

We formulate the control as the sum of two components $u = u_e + u_F$, where, as before, we choose the first component u_e with a discontinuous coefficient, i.e.,

$$u_e = -\psi_e e, \quad \psi_e = \begin{cases} k_1, & e\sigma \geq 0, \\ k_2, & e\sigma < 0. \end{cases}$$

Substituting these expressions into (2.58), we obtain

$$\dot{\sigma} = (c - a_2)\sigma + \dot{\sigma}_0 + bu_F + F, \tag{2.59}$$

where

$$\dot{\sigma}_0 = -[c(c - a_2) + a_1]e + b\psi_e e.$$

If the parameters of the ψ_e-cell are chosen such as in Example 17, namely,

$$k_1 < \min_{a_i, b} \frac{c(c - a_2) + a_1}{b}, \quad k_2 > \max_{a_i, b} \frac{c(c - a_2) + a_1}{b},$$

then the inequality $\sigma\dot{\sigma}_0 \leq 0$ is satisfied. If we now choose the second component u_F of the control such that the condition

$$\sigma(bu_F + F) \leq 0 \tag{2.60}$$

is fulfilled, then, as can be seen from the expression

$$\sigma\dot{\sigma} = (c - a_2)\sigma^2 + \sigma\dot{\sigma}_0 + \sigma(bu_F + F), \tag{2.61}$$

the condition $\sigma\dot{\sigma} \leq 0$ is fulfilled on the surface $\sigma = 0$ and we have a sliding mode. Moreover, inequality (2.60) guarantees the hitting of the surface $\sigma = 0$ by the representative point. Indeed, if, for $F \equiv 0$, the conditions under which the point hits the surface $\sigma = 0$ are fulfilled, then the third term in (2.61) strengthens these conditions.

Thus, if the polynomials

$$\varphi^{\pm}(s) = s^2 + a_2^{\pm}s + a_1^{\pm} - b^- k_1$$

have no real positive zeros, then the hitting conditions are ensured by inequality (2.60).

Inequality (2.60) is satisfied by the set of functions u_F. Here are some of them which are typical of VSS. In the first version, when the disturbance F is known, it is the function

$$u_F = \psi_F F, \quad \psi_F = \begin{cases} l_1, & \sigma F \geq 0, \\ l_2, & \sigma F < 0, \end{cases}$$

where $b^- l_1 < -1$, $b^- l_2 \geq 1$. In the second version, when the disturbance F is unknown, the form of the law is preserved, but, instead of the function F, we use its majorant F_m,

$$u_F = \psi_F F_m, \quad \psi_F = \begin{cases} l_1, & \sigma \geq 0, \\ l_2, & \sigma < 0, \end{cases}$$

where $b^- l_1 \leq -1$, $b^- l_2 \geq 1$.

In the structure of the synthesized control system for the case of the measured disturbance, shown in Fig. 2.78, we see two ψ-cells and this characterizes this

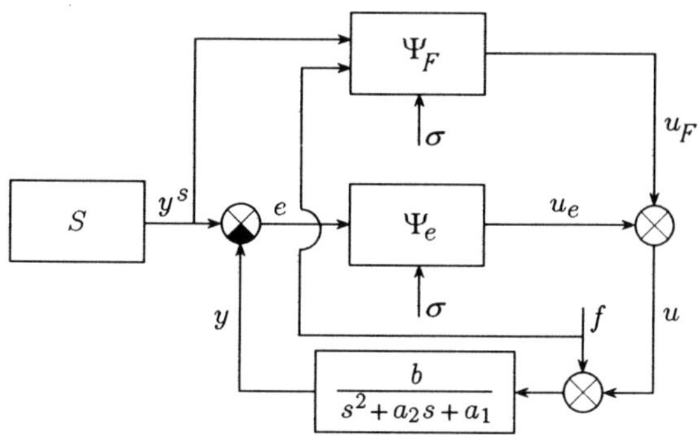

Figure 2.78.

system as a variable structure system. In a VSS, not only the feedback factor ψ_e varies jumpwise, but also the disturbance feedforward factor.

Note that the ψ-cell is a functional element for two inputs and can be represented as a relay with a varying value of the signal being commuted. This is clearly seen in the so-called quasirelay representation of the ψ-cell.

2.3.8. Quasirelay representation of a ψ-cell

The standard representation of a ψ-cell is given in Fig. 2.79. The input and output of the ψ-cell are related by the expressions

$$u = \psi e, \qquad \psi = \begin{cases} k_1, & e\sigma \geq 0, \\ k_2, & e\sigma < 0. \end{cases}$$

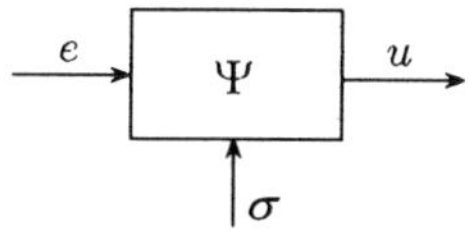

Figure 2.79.

Without loss of generality, we set $k_1 = -k_2 = -k$, and then we have an expression $\psi = -k\,\mathrm{sgn}(e\sigma)$ for the ψ-cell and, consequently, the VSS law of feedback has the form

$$u = \psi e = -k|e|\,\mathrm{sgn}\,\sigma.$$

The last expression is called a quasirelay representation of the ψ-cell. Its block diagram is given in Fig. 2.80, and, in greater detail, in Fig. 2.81. The sign ⊠ in Fig. 2.81 is used to denote the multiplicator, i.e., the multiplication operator for the signals, and abs $(\cdot)$ denotes the operations of subtraction of the absolute value.

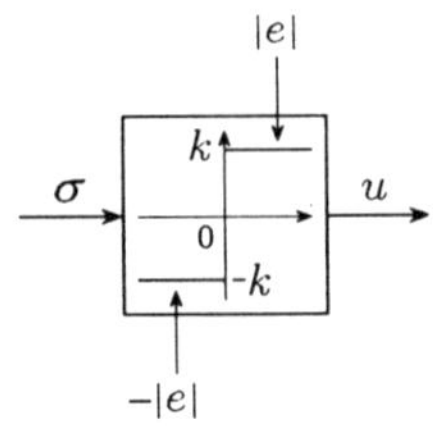

Figure 2.80.

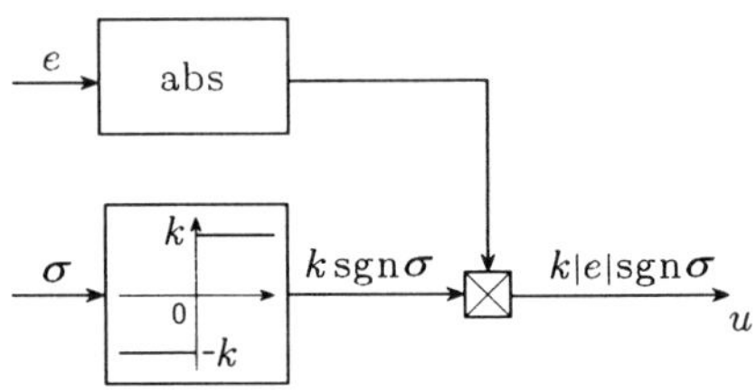

Figure 2.81.

Example 18 (VSS under a singular perturbation). Let us investigate the effect produced by the time delay in switchings on the properties of the VSS synthesized in Example 17, i.e., analyze the qualitative behavior of the solutions of the equation

$$\ddot{y} + a_2\dot{y} + a_1 y = -bk|y|\,\mathrm{sgn}_\tau\,\sigma,$$
$$\sigma = \dot{y} + cy, \quad \mathrm{sgn}_\tau\,\sigma = \mathrm{sgn}\,\sigma(t-\tau),$$

where all constants and parameters satisfy the standard conditions and τ is a sufficiently small time delay constant. The index τ signifies, as usual, that the ψ-cell "operates" not at the

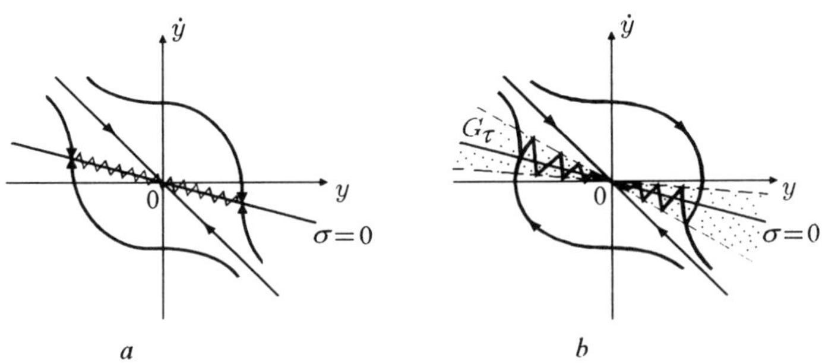

Figure 2.82.

moment of sign change by the input signal but in time τ. It follows that instead of an ideal sliding mode that we have for $\tau = 0$ (Fig. 2.82*a*) a switching mode known as a real sliding mode occurs (Fig. 2.82*b*). Whereas in the ideal sliding the phase point belongs to the line $\sigma = 0$,

$$(y, \dot{y}) \in G_0 = \{(y, \dot{y}) \mid \sigma = 0\},$$

for a sufficiently small τ, when the sliding is real, the phase point is in a certain neighborhood of the line $\sigma = 0$, i.e.,

$$(y, \dot{y}) \in G_\tau = \{g(y, \dot{y}) \mid |\sigma| \leq \delta(\tau)|y|\},$$

where $\delta = O(\tau)$, i.e., δ is a quantity of order τ.

It follows from this example that the asymptotic stability is preserved under such a singular perturbation, in contrast to a relay system (Fig. 2.82*b*). Formally, the convergence of the solution to the origin in the real sliding mode follows from the differential equation and the inequality that describe this mode, namely,

$$\dot{y} + cy = \sigma, \quad |\sigma| \leq \delta|y|.$$

It is obvious that for a sufficiently small δ (i.e., small τ) we have an exponential convergence to zero. In other words,

- the VSS is robust relative to singular perturbations.

Example 19 (VSS under a functional perturbation). Let us consider the effect produced by a functionally nonideal mode, say, a positive hysteresis of value $\Delta > 0$ in the switchings of the ψ-cell on the qualitative behavior of a VSS. In this case, we deal with an equation

$$\ddot{y} + a_2\dot{y} + a_1 y = -bk|y| \operatorname{sgn}_\Delta \sigma,$$

in which the switchings occur on the surface

$$\sigma = \dot{y} + cy,$$

and all parameters and their choice are described in Example 17.

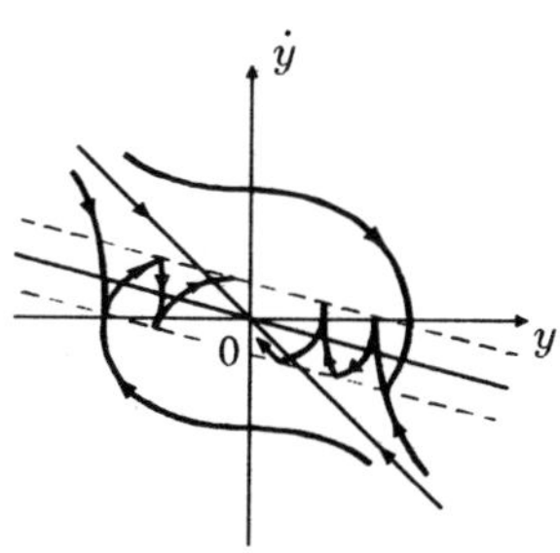

Figure 2.83.

The properties of the discontinuous element $\operatorname{sgn}_\Delta \sigma$ are such that a switching occurs only on the lines $|\sigma| = \Delta$ after the signal passes $\sigma = 0$, and therefore, instead of the standard phase portrait of a VSS with an ideal sliding mode, we have a real sliding mode in the strip $|\sigma| \leq \Delta$ (Fig. 2.83). Of course, this leads to dissipativity, i.e., to the loss of the property of asymptotic stability. Formally, this fact follows from the relations

$$\dot{y} + cy = \sigma, \quad |\sigma| \leq \Delta.$$

Consequently,

- the VSS is not robust relative to functional perturbations.

2.3.9. Limitations and drawbacks of the VSS theory and the related problems

The examples considered above and many results from the theory and practice of the variable structure systems allow us to make conclusions which are important for the development of the feedback theory. Here are some of them.

- The parameters of a real sliding mode depend on the hidden parameters τ, Δ in the examples considered above. The "beatings" that arise, i.e., modes of high-frequency oscillations in the neighborhood of $\sigma = 0$, affect the operation of mechanical and electromechanical drives which are often used in practice as power plants. This circumstance makes the practical use of classical VSS impossible.

- The standard sliding mode can be simulated on a discrete computer only with the use of the first-order methods (the methods of Euler, Adams, etc.) since higher-order methods require the existence of derivatives of the right-hand side of the differential equation. This condition is not fulfilled for equations describing VSS. For a "discrete" sliding we have relations

  $$\sigma \sim O(h), \quad \dot{\sigma} \sim O(h),$$

  where h is a discretization step. These relations mean that the deviation of the trajectories from the line of sliding is proportional to the step of discretization. In order to increase the precision of sliding, we must diminish the discretization step h, and this immediately leads to an increase in the time of computations since the latter are of order $1/h$. It is precisely this circumstance that makes impossible a direct use of VSS with respect to sliding modes in systems of direct digital control.

- The difficulties encountered in the stabilization of an uncertain object of relative order r such that $1 < r < n-1$. To a large degree, these difficulties lie in obtaining information about the state of the uncertain system, in particular, in obtaining "good" estimates for the derivatives of the output signal of the object.

In connection with these circumstances we can formulate the following problem: *it is required to construct a control system with the properties of an ideal VSS but with a smooth feedback which is robust relative to regular, singular, functional, and structural disturbances.* It is this problem that is considered in the second part of the monograph.

2.4. Bibliographical Comments

The researchers began to be interested in discontinuous systems long ago. For instance, in classical mechanics this interest was connected with the analysis of the influence of the Coulomb friction and hysteresis phenomena and in the electrical engineering and radio engineering it was connected with the appearance of stabilizers, various generators, economical power amplifiers, etc.

It was noted that relay systems were prone to self-excited oscillations, but, at the same time, certain motions in relay systems were hardly sensitive to parameters and therefore proved to be useful for compensating for the influence of uncertainty factors on the properties of a control system. The need of a purposeful use of this property as well as the need of means of damping self-excited oscillations led to the creation of the theory of relay systems. At the first stages, a considerable contribution to this theory was made by Hazen [91], Flugge-Lotz [90], and Tsypkin [75]. The reader can get acquainted with the contemporary state of the theory of relay systems by reading the books [9, 14].

The interest that the contemporary control theory shows to relay systems is connected, first of all, with problems of time-optimal action under a limited amount of resources [60, 94]. Some aspects of relay stabilization under uncertainty conditions were considered in [7]. Mathematical methods of analysis of sliding modes arising in relay systems can be found in [71]. Many works are devoted to problems of control of uncertain objects, see, for instance, [46, 55, 69, 73, 74, 76]. The fundamentals of the theory of variable structure automatic control systems are exposed in the monograph [15].

Part II

New Types of Feedback

The second part of the monograph is devoted to the exposition of the theoretical foundations of designing the new class of nonlinear control systems. The mathematical basis of designing these control systems is the binarity principle in accordance with which the coordinates and operators of a nonlinear control system are regarded as the totality of its state variables and they are not radically distinguished. Moreover, every element of this totality can be a coordinate or operator, and the substantive interpretation of a state variable is predetermined by the role it plays in a specific local transformation, and this role can certainly vary from one transformation to another. This means that operators can be subjected to transformations similar to those used for the transformation of coordinates. This circumstance immediately leads to the necessity of introducing new types of feedback when the aim of feedback is the formation of an operator rather than the formation of a variable as it is done in classical control theory.

The combination of the binarity principle and the deviation control principle makes it possible to pass to an automatic formation of control laws under the conditions where the a priori information is insufficient for a direct synthesis of feedback which would provide the feedback control system with the required set of properties.

In the second part of the monograph we also speak about the development of the conceptual apparatus of the new theory and about the synthesis of generalized structures of feedback systems with new types of feedback. A considerable attention is given to the analysis of specific examples.

Bibliographical note. The general principles of the theory of new types of feedback are formulated in [18, 19, 79]. The first systematic exposition of this theory can be found in [17]. The works [20, 21, 37, 38, 39, 40, 41, 42, 85, 86, 87, 88] are devoted to its different aspects. The stabilization of uncertain systems with the use of nonstandard feedback is considered in [26, 27, 28, 31, 32, 33, 34, 35, 36, 44, 45, 81, 82, 83, 84].

The mathematical theory of discontinuous systems was developed by Filippov in [72], and different applied aspects of standard sliding modes can be found in [9, 14, 68]. The works [22, 23, 80] are devoted to higher-order sliding modes. The article [25] contains the detailed exposition of this theory.

The reader can get acquainted with the problem of obtaining high quality derivatives of signals by reading [1, 6, 49]. Mathematical problems of differentiation are considered in the monograph [66], different discrete approximations can be found in [65], and the contemporary interpretation of the differentiation problem is given in [2].

Thanks to the works by Pontryagin (the maximum principle) [63] and Bellman (the optimality principle) [4] the optimal control problem assumed its contemporary form. Optimal control is the most thoroughly worked out and the most rapidly developing division of control theory. The classical theory can be found in [3, 5, 8, 10, 50, 54, 70, 94], the numerical aspects of optimal control are discussed in [59]. The contemporary interpretation of the main ideas of optimal control is given in [51]. The generalization of the method of suboptimal control can be found in [29, 30, 43].

Chapter 3

General Aspects of the Theory of New Types of Feedback

The new theory is based on the binarity principle which reveals the dual nature of signals in nonlinear dynamical systems, namely, the fact that signals can play either the part of variables subjected to transformations or the part of operators that define these transformations. The binarity principle makes it possible to lay the synthesis of the operator of stabilizing feedback on an auxiliary nonlinear system with feedback. The development of this idea and the corresponding methods of synthesis led to the necessity of considering three new types of feedback, namely, the operator, the operator-coordinate, and the coordinate-operator feedback.

In this chapter we give the main concepts, definitions, and principles of design of automatic control systems with new types of feedback.

3.1. Introductory Remarks

The material exposed in the preceding chapter leads to the conclusion that the traditional theory of automatic control systems was developed, in the main, in the framework of block diagrams shown in Fig. 3.1. In the framework of these diagrams, the problem of synthesis of a control system reduces to the choice of a priori information about the object, the disturbance, and the purpose of adjustment of the controller operator R which solves the stabilization problem, i.e., which ensures the following condition: $e \to 0$ as $t \to \infty$.

If the a priori information is sufficient for solving the synthesis problem, i.e., the information is exact and only slight deviations ΔP from the model of linear operator P and of the external forces f from their wave model are admissible, and, in addition, if the purpose of adjustment is defined not very "strictly," then the classical methods of synthesis described in the preceding chapter are quite suitable for synthesizing the stabilizing operator R.

When the deviation ΔP is not very small and the necessary information about the variation of characteristics of the control object can be obtained in

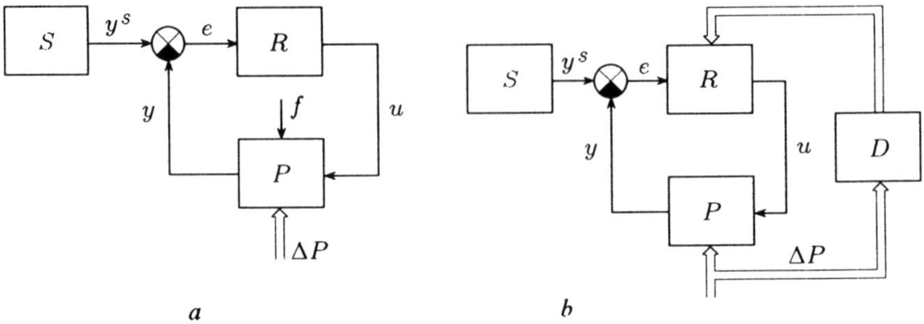

Figure 3.1.

the on-line mode, then we can organize, also in the on-line mode, an additional adjustment of the feedback operator R according to the block diagram shown in Fig. 3.1*a* which is accepted in the adaptive control theory.

Now if the variation of the object operator ΔP is intensive and, in addition, in a manner that cannot be controlled, the disturbance f is not a wave disturbance, and, moreover, "strict" requirements to the quality of control are formulated, and there are constraints on the phase variables and control, then the *traditional methods of synthesizing the stabilizing controls are not suitable and new approaches must be sought to the synthesis of feedback under these difficult conditions.*

When defining the direction of search, it is useful, as usual, to use the experience and analogies. Recall that the feedback principle has replaced the disturbance control principle when this disturbance became unknown. And whereas in the case where the disturbance was known the aim of control synthesis was the obtaining of the $u^s(t)$ "program", in the case where the information about disturbance is absent the aim of synthesis of the control is the finding of its operator R which generates the required control signal $u(t)$ with the aid of feedback. As to the values of the signal $u(t)$, they became inessential, of importance is only the fact that when the choice of the operator R is correct, the feedback control system has the required property, namely,

$$e(t) \to 0, \quad t \to \infty.$$

The transition from the formation of a signal to the formation of the operator that generates the needed signal, indicated above, must be made a principle and formulated as follows, for instance:

- when there is a deficit of information, we must pass from the program formation of a function (element) to the synthesis of an algorithm (operator) that generates the function.

Just as the time function $u(t)$ is an element of a certain set of admissible controls, the operator R is only an element of the set $\mathcal{R}$ of stabilizing feedbacks. Therefore, in order to obtain the required operator R under the deficit of information, we use the principle formulated above, namely, *we synthesize not the feedback operator R itself but the algorithm of its formation, and since all this is done under the conditions of uncertainty, we cannot do without the feedback mechanism.* Note that this idea is an alternative of the idea of adaptive control (Fig. 3.1*b*) where the procedure of identification is used to realize the "program" generation of the required operator R.

The principle formulated above will be called the *generation principle*. It stands to reason that the generation principle can be generalized to the formation of an algorithm which generates the operator R, and so on and so forth. When we apply the generation principle, the following urgent questions naturally arise:

- in what way should the generation loop be synthesized,
- what is the error or the controlled coordinate of this loop,
- what should be used as the control device in this loop,
- can we gain any practical benefit from this idea since, as a matter of fact, the generation loop can be regarded as a part of the control, and it follows that this synthesis of the control "by parts" is simpler than the synthesis of the control "as a whole"?

Whereas it is easy to answer some of these questions, some other questions have no unambiguous answer at all. This means that there is much heuristic reasoning and arbitrariness in the realization of the outlined plan of automatic synthesis of stabilizing controls under the conditions of unremovable uncertainty based on a priori and current information. This arbitrariness can be partly removed with the aid of the following system of basic concepts.

3.2. System of Basic Concepts

In order to make the theory that we are developing constructive, we shall need a number of concepts which are nonstandard for the classical control theory, and we introduce the key concept in the following section.

3.2.1. Operator signal

To show that the operator is subjected to a change, we shall use a double-lined arrow (Fig. 3.2). The corresponding variable will be denoted by a letter of the Greek alphabet, say, μ, and called an operator signal, or O-signal for brevity

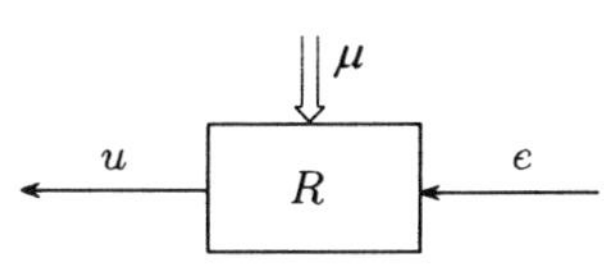

Figure 3.2.

(Fig. 3.3a), in contrast to ordinary variables which are denoted by Latin letters and ordinary arrows (Fig. 3.3b) and called coordinate signals or C-signals. Since

Figure 3.3.

every O-signal, just as a C-signal, has a physical nature, it is obvious that the difference between O- and C-signals is conventional and is predetermined by the interpretation of the role played by the signal in the local transformation $y = P_1(\mu, x) = P_2(x, \mu)$ (Fig. 3.4). In the first case (Fig. 3.4a), the signal x is

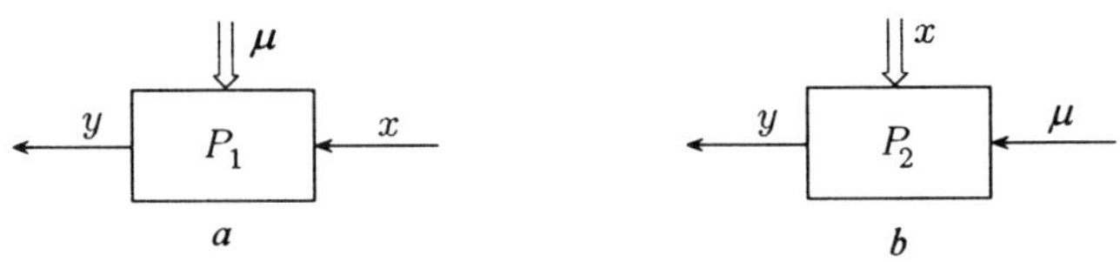

Figure 3.4.

transformed and the signal μ defines the transformation operator. In the second case (Fig. 3.4b), everything is the other way round. Thus, the O- and C-signals form a common set of state variables of a nonlinear system, and therefore it is convenient to introduce similarity transformers whose actions can be understood by analyzing Fig. 3.5. The typization of signals that we have introduced naturally entails the typization of the main structural elements of an automatic control system.

Figure 3.5.

3.2.2. Types of dynamical objects

Depending on the type of the input and output signals of a dynamical element, we can distinguish four main types of dynamical objects shown in Fig. 3.6. It should be pointed out that objects of these types have been used for a long time in different branches of science, but their classification in accordance with the criteria indicated above have not been carried out. We shall only give some examples of different types of objects.

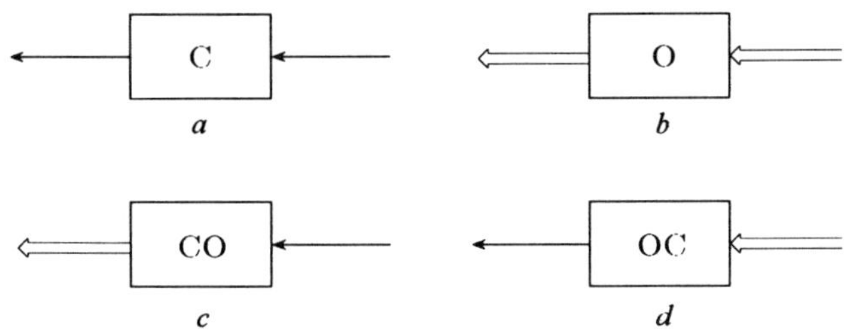

Figure 3.6.

C-object, a standard oscillatory radio circuit which converts input voltage into output voltage,

O-object, an oscillatory circuit with a varicond which converts the capacity of the circuit into the frequency of oscillations,

CO-object, a standard oscillatory circuit whose input is voltage and whose output is the characteristic,

OC-object, it establishes the relationship between the parameters of the oscillatory circuit and the output voltage.

3.2.3. Binary operation

From what was said above we can see the importance of the part played by the element with two independent inputs shown in Fig. 3.7. This element is described by the equation $y = \beta(\mu, x)$ and is called a binary element. A binary element may be

linear with respect to x: $y = \beta_1(\mu)x$,

linear with respect to μ: $y = \mu\beta_2(x)$,

linear with respect to x and μ, and then it is said to be bilinear.

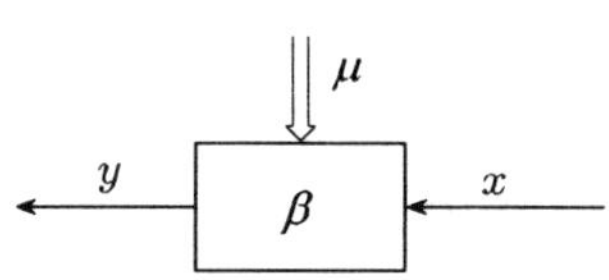

Figure 3.7.

The simplest bilinear element is a multiplier with scaling described by the equation

$$y = k\mu x.$$

A binary element is said to be separable if

$$y = \beta_1(\mu)\,\beta_2(x).$$

A binary element generalizes the concept of a relay element and a ψ-cell. For instance, a relay element can be obtained from a binary one for $\mu = \text{const}$ and a ψ-cell can be obtained from it for $\beta_1 = \operatorname{sgn}\mu$, $\beta_2(x) \geq 0$.

3.2.4. Types of control elements

The element of a control system whose output variable exerts a direct influence on the input of the object is called a control element. Control elements may be statical and dynamical.

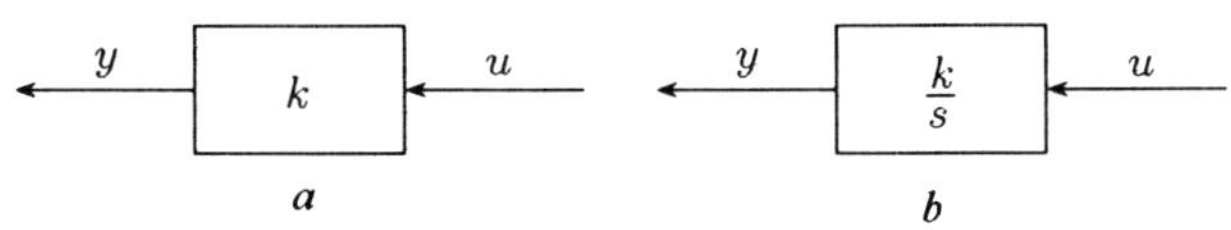

Figure 3.8.

In the classical control theory, most frequently a statical amplifier (Fig. 3.8*a*) or an integrator (Fig. 3.8*b*), i.e., a linear element, plays the part of a control element. When a binary element is used, control elements become nonlinear, and additional possibilities appear to control their properties. Examples of a statical and a dynamical binary control element are given in Fig. 3.9.

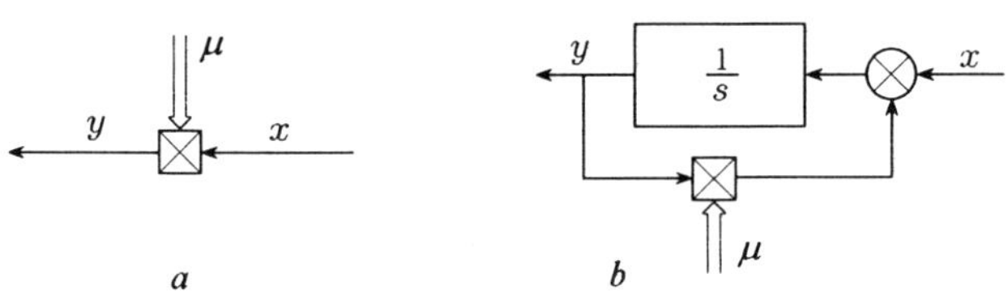

Figure 3.9.

3.2.5. New types of feedback

When we have an extended collection of types of dynamical elements, it is expedient to introduce, in addition to ordinary or, to put it otherwise, coordinate feedback (CF, Fig. 3.10*a*), three new types of feedback, namely, coordinate-operator feedback (COF, Fig. 3.10*b*), operator-coordinate feedback (OCF, Fig. 3.10*c*), and operator feedback (OF, Fig. 3.10*d*). We now have the necessary minimum of basic concepts and can begin the exposition of the main facts of the theory.

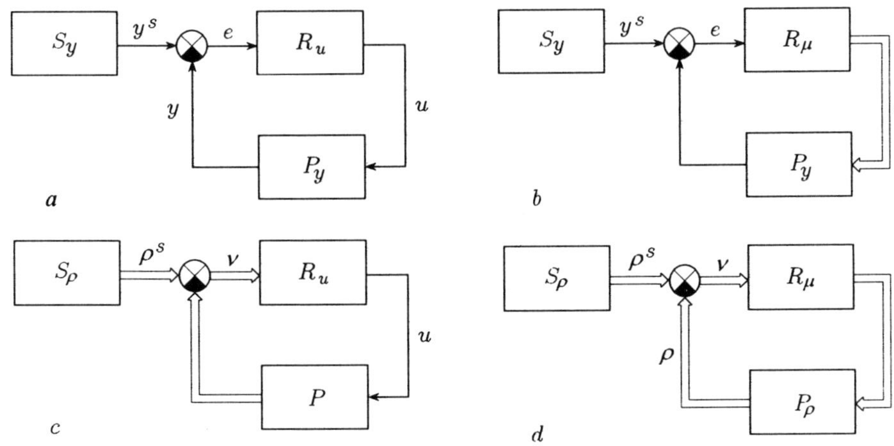

Figure 3.10.

3.3. Structural Synthesis of Binary Systems

3.3.1. Stabilization problem

Let us consider a standard problem of stabilization of an uncertain object P (Fig. 3.11) under the conditions where, using the a priori information about the coordinate $f \in F$ and operator $a \in A$ disturbances, we must choose an operator R_u of stabilizing feedback (Fig. 3.12). Assume that by the hypothesis we have not only to achieve stabilization, i.e., the "stability" of the control error, but also to ensure the required quality of the transient process. Suppose, for instance, that the relation

$$S_e(c)e = e$$

must be satisfied in the transient processes. Here $S_e(c)$ is the known operator and c is its parameter, $c \in C$.

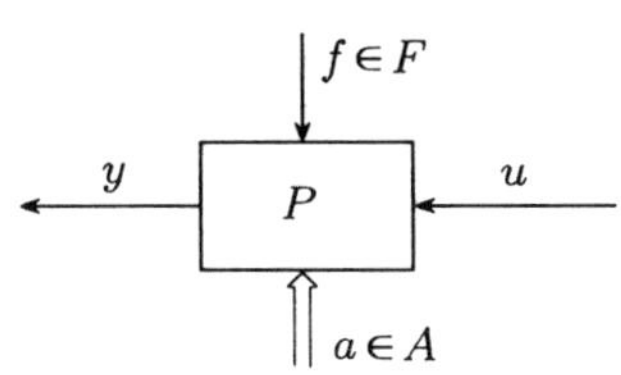

Figure 3.11.

Under these natural assumptions, it is possible to carry out a structural synthesis of a binary stabilization system. Indeed, following the general conception of binary control given above, we refuse the a priori choice of the feedback operator R_u according to the diagram shown in Fig. 3.12 and pass to its automatic choice with the aid of operator signal μ according to the diagram shown in

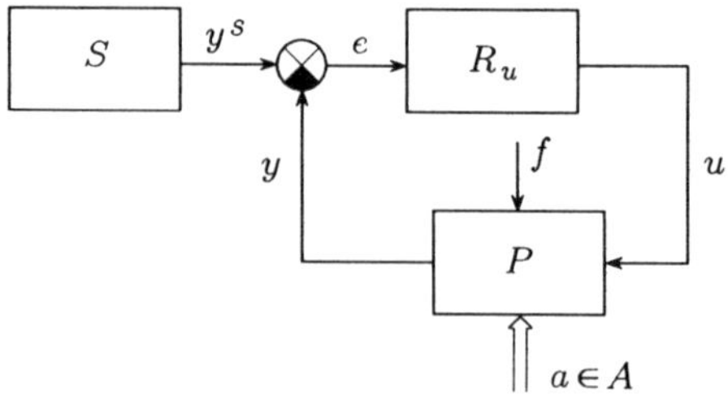

Figure 3.12.

Fig. 3.13. As a result of this transition, the problem of the choice of specific operator R_u is replaced by a simpler problem of the choice of a family of stabilizing operators R, which are marked by the operator signal μ, and of the formation of the required value of the operator signal μ itself.

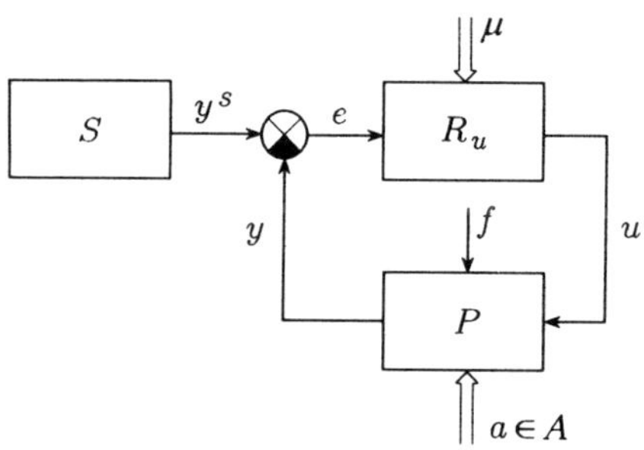

Figure 3.13.

We shall put off the discussion of the problem of choosing the family of operators R and shall discuss the principles of formation of an appropriate μ. Since the quality of the control system in the problem that we have posed is expressed in terms of the control error, we can represent the system shown in Fig. 3.13 in the form of a new (generalized) object P_e (Fig. 3.14) which has a

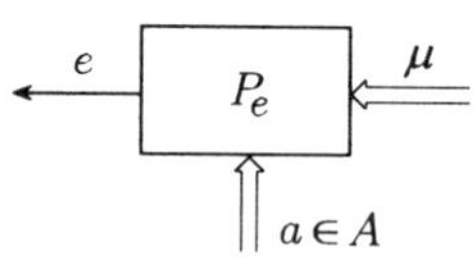

Figure 3.14.

control error e as its output and an operator signal μ as its input, i.e., P_e is a nonlinear object of the OC-type subjected to an unknown perturbation $a \in A$. For the control theory this is a standard statement of the stabilization problem which can be solved with the aid of control principles considered in the preceding chapter. In particular, if we introduce into consideration a signal

$$e^s = S_e e,$$

where S_e is an operator that reflects the requirements imposed on the standard behavior of the error, then we can introduce into consideration a new error of control $\sigma = e^s - e$ for whose stabilization at zero (and this is given by the solution of the original problem since $\sigma = 0$ implies the required relation $S_e e = e$) it is expedient to use feedback according to the block diagram shown in Fig. 3.15,

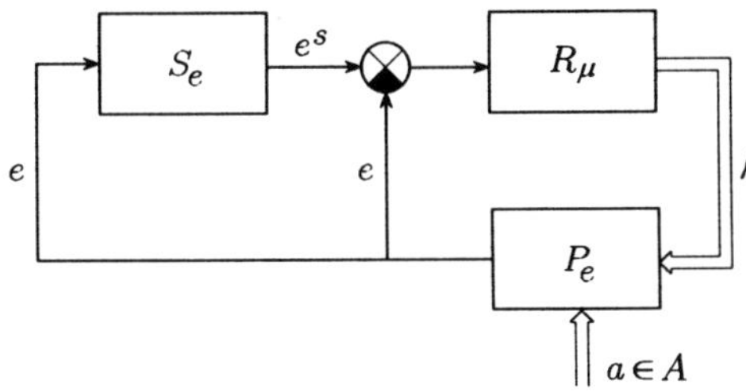

Figure 3.15.

where R_μ is an operator of the coordinate-operator feedback whose synthesis can be carried out with the aid of all known techniques of synthesis of a stabilizing feedback when there exists coordinate perturbation $a \in A$ which cannot be measured.

It stands to reason that certain peculiarities arise due to the nonlinearity of the object P_e, but at least no considerable difficulties are encountered. And this means that

- the complicated problem of compensating for the unknown operator perturbation reduces to the traditional and well-studied problem of compensating for coordinate perturbation.

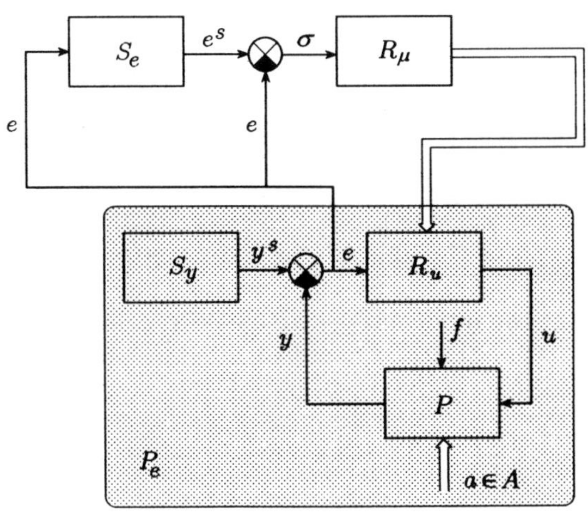

Figure 3.16.

We can put the block diagram shown in Fig. 3.15 into correspondence with a more detailed diagram of a control system with two types of feedback, a coordinate feedback and a coordinate-operator one. Figure 3.16 shows a block diagram with two local controls, a C-control R_u and a CO-control R_μ.

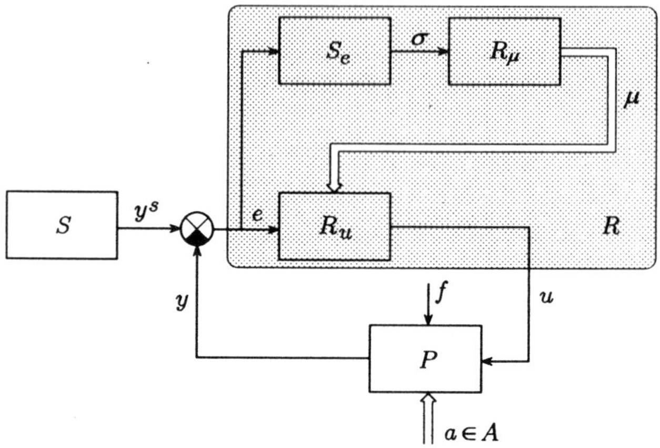

Figure 3.17.

We can interpret the exposed technique of synthesizing the structure of a control as a technique of obtaining a nonlinear controller R in the block diagram in Fig. 3.17, and, generally speaking, it is not a priori obvious why this linear controller must have a structure of exactly this type. In the preceding diagram, the operator $S_e(c)$, which defines the standard property of a control system, depends on the parameter c which is often not exactly defined but only with an accuracy to within the inclusion into a certain set C, i.e., $c \in C$. It is natural to suppose that by varying the parameter c in the range of the set C, we can improve some characteristics of the system.

It is well known, for instance, that by lowering the gain factors in feedback we can make the system disturbance-proof, decrease the effect produced by constraints, etc. On the other hand, if the operator R_u in the block diagram shown in Fig. 3.16 realizes a statical binary operation, i.e.,

$$R_u = k\mu x, \quad c = \text{const},$$

then the O-signal μ plays the role of the gain factor and therefore must be bounded, i.e., $|\mu| \leq 1$.

Let us now analyze the mode of operation of the structure shown in Fig. 3.15 when the required relation $\sigma = 0$ is satisfied. This relation is equivalent to two relations

$$S_e(c)\, e = e, \quad e = P_e(a)\, \mu.$$

For the defined parameter c, the first relation yields a function $e(c)$ and, then, the second equation yields the required value of the O-signal

$$\mu = P_e^{-1}(a)\, e(c), \tag{3.1}$$

which, naturally, depends on the unknown parameter a, and therefore we could not calculate it with the use of the a priori data.

The second inference from (3.1) is that when the parameter a varies in a wide range, then the range of variation of the signal μ must also he sufficiently wide, and this may turn out to be incompatible with the constraint

$$|\mu| \le 1. \tag{3.2}$$

There is one possibility of satisfying relation (3.1) under constraint (3.2). It is obviously connected with the variation of the parameter $c \in C$ as a function of the O-signal μ. The realization of this plan means a transition from the diagram in Fig. 3.15 to the diagram of the system shown in Fig. 3.18, where we now have to carry out the variation of the parameter c with the aid of the O-signal ρ, i.e., $c = c(\rho)$. Since, in this case, μ is an output signal and ρ is an input signal, we deal with the control of the O-object (Fig. 3.19) which replaces the diagram shown in Fig. 3.18

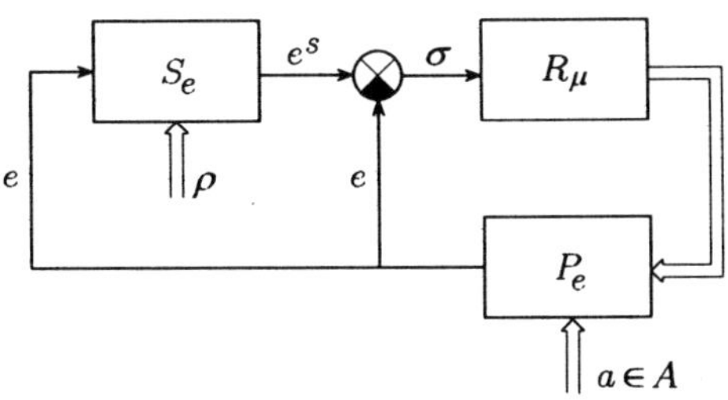

Figure 3.18.

We again arrive at the standard problem of controlling an uncertain object for whose solution it is natural to use feedback, in this case, of course, of the operator type. Since, by the hypothesis, the aim of the control is the fulfilment of the constraint $|\mu| \le 1$, the problem can be interpreted (without loss of generality) as a stabilization problem of an O-signal μ at zero. Therefore there is no need to use a reference action signal, and, after the application of the controlled coordinate feedback, we obtain the structure of the system shown in Fig. 3.20, where R_ρ is the operator of the O-control.

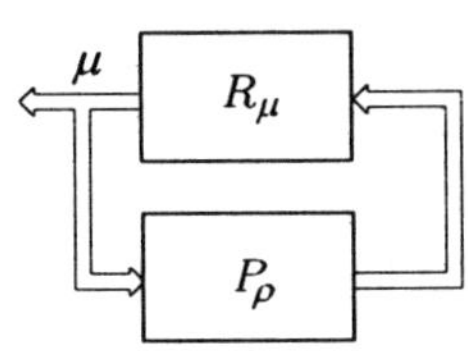

Figure 3.19.

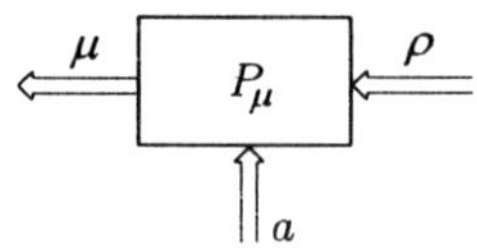

Figure 3.20.

Considering the structure of the object P_e with the use of Fig. 3.18, we first find a system with two types of feedback (Fig. 3.21) and then, analyzing it in detail (Fig. 3.22), we reveal that, in fact, three types of feedback are used in the control system.

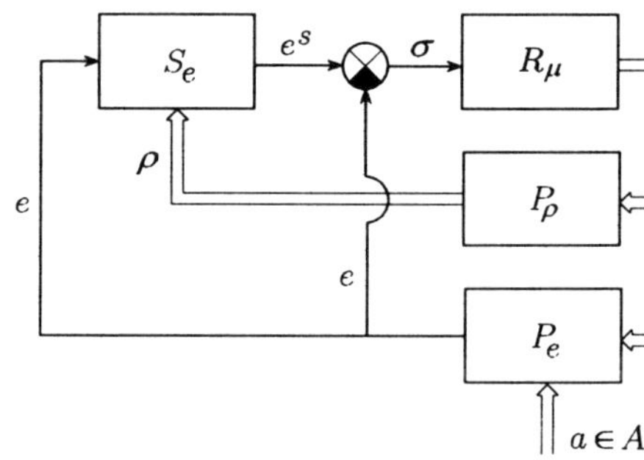

Figure 3.21.

It is easy to see from the block diagrams shown in Figs. 3.21 and 3.22 that the O-feedback is consistent with the OC-feedback and acts with the aim of stabilizing σ at zero.

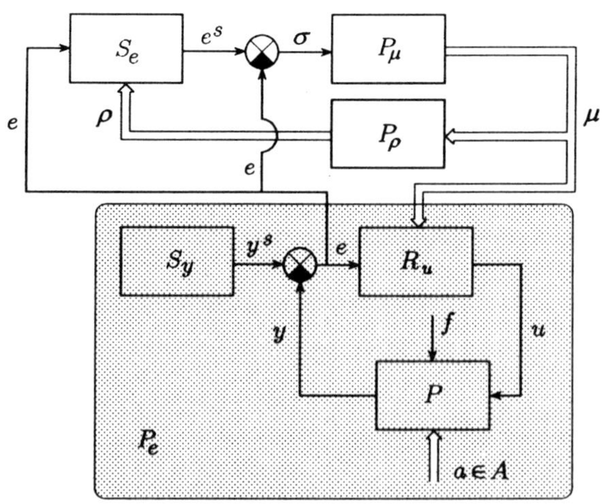

Figure 3.22.

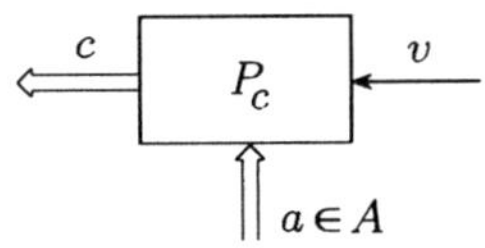

Figure 3.23.

In the preceding diagram, the choice of the feedback operator R_ρ is arbitrary, and therefore there is a sufficiently large arbitrariness in this diagram which allows us to endow the trajectory of the shift of the parameters of the operator $S_e(c)$, $c(t) \in C$, with the required properties. This may prove to be the right thing since the operator $S_e(c)$ is responsible for the quality of the control system. Suppose, for instance, that we must make the function $c(t)$ close to the function $c^s(t) \in C$ with the use of an additional action v at the control object P which is summed up with the basic signal $u(t)$. To synthesize the controller which would solve this problem, we can represent the object shown in Fig. 3.22 from

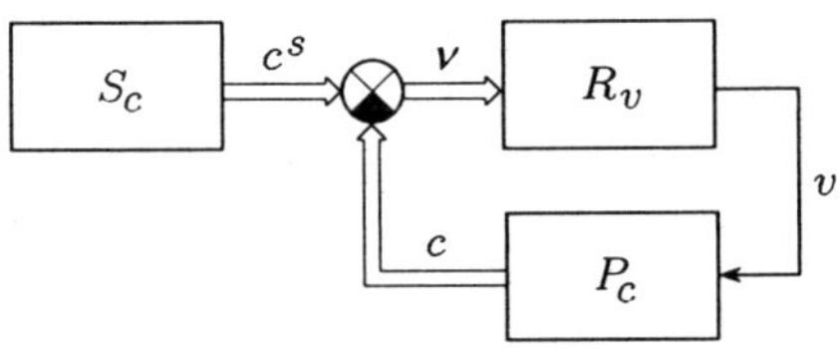

Figure 3.24.

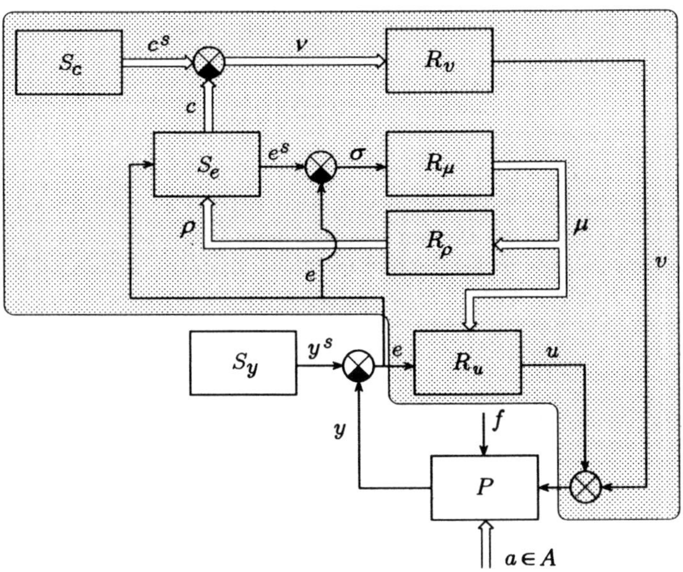

Figure 3.25.

the input v to the output $c(t)$ as a CO-object P_c (Fig. 3.23). Again following the classical tradition, in order to solve the problem of tracking the signal $c^s(t)$ we use feedback, this time of an operator-coordinate type, and obtain a block diagram shown in Fig. 3.24, where S_c is the reference device for the O-signal c^s, v is the O-error of control, and R_v is an OC-controller. Revealing the inner structure of the P_c-object, we obtain a control system with all possible four types of feedback (Fig. 3.25). It should be emphasized once more that the contoured part in Fig. 3.25 is, in fact, a nonlinear controller whose a priori synthesis is impossible with the use of traditional means, whereas the proposed theory based on the binarity principle gives the desired result.

3.3.2. Nonlinear feedback as a means of suppressing uncertainty

The main idea of the approach to the structural synthesis of dynamical systems demonstrated above consists in reducing a complicated new problem to a sequence of traditional problems which are solvable, at least in principle, by well-known means.

For simplicity, we have used only the deviation control principle which, in conjunction with the binarity principle, has allowed us to obtain a sufficiently large variety of structures of nonlinear systems. It stands to reason that this variety will be increased if we also apply other well-known principles of control to the structural synthesis which, without any stipulations but only due to their generality, can be applied to the stabilization problem in question. We omit the details here and only point out the aspects of the approach which are fundamental.

The idea of the new approach itself can be formulated as a principle of generation of structures, i.e., the procedure of synthesis of a complicated controller should be represented as a sequence of synthesis problems solvable by traditional methods.

It stands to reason that not every problem admits a decomposition of this kind, but the number of problems admissible to decomposition is sufficiently large. Let us consider, for instance, a problem on lowering the dimension of a feedback system. We can see from the block diagram shown in Fig. 3.25 that when the error of the OC-feedback vanishes, i.e., when the relation

$$\sigma = 0 \tag{3.3}$$

is satisfied, then an additional dependence of the state variables on one another appears,

$$S_e e = e, \tag{3.4}$$

and, hence, the order of the system becomes lower. Therefore the natural question is whether it is possible to arrange a regular degeneration of the order of the system by establishing relationships of form (3.3). But how can we do it? Whereas in the diagram shown in Fig. 3.26 a natural control input is used for

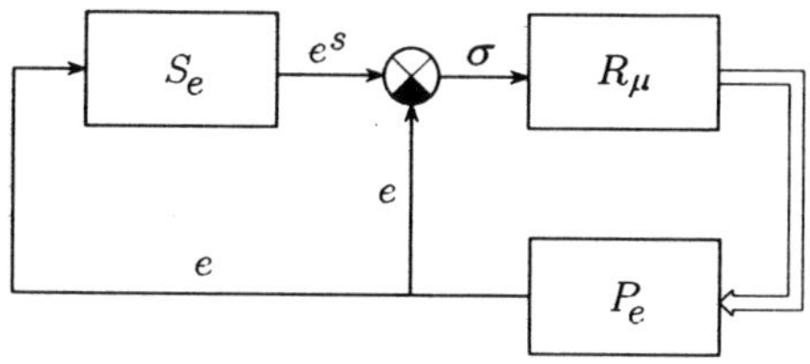

Figure 3.26.

degeneration, there is no control input of this kind in the diagram of the system in Fig. 3.27 which corresponds to relation (3.4). Consequently, this problem cannot be solved by standard techniques, but if we recollect that the operator S_e depends on the parameter c, i.e., $S_e(c)$, then we can consider precisely this

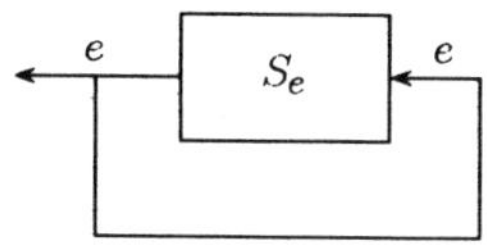

Figure 3.27.

parameter or, what is the same, the O-signal ξ since $c = c(\xi)$, to be a control and regard it as an OC-object with operator P_e' (Fig. 3.28). Following now exactly the approach described above, we synthesize the structure represented in Fig. 3.29. Next we choose an operator S_e' and a controller R_ξ such that the rela-

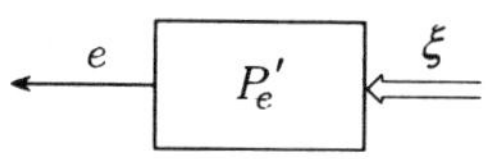

Figure 3.28.

tion $\sigma' = e_1^s - e = 0$ is satisfied from some time moment. Together with relation (3.4), this relation establishes dependences $S_e' e = e$, $S_e e = e$, and, consequently, the order of the system is lowered again.

Acting successively in the way indicated above, we can reduce an n-dimensional object, in a finite number of iterations, to a first-order object and, as

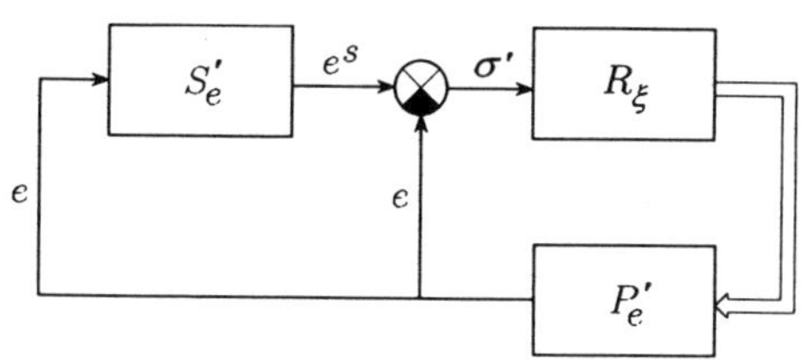

Figure 3.29.

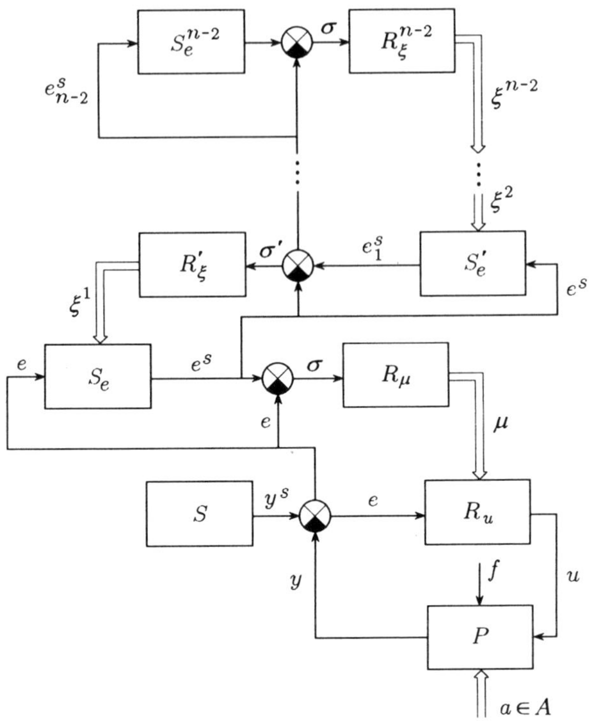

Figure 3.30.

a result, obtain structures represented in Fig. 3.30. We can give the following substantive interpretation to this fact: by adding a nonlinear controller to a complicated and uncertain object we can considerably simplify the behavior of a feedback system and make it predictable. In other words, *the controller that we propose is a "swallower" of complexity and uncertainty of an object.*

3.3.3. Filtration problem

The filtration problem is one of the main problems of control theory. Here is a typical statement of this problem. We are given an operator P of the desired transformation of the basic signal x into an output signal y^x (Fig. 3.31). Actually, an interference ξ is usually added to the basic signal (Fig. 3.32), and this leads

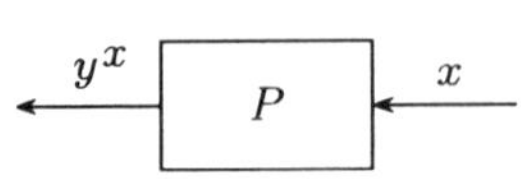

Figure 3.31.

to a deviation of the value of the output from the required value. Therefore the operator P is no longer the best and a problem arises of defining the best operator in the block diagram in Fig. 3.32 which would minimize the "distance" between y^x and y, i.e.,

$$P_{\text{opt}} = \operatorname{argmin} r(y^x - y),$$

where $r(\cdot,\cdot)$ is an appropriately determined distance, say, the mean square deviation, when random signals are meant.

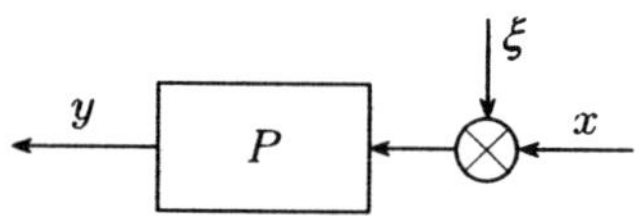

Figure 3.32.

It is customary to consider the signals x and ξ to be random processes with known statistical properties. Most commonly the filters forming these signals are known with operators P_x and P_ξ respectively (Fig. 3.33), where δ is white noise. With this representation of the signals x, ξ the block diagram of the filter

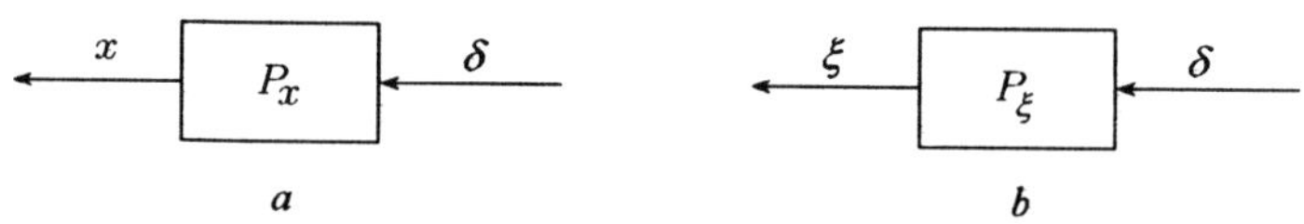

Figure 3.33.

assumes the form shown in Fig. 3.34. If the forming filters P_ξ, P_x are known and stationary, then the filter P is also stationary and is called a Wiener filter. Now if P_ξ, P_x are known but not stationary, it is called a Kalman-Bucy filter. The methods of synthesizing these filters are well known.

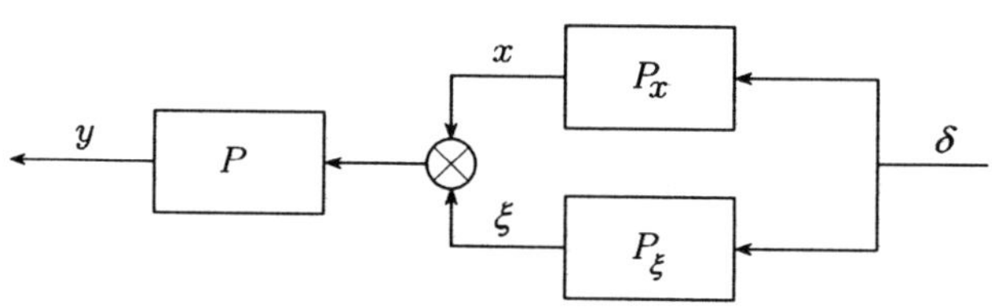

Figure 3.34.

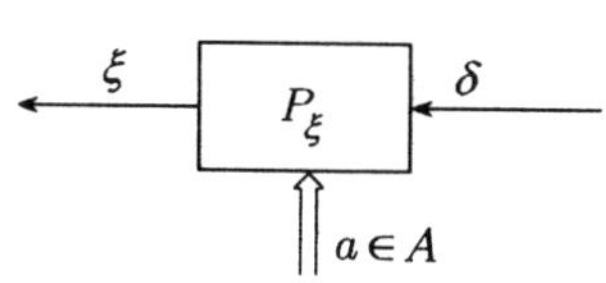

Figure 3.35.

A problem arises when the interference operator P_ξ is unknown. This situation can be simulated by the action of perturbation $a \in A$ (Fig. 3.35). Since we have an operator perturbation in this case, it should be compensated also by an operator perturbation μ (Fig. 3.36). As a result, the choice of the optimal

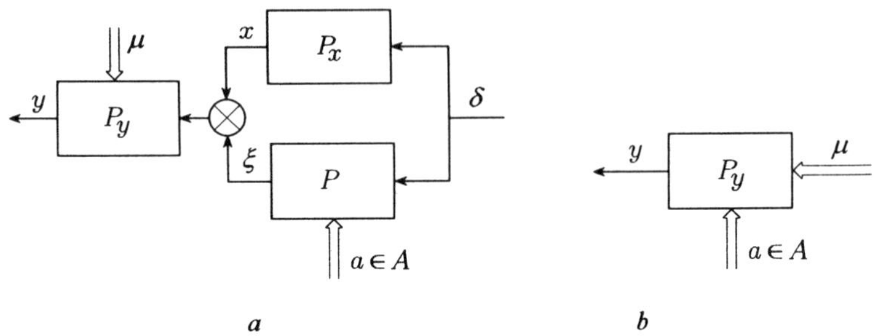

Figure 3.36.

filter operator can be laid on the CO-feedback according to the scheme which is standard for the theory of binary control (Fig. 3.37).

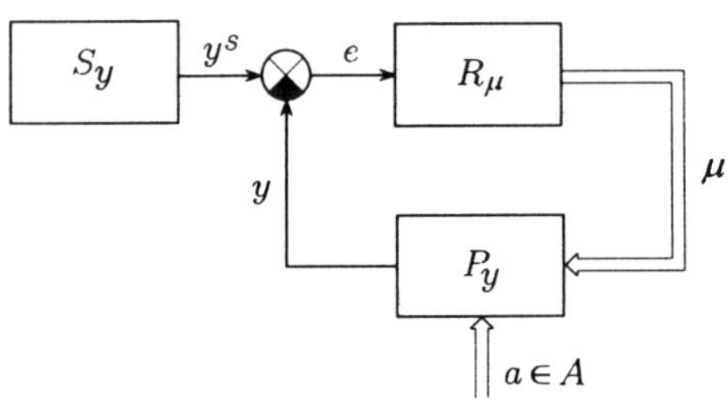

Figure 3.37.

The block diagram of the reference action S_y is shown in Fig. 3.38. It stands to reason that statistical methods should be used to calculate the CO-operator R_μ of feedback.

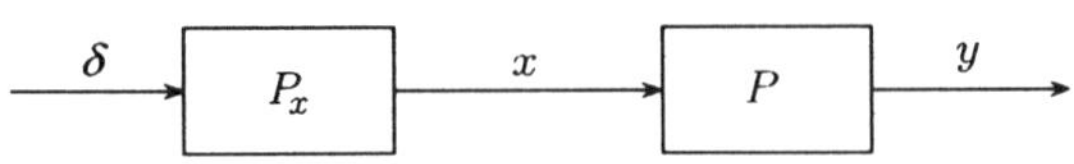

Figure 3.38.

Let us now consider simple examples which will illustrate the specific features of synthesis of new types of controllers. We shall preface the consideration with a table entitled "Problem—a type of a nonlinear operation" in which we shall give the plan of verification of the usefulness of the construction exposed above by way of examples of standard problems of control theory.

Problem	Type of operation	
	Unary	Binary
Stabilization		
Filtration and differentiation		
Optimality		
Invariance		

The hatched column of the table reflects the achievements of the classical control theory in which the binary operation was not actively used. The merits and drawbacks of the results that were obtained were mentioned above. In the second column of the table the squares are not hatched, and we have to find out whether the binarity principle gives anything new for solving these problems.

Chapter 4

Theory of Coordinate-Operator Feedback

In this chapter, using a simple example, we consider in detail the principles of synthesis of stabilization systems with coordinate-operator feedback as well as the principal properties and peculiarities of these control systems. For convenience, we call the theory exposed in the sequel a CO-theory. The CO-theory contains regular methods of analysis and synthesis of control systems with CO-controllers in the block diagram shown in Fig. 4.1 or, what is the same, the methods of control of an CO-object shown in Fig. 4.2.

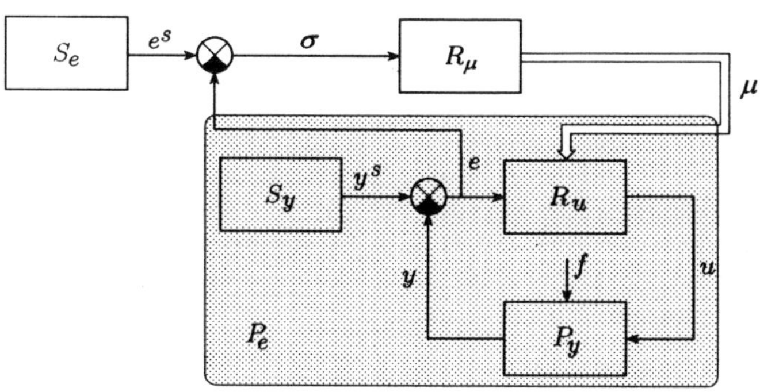

Figure 4.1.

The peculiarities and the features that distinguish this theory from the classical stabilization theory can be seen from Figs. 4.1 and 4.2. First of all, it is the fundamental nonlinearity of the control object P_e. Therefore the standard methods of synthesis of feedback are not applicable here and we have to look

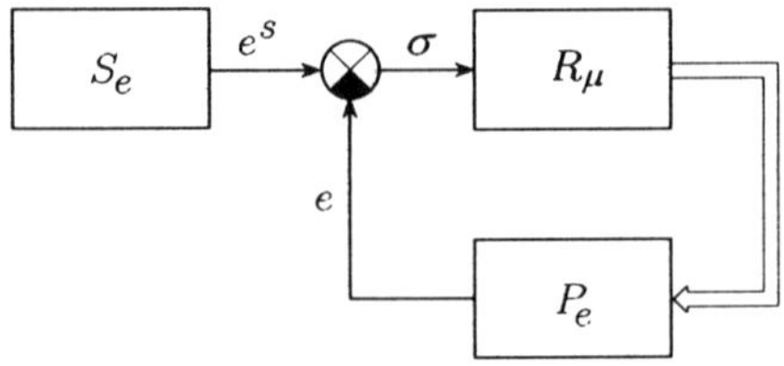

Figure 4.2.

for new methods. We shall demonstrate that the *principles* discussed in the preceding chapter are useful and constructive for this purpose.

4.1. Stabilization of a Second-Order Object with Unknown Parameters and an External Action

Consider a simple second-order object Σ^p:

$$\begin{aligned} \dot{x}_1 &= x_2, \\ \dot{x}_2 &= ax_1 + bu + f, \\ y &= cx_1 + x_2 \end{aligned} \tag{4.1}$$

with unknown parameters a, b and disturbance f which satisfy the inclusions

$$a \in A \stackrel{\text{def}}{=} \{a \mid |a| \leq a_0\}, \quad b \in B \stackrel{\text{def}}{=} \{b \mid 0 < b^- \leq b \leq b^+\},$$
$$f \in F \stackrel{\text{def}}{=} \{f \mid |f| \leq f_m\},$$

where the numbers a_0, $b^\pm$ and the function f_m are known. This object, called a Σ^p-system in what follows, allows us to give a substantive outline of the CO-theory which reflects, practically without exclusions, its main aspects and results.

In the *problem of stabilization* of a Σ^p-system we have to indicate a robust smooth state feedback which stabilizes the Σ^p-system under an arbitrary variation of the parameters $a \in A$, $b \in B$ and any disturbance $f \in F$.

It is clear from what was said above that the standard approaches to this problem do not guarantee its solution since

- feedback with high gain is not robust,
- the adaptive control is not applicable since the main conditions, namely, the quasistationarity of the parameters and the condition of vanishing disturbance, are not fulfilled,
- the methods of VSS are oriented to the use of discontinuous control when the robustness is not achieved.

When synthesizing a CO-controller, we follow the recommendations of the general theory of systems with new types of feedback but begin with indicating two useful observations which simplify the matter.

4.1.1. The scalarization principle and the equation of an object in the error space

This principle makes it possible to reduce, under natural conditions and upon the fulfilment of the matching condition (i.e., the condition of disturbance matching), the problem of stabilization of a multidimensional object to the problem of stabilization of a scalar object.

Indeed, it is easy to make sure that the output of the Σ^p-system exponentially tends to zero (and this means solving the stabilization problem) if the feedback relation

$$\dot{x}_1 + dx_1 = 0 \tag{4.2}$$

is satisfied for a certain number $d > 0$. We call the quantity

$$\sigma = \dot{x}_1 + dx_1 = x_2 + dx_1$$

the error of realization of the desired feedback (4.2).

It is obvious now that the stabilization of the error σ at zero means the solution of the initial problem. However, it does not yet follow that the problem has become scalar. In order to verify this fact, we pass from the original coordinate space (x_1, x_2) to the error space, i.e., coordinates (x_1, σ). To this end, we first find that

$$\dot{\sigma} = \ddot{x}_1 + d\dot{x}_1 = \dot{x}_2 + dx_2 = ax_1 + bu + f + dx_2.$$

From the equation $\sigma = x_2 + dx_1$ we express the variable x_2,

$$x_2 = \sigma - dx_1.$$

In the equations for the Σ^p-system we replace x_2 by the expression that we have found. As a result, we obtain the required motion equations in the form

$$\dot{x}_1 = -dx_1 + \sigma, \tag{4.3}$$

$$\dot{\sigma} = d\sigma + (a - d^2)x_1 + bu + f. \tag{4.4}$$

We shall call Eq. (4.3) a Σ_1^p-system and Eq. (4.4) a Σ_2^p-system. We can see from Eqs. (4.3), (4.4) that as $\sigma \to 0$, it automatically follows that $x_1 \to 0$ as well. Consequently, we can restrict ourselves to a scalar Σ_2^p-equation in order to solve the stabilization problem.

If we introduce a notation

$$\tilde{f} = f + (a - d^2)\, x_1,$$

then we can use the indicated transformations to reduce the initial problem to the stabilization of a scalar Σ_2^p-object of the form

$$\dot{\sigma} = d\sigma + bu + \tilde{f}, \tag{4.5}$$

which is acted upon by the unknown disturbance $\tilde{f} \in \tilde{F}$ for which only the majorant

$$\tilde{F} = \left\{\tilde{f} \mid |\tilde{f}| \leq (d^2 + a_0)|x_1| + f_m = \tilde{f}_m\right\}$$

is known. Since the equation

$$\sigma = x_2 + dx_1 = 0 \tag{4.6}$$

defines a straight line in the plane (x_1, x_2), geometrically the scalarization principle means choosing a control that would create conditions under which the straight line (4.6) is an attractor, i.e.. an attracting invariant set (Fig. 4.3a). In the error space, this attractor turns into one of the coordinate axes (Fig. 4.3b).

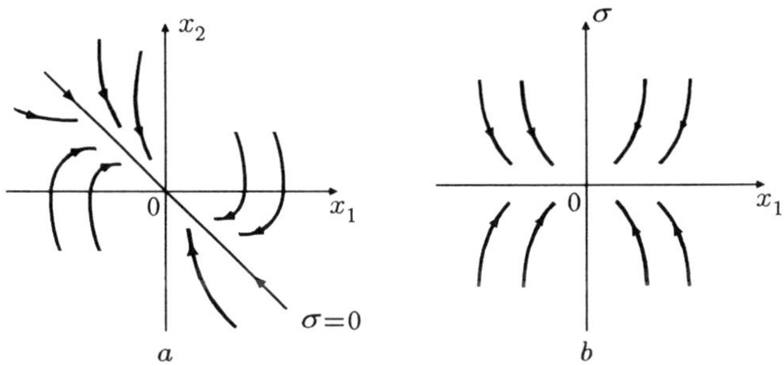

Figure 4.3.

Some remarks are due concerning the reduction of the general stabilization problem to a scalar problem and the related problems of control of a scalarized object Σ_2^p of the form (4.5).

4.1.2. Some remarks concerning the statement of the problem and its generalizations

When carrying out the scalarization of a stabilization problem, we must bear in mind the following.

Remark 1. The consideration of an object with only two unknown parameters $a \in A$, $b \in B$

$$\dot{x}_1 = x_2,$$
$$\dot{x}_2 = ax_1 + bu + f$$

does not lead to a loss of generality of the result since we can use a nonstationary change of variables

$$x_1 = c(t)\, z_1, \quad z_2 = \dot{z}_1$$

in order to impart to the equations of the arbitrary object with three unknown parameters

$$\dot{x}_1 = x_2,$$
$$\dot{x}_2 = a_1x_1 + a_2x_2 + bu + f,$$

where $a_1 \in A_1$, $a_2 \in A_2$, $b \in B$, the form

$$\dot{z}_1 = z_2,$$
$$\dot{z}_2 = \tilde{a} z_1 + \tilde{b} u + \tilde{f}$$

required by the statement of the problem in question. Here

$$\tilde{a} = \frac{-\ddot{c} + a_2 \dot{c} + a_1 c}{c}, \quad \tilde{b} = \frac{b}{c}, \quad \tilde{f} = \frac{f}{c},$$

only if $c(t)$ is a solution of the differential equation

$$2\dot{c} - a_2 c = 0.$$

Remark 2. The scalarization principle is applicable to arbitrary objects of the general position with one input and one output, i.e., to objects of the form

$$\dot{x}_i = x_{i+1}, \quad i = 1, \dots, n-1,$$
$$\dot{x}_n = \sum_{i=1}^{n} a_i x_i + u + f, \quad y = cx.$$

Indeed, in this case it suffices to choose a new error in the form

$$\sigma = x_n + \sum_{i=1}^{n-1} d_i x_i, \quad d_i = \text{const} > 0,$$

and, in the $(n-1)$th equation of the Σ-system, replace x_n by the expression

$$x_n = -\sum_{i=1}^{n-1} d_i x_i + \sigma$$

and write the equation

$$\dot{\sigma} = (d_{n-1} + a_n)\sigma + \sum_{i=1}^{n-1} \alpha_i x_i + u + f$$

instead of the last equation in the Σ-system.

As a result, we get two subsystems, namely, a Σ_1-system of order $n-1$,

$$\dot{x}_i = x_{i+1}, \quad i = 1, \dots, n-2,$$
$$\dot{x}_{n-1} = -\sum_{i=1}^{n} d_i x_i + \sigma,$$

and a scalar Σ_2-system

$$\dot{\sigma} = d\sigma + u + \tilde{f}, \quad y = (c' - d')x' + c_n \sigma,$$

where $x' = \mathrm{col}(x_1, \ldots, x_{n-1})$, $c = (c', c_n)$, $d' = (d_1, \ldots, d_{n-1})$. In this case, if the parameters d_i are such that the Σ_1-system for $\sigma = 0$

$$\dot{x}_i = x_{i+1}, \quad i = 1, \ldots, n-2,$$

$$\dot{x}_{n-1} = -\sum_{i=1}^{n} d_i x_i$$

is exponentially stable, then, again, it follows from the condition $\sigma \to 0$ that $x_i \to 0$, and this is a formal expression of the scalarization principle.

Remark 3. The transformation of coordinates indicated above, which serves as the basis of the scalarization principle, is expressed by the block diagram in Fig. 4.4. The peculiarity of the decomposition shown in the figure is that the Σ_1-system is completely defined and asymptotically stable and the Σ_2-system is scalar, subject to the influence of uncertainty factors $\{a, f\}$, and can be efficiently controlled.

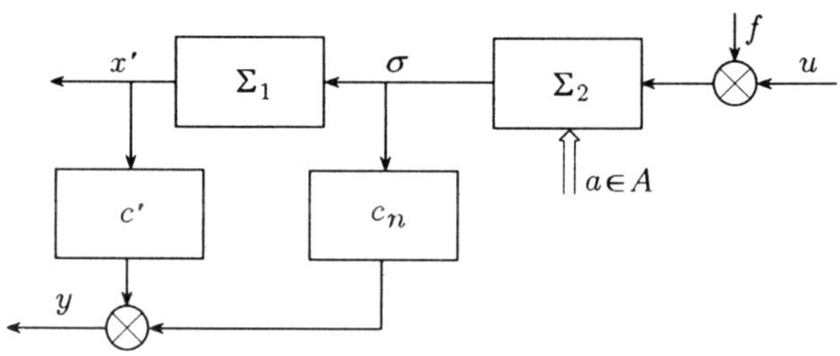

Figure 4.4.

Note that, in the general case, the scalarization principle also holds for systems with many inputs and many outputs described by equations

$$\dot{x} = Ax + Bu, \quad u \in R^m,$$

$$y = Cx, \quad y \in R^r.$$

However, in this case, the Σ_2-system is of dimension m which coincides with the dimension of the control vector u

$$\dot{\sigma} = D\sigma + CBu + \tilde{f},$$

and if $\det CB \neq 0$, then, upon the change $u = (CB)^{-1}v$, we get m one-dimensional stabilization problems instead of one m-dimensional problem.

Remark 4. Assuming that there is no disturbance in the Σ^p-system, i.e., that $f \equiv 0$, and that all parameters are constant,

$$\dot{x}_1 = x_2, \quad \dot{x}_2 = ax_1 + bu, \quad y = cx_1 + x_2,$$

we find transfer functions from the input u to the outputs y and $\sigma = x_2 + dx_1$ respectively. It is easy to obtain expressions

$$W_y(s) = b\frac{s+c}{s^2+a}, \quad W_\sigma(s) = b\frac{s+d}{s^2+a}, \tag{4.7}$$

comparing which we find the following: for $c > 0$, instead of the variable σ we can use the output y of the object, and then, actually, the scalarization principle leads to the problem of stabilization of the output

$$\dot{y} = cy + bu + \tilde{f},$$

i.e., we can replace a second-order object by a first-order object. We can see from (4.7) that such an object has a relative order $r = 1$ and a minimal phase.

Remark 5. The choice of feedback stabilizing the Σ_2^p-object,

$$\dot{\sigma} = d\sigma + bu + \tilde{f},$$

is especially simple when we have full information about the parameters and the disturbance. Indeed, under this assumption, the control $u = v - \tilde{f}/b$ reduces the stabilization problem to a trivial problem

$$\dot{\sigma} = d\sigma + v,$$

where the choice of the control v is obvious. Now if the parameters or the disturbance are unknown, the problem becomes more complicated since the stabilization methods from the classical control theory, namely, feedback with high gain ($u = -k\sigma$, $k \to \infty$), adaptive feedback ($u = -k\sigma$, $\dot{k} = \gamma\sigma^2$, $\gamma = \text{const} > 0$), a variable structure system ($u = -\tilde{f}_m \operatorname{sgn} \sigma$), have certain merits but also have some drawbacks and restrictions in applications.

Remark 6. We shall show that the linear feedback

$$u = -k_2\sigma - k_1x_1 \tag{4.8}$$

with bounded coefficients is suitable for stabilization of free motions of an uncertain object, with constant a priori bounded parameters (called earlier an interval object)

$$\dot{\sigma} = d\sigma + bu + (a - d^2)x_1, \tag{4.9}$$

but, generally speaking, does not solve the problem when the parameters a, b are changed or there exists an external disturbance

$$\dot{\sigma} = d\sigma + bu + (a - d^2)x_1 + f, \quad f \in F. \tag{4.10}$$

Indeed, after substituting (4.8) into (4.9) and when $f \equiv 0$, we obtain

$$\dot{\sigma} = (d - bk_2)\sigma + (a - d^2 - bk_1)x_1, \tag{4.11}$$

and, to analyze the asymptotic behavior of $\sigma(t)$, we must complete this equation by an equation for the error x_1, i.e.,

$$\dot{x}_1 = -dx_1 + \sigma. \tag{4.12}$$

The characteristic polynomial of the feedback system (4.11), (4.12) is defined by the expression

$$\det \begin{bmatrix} s+d & -1 \\ bk_1 - a + d^2 & s + bk_2 - d \end{bmatrix} = s^2 + bk_2 s + bk_1 - a + dbk_2 = 0,$$

and it is clear that when the conditions

$$k_2 > 0, \quad b^- k_1 + db^- k_2 > a^- \tag{4.13}$$

are fulfilled, the feedback system is asymptotically stable.

When the parameters are changed, conditions (4.13) do not longer guarantee the stability of the system and the action of the disturbance does not guarantee stabilizability even when the free motions are stable.

Even for constant parameters and $f \equiv 0$ the quality of transient processes may vary considerably with the variation of the parameters and do not satisfy the requirements imposed upon the system.

We can thus formulate the following problem:

- how can we achieve the independence of the properties of a feedback system of the uncertainty factors when the gain factors in the feedback channels are restricted?

4.1.3. The coordinate-operator phase space

The scalarization principle is closely connected with new types of feedback and the binarity principle, namely, if we take the variable σ as the error of the CO-loop of control and use the CO-feedback, then the structure shown in Fig. 4.5, which is standard for the theory of binary control, naturally arises. In this case, the following problems of choice become key problems: the choice of the operator R_u, the choice of the operator R_μ, and the choice of the type of binary operation.

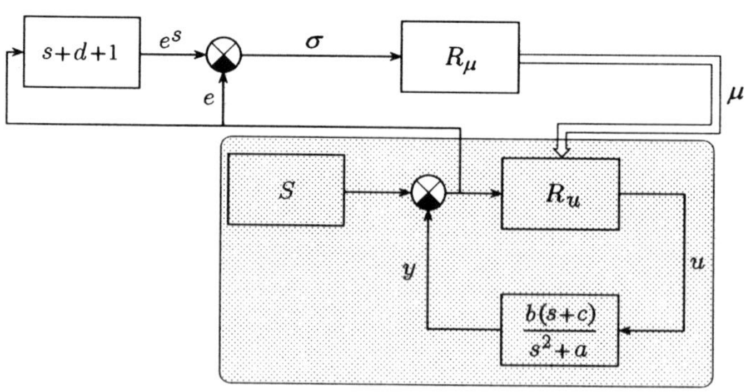

Figure 4.5.

Let us solve these problems recalling, however, that the following transformations are of importance here.

It is obvious that the motion in the system

$$\dot{x}_1 = -dx_1 + \sigma,$$
$$\dot{\sigma} = d\,\sigma + a^*\,x_1 + b\,u + f, \quad a^* = a - d^2$$

is close to the required motion defined by the equation $\dot{x}_1 = -dx_1$ and only slightly depends on the uncertainty factors $\{a, f\}$ if the following inequality is satisfied for a sufficiently small number $\delta > 0$:

$$|\sigma| \leq \delta |x_1|. \tag{4.14}$$

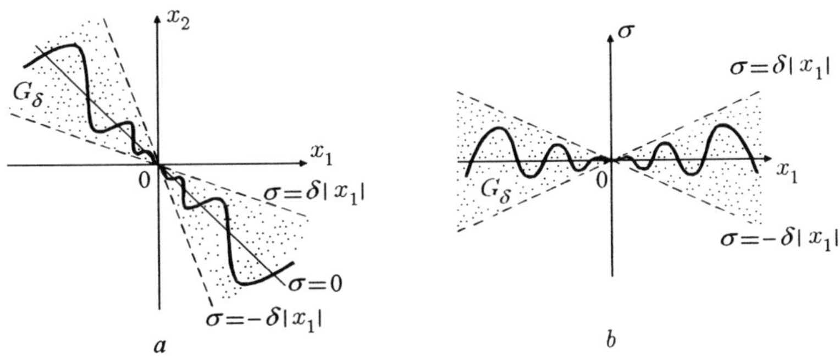

Figure 4.6.

This statement is sufficiently clear from geometrical representations since the fulfilment of condition (4.14) means that the phase trajectory does not leave the sectoral set G_δ (Fig. 4.6) which surrounds the straight line $\sigma = 0$. Therefore the transfer of the phase point into G_δ and confining it there may be the purpose of control. To analyze the motion and to synthesize the controller in the set $G_\delta = \{x \mid |\sigma| \leq \delta |x_1|\}$, it is convenient to use the nonlinear change of coordinates $\xi = \sigma/x_1$. Figure 4.7 explains the geometrical meaning of this change, namely, ξ defines the slope of the line $\sigma = \xi x_1$. Therefore ξ can be regarded as a parameter, or, more generally, an operator variable, if we employ the following relations. The variables σ and x_1 are related by the differential operator $\pi\left(\frac{d}{dt}\right)$:

$$\sigma = \dot{x}_1 + dx_1 = \left[\frac{d(\cdot)}{dt} + d\right] x_1 = \pi\left(\frac{d}{dt}\right) x_1.$$

A similar relationship can be established between the variables σ_ξ and x_1:

$$\sigma_\xi = \sigma - \xi x_1 = \pi\left(\frac{d}{dt}\right) x_1 - \xi x_1 = \pi_\xi\left(\frac{d}{dt}\right) x_1,$$

where

$$\pi_\xi\left(\frac{d}{dt}\right) = \frac{d}{dt} + d - \xi.$$

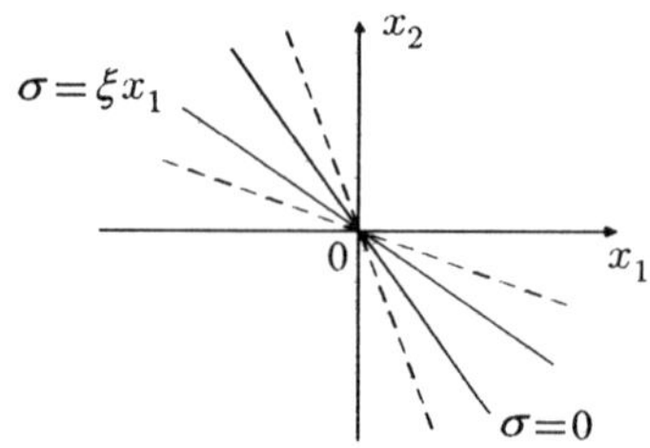

Figure 4.7.

Since ξ defines the operator π_ξ, it is expedient to call it an operator variable. Consequently, the space (x_1, ξ) can be called a coordinate-operator phase space, or, briefly, a CO-space. What advantage can be gained from this change of variables?

In order to answer this question, it suffices to find an equation for the change of the new variable $\xi = \sigma/x_1$. For the time being, we set $f \equiv 0$, and then we successively obtain

$$\begin{aligned}\dot{\xi} &= \frac{\dot{\sigma}}{x_1} - \frac{\sigma}{x_1}\frac{\dot{x}_1}{x_1} = \frac{d\sigma + bu + a^* x_1}{x_1} - \frac{\sigma}{x_1}\frac{\sigma - dx_1}{x_1} \\ &= d\xi + b\frac{u}{x_1} + a^* - \xi(\xi - d).\end{aligned}$$

If we introduce a notation $\mu = u/x_1$ and call μ a new control, then the problem of stabilizing the free motion of a parametrically uncertain object

$$\dot{\sigma} = d\sigma + a^* x_1 + bu$$

reduces to the problem of stabilizing a definite object which is under the action of the coordinate perturbation a^* since we now deal with an equation for Σ_ξ of the form

$$\dot{\xi} = d\xi - \xi(\xi - d) + b\mu + a^*.$$

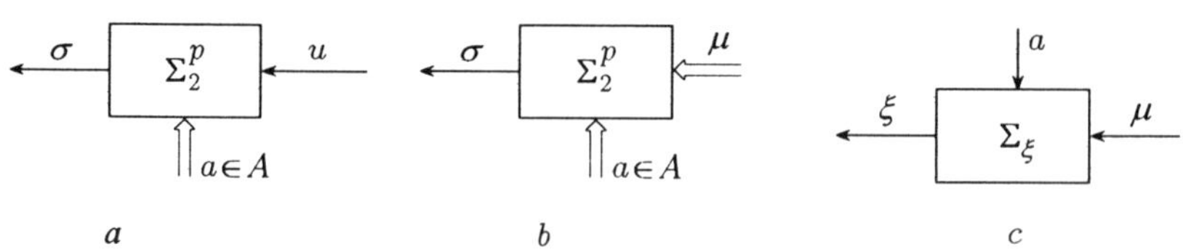

Figure 4.8.

In other words, by means of the nonlinear change $\xi = \sigma/x_1$, we have transformed the complicated problem of stabilization of the uncertain Σ_2^p-object (Fig. 4.8a, b) into the well-studied problem of compensation for the coordinate perturbation for the Σ_ξ-object (Fig. 4.8c). The remaining part of this subsection is devoted to the discussion and comparison of versions of the choice of control μ

on the basis of standard techniques of compensation, and here we want to point out that although the problem of synthesis of stabilizing feedback is solved, in principle, certain complications are inevitable since the known principles must be applied under new conditions because the object is essentially nonlinear:

$$\begin{aligned} \dot{x}_1 &= -dx_1 + \xi\, x_1, \\ \dot{\xi} &= d\xi - \xi\,(\xi - d) + b\mu + a^*, \quad b \in B,\ a^* \in A^*. \end{aligned}$$

Since these equations act on the set G_δ and δ is small, the second equation can be linearized, and, without loss of generality, we can restrict the consideration to a first-approximation equation

$$\dot{\xi} = 2d\xi + b\mu + a^*.$$

Concluding this section, we want to point out that when transforming the coordinates, we introduced the notation

$$\mu = \frac{u}{x_1},$$

which, actually, defines the statical binary operation

$$u = \beta(\mu, x_1) = \mu x_1$$

which defines the operator R_u of the coordinate feedback in the block diagram in Fig. 4.5.

This is, certainly, not the only possibility, but the simplest one. We could have used a dynamical binary operation, say, an operation of an integral type,

$$\frac{u}{x_1} = \int \mu\, dt,$$

or that of an inertial type,

$$\frac{u}{x_1} = \tau\mu + \int \mu\, dt, \qquad \tau = \text{const}.$$

In any case, the general conclusions remain valid, of course, but, instead of a scalar object, we should have to deal with a higher-order object. For instance, in an integral binary operation the system being stabilized is described by equations

$$\begin{aligned} \dot{\xi} &= 2d\,\xi + b\mu_1 + a^*, \\ \dot{\mu}_1 &= \mu. \end{aligned}$$

4.2. CO-Algorithms of Stabilization

Let us use the standard methods of the classical control theory for synthesizing a control which stabilizes the scalar object:

$$\dot{\xi} = 2d\,\xi + b\mu + a - d^2, \quad a \in A,\ b \in B,\ d > 0. \tag{4.15}$$

4.2.1. Direct compensation

In Eq. (4.15) the parameter d^2 can be interpreted as a known perturbation, and, for a known parameter b, the effect produced by this "perturbation" can be annihilated with the aid of a direct compensation

$$\mu = \mu_1 + \frac{d^2}{b} \tag{4.16}$$

since

$$\dot{\xi} = 2d\xi + b\mu_1 + a.$$

Now if the parameter b is unknown, then the direct "frontal" compensation does not produce any effect, but it can be successfully combined with the identification of the parameters. We shall speak about it somewhat later and now will only point out that control (4.16) in the initial variables is associated with feedback of the form

$$u = \mu x_1 = \mu_1 x_1 + \frac{d^2}{b} x_1,$$

which has a binary and a linear component.

Note that if the parameter a is also known, then there is no need to use the binary component.

4.2.2. Asymptotic estimation or an indirect measurement of the O-perturbation

Let us consider a version of the problem of stabilization of an object[1]

$$\dot{\xi} = 2d\xi + a + b\mu, \tag{4.17}$$

where the parameters a, b are fixed, b is known, and a is an arbitrary element from A. To find the estimate $\tilde{a}$ of the unknown parameter a, we use the observer

$$\begin{aligned} \dot{\varphi} &= (2d - k_1)\varphi + \tilde{a} + k_1\xi + b\mu, \\ \dot{\tilde{a}} &= -k_2(\varphi - \xi), \quad k_1, k_2 = \text{const}. \end{aligned} \tag{4.18}$$

Subtracting (4.17) from (4.18) and introducing a notation

$$e = \varphi - \xi, \qquad \alpha = \tilde{a} - a,$$

with due account of the fact that the parameter a is fixed, i.e.,

$$\dot{a} = 0,$$

we obtain an observer equation for the errors (e, α)

$$\begin{aligned} \dot{e} &= (2d - k_1)e + \alpha, \\ \dot{\alpha} &= -k_2 e. \end{aligned} \tag{4.19}$$

[1]Here and in the sequel, we shall write, for simplicity, a instead of $a^* = a - d^2$ assuming that the component d^2 has already been formed.

The characteristic polynomial of observer (4.19) has the form

$$\det \begin{bmatrix} s-(2d-k_1) & -1 \\ k_2 & s \end{bmatrix} = s^2 + s(k_1 - 2d) + k_2 = 0$$

and is a Hurwitz polynomial when the inequalities $k_2 > 0$, $k_1 > 2d$ are satisfied. Therefore the estimate $\tilde{a}$ asymptotically (exponentially) converges to the number a. If we now give the control μ the form

$$\mu = \mu_1 - \frac{\tilde{a}}{b}, \tag{4.20}$$

then the perturbation a will be asymptotically compensated since

$$\dot{\xi} = 2d\xi + b\mu_1 + a - \tilde{a},$$

and, consequently, $a - \tilde{a} \to 0$.

Thus, when choosing the control μ_1, it suffices to deal with the free motion of the object,

$$\dot{\xi} = 2d\xi + b\mu_1.$$

In the initial variables, the control algorithm (4.20) is associated with an algorithm of the form

$$u = \mu x_1 = \mu_1 x_1 - \frac{\tilde{a}}{b} x_1,$$

where it is natural to say that the second component is additive. In other words,

- the technique of asymptotic evaluation of the constant perturbation a described above realizes the standard procedure of adaptive control.

If, however, $a = a(t)$, then the *theory of adaptive control does not give any recommendations concerning the synthesis of stabilizing control whereas the theory that we expose here can be easily generalized to this case.*

4.2.3. Compensation for a wave O-perturbation

Suppose that the differential operator $K\left(\frac{d}{dt}\right)$ that annihilates the O-perturbation is known, i.e.,

$$K\left(\frac{d}{dt}\right) a \equiv 0,$$

for instance, the numbers r, p are known such that

$$\dot{a} + pa + r \equiv 0, \tag{4.21}$$

but the initial condition $a(0)$ is unknown. This makes the perturbation a unknown.

According to the recommendation of the classical control theory, when stabilizing the object

$$\dot{\xi} = 2d\xi + b\mu + a, \tag{4.22}$$

we must use Eqs. (4.21), (4.22) to form an observer equation which gives an asymptotic estimate $\tilde{a}$ of the perturbation a. Such an observer can be constructed in the standard way and has the form

$$\begin{aligned} \dot{\varphi} &= 2d\varphi + b\mu + \tilde{a} - k_1(\varphi - \xi), \\ \dot{\tilde{a}} &= -p\tilde{a} - r - k_2(\varphi - \xi). \end{aligned} \tag{4.23}$$

When we subtract termwise (4.21) and (4.22) from (4.23) and pass to the error of evaluation

$$e = \varphi - \xi, \quad \alpha = \tilde{a} - a,$$

the observer equation assumes the form

$$\begin{aligned} \dot{e} &= (2d - k_1)e + \alpha, \\ \dot{\alpha} &= -p\alpha - k_2 e. \end{aligned} \tag{4.24}$$

The characteristic polynomial of system (4.24) is given by the expression

$$\det \begin{bmatrix} s + (k_1 - 2d) & -1 \\ k_2 & s + p \end{bmatrix} = s^2 + (p + k_1 - 2d)s + k_2 + p(k_1 - 2d) = 0,$$

and it is clear that the observer is exponentially stable when

$$k_1 > 2d - p, \quad k_2 > p(k_1 - 2d).$$

The last condition can be easily fulfilled, and, consequently, observer (4.24) gives an asymptotic estimate

$$\tilde{a} \xrightarrow{\text{exp}} a.$$

By virtue of the obtained asymptotic estimate, the equation

$$\mu = \mu_1 - \frac{\tilde{a}}{b}$$

solves the compensation problem for the variable perturbation $a(t)$ and reduces the original stabilization problem to a trivial stabilization problem of the object

$$\dot{\xi} = 2d\xi + b\mu_1.$$

Note that it is impossible to obtain a similar result acting in the framework of the standard conception of adaptive control.

4.2.4. Relay CO-stabilization

When we do not have a wave model of O-perturbation but know the majorant a^0, i.e., a function or a constant such that

$$|a(t)| \leq a^0,$$

it is possible to use a discontinuous feedback, in particular, a relay feedback

$$\mu = -k \operatorname{sgn} \xi$$

for $a^0 = \text{const}$, in order to stabilize the object

$$\dot{\xi} = 2d\xi + \mu + a.$$

In this case, the feedback system can be described by the equation

$$\dot{\xi} = 2d\xi - k \operatorname{sgn} \xi + a, \quad k = \text{const}, \tag{4.25}$$

and, when the condition

$$k > a^0/b^-$$

is fulfilled, there exists a neighborhood in which the zero of Eq. (4.25) is asymptotically stable.

We can get an idea of the qualitative behavior of solutions of the equation

$$\dot{\xi} = 2d\xi - k \operatorname{sgn} \xi + a \tag{4.26}$$

from Fig. 4.9a, which shows the variety of solutions of Eq. (4.26), or from Fig. 4.9b, which shows the phase trajectories in the CO-space (x_1, ξ). The sliding mode in a relay system is not robust and is "washed out" to the real sliding mode when a space delay Δ or a small delay in switchings is introduced, i.e., when instead of (4.26) we deal with equations of the form

$$\dot{\xi} = 2d\xi - k \operatorname{sgn}_\Delta \xi + a, \quad \dot{\xi} = 2d\xi - k \operatorname{sgn}_\tau \xi + a. \tag{4.27}$$

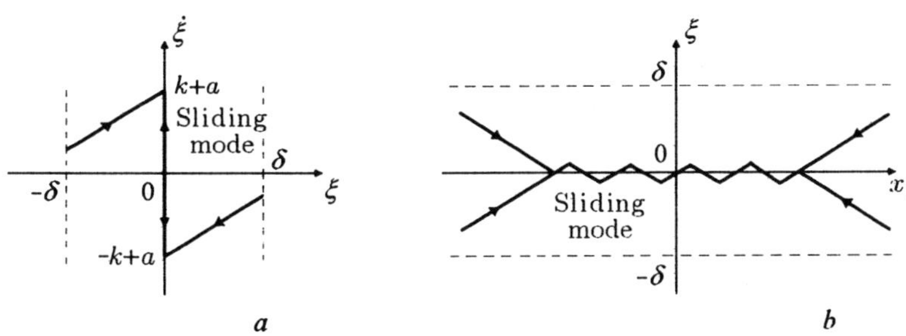

Figure 4.9.

Figure 4.10 illustrates what we have said, where $\Delta(\tau)$ is the "amplitude" of the real sliding mode. All the indicated properties of the real system are well known and have been pointed out, and now it is more interesting to see what motions in the initial coordinate space (x_1, x_2) correspond to the indicated motions in the CO-space (x_1, ξ).

We shall first do it formally, using the first equation of the object in the variables (x_1, ξ), i.e., the equation

$$\dot{x}_1 = -dx_1 + \xi x_1 = -(d - \xi)x_1.$$

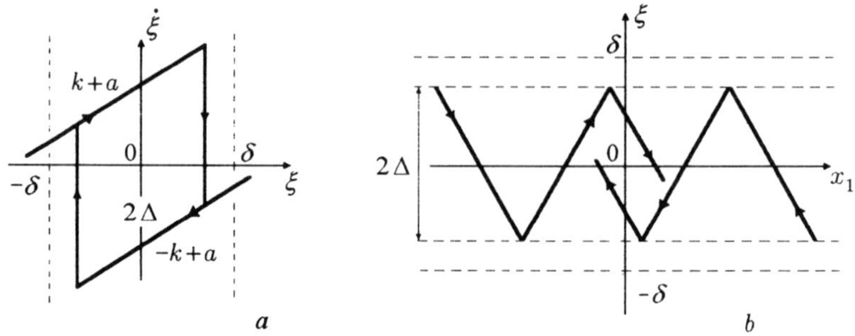

Figure 4.10.

When an ideal sliding mode appears, we have a relation $\xi = 0$, and, consequently, the variable x_1 and, together with it, the output y of the object exponentially decrease to zero since $\dot{x}_1 = -dx_1$. The condition $|\xi| \le \Delta < \delta$ is fulfilled in the mode of real sliding and we also have an exponential stability if $\delta < d$, and this, of course, is fulfilled. Therefore the obtained system of binary control with relay CO-algorithm is exponentially stable both upon ideal and upon real switchings. To put it otherwise,

- it is robust relative to nonideal switchings, and not only of a temporal type but also (!) of a spatial type.

We may say that the properties of robustness of a nonlinear discontinuous system depend on the position of the discontinuous (relay) element in its structure.

Let us now consider the constructed binary system with a relay CO-algorithm of stabilization in the initial coordinate space (x_1, x_2). Since $u = \mu x_1$, $\sigma = \xi x_1$, $\mu = -k \operatorname{sgn} \xi$, we successively have

$$u = -kx_1 \operatorname{sgn} \xi = -kx_1 \operatorname{sgn} \frac{\sigma}{x_1} = -k|x_1| \operatorname{sgn} \sigma. \tag{4.28}$$

Let us extend the action of this algorithm of control beyond the limits of the set G_δ to the whole plane (x_1, x_2), and then this algorithm becomes a standard control algorithm for variable structure systems. Indeed, the last expression in (4.28) defines the standard ψ-cell,

$$u = \psi x_1, \quad \psi = \begin{cases} -k, & x_1\sigma > 0, \\ k, & x_1\sigma < 0, \end{cases}$$

and, consequently, the phase portrait of the system in question in the coordinates (x_1, x_2) has the form that we already know (Fig. 4.11). The change in the structure of the system takes place on the straight lines $x = 0$ and $\sigma = 0$. The block diagram of the binary system with relay CO-feedback synthesized in this way is shown in Fig. 4.12. For the sake of comparison, we give in Fig. 4.13 the

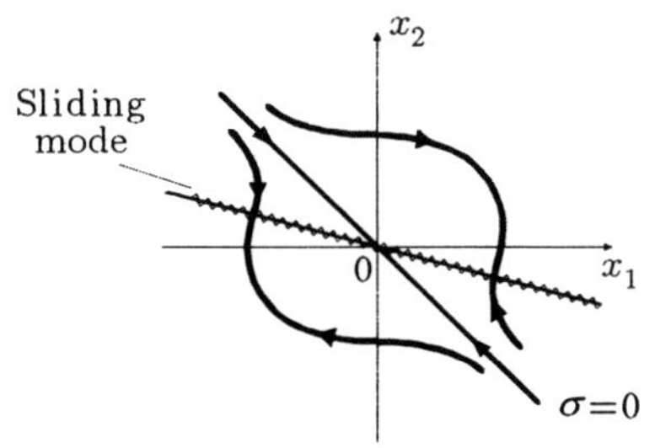

Figure 4.11.

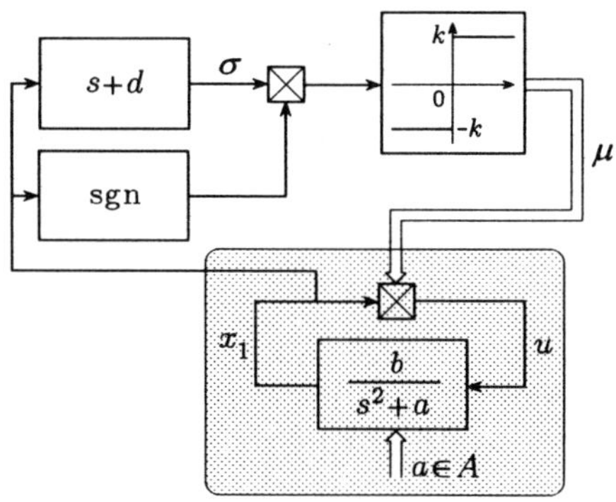

Figure 4.12.

structure of the same system in the standard version of a VSS. Comparing the figures, we can see that the binary control theory allows us to reveal the delicate construction of a discontinuous feedback and, consequently, makes it possible to synthesize regularly a feedback of this type whereas in the VSS theory we have guessed the type of feedback. One more principal peculiarity of CO-feedback is

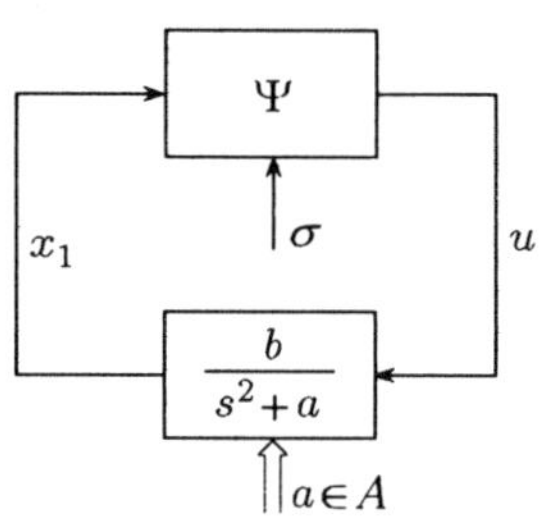

Figure 4.13.

its sign alternation depending on the position of the phase point in the plane. Formally, this follows from the relation

$$u = \mu x_1 = -k|x_1| \operatorname{sgn} \sigma,$$

and, consequently, $\mu = -k\,\mathrm{sgn}(x_1\sigma)$. To put it otherwise,

- the efficiency of systems with CO-feedback is achieved only when the structure of the system is variable.

4.2.5. Remark concerning the robustness of systems with relay CO-feedback

There exists, however, an essential difference between the classical VSS and the binary control system under consideration which becomes obvious when switchings are nonideal. Namely, in the binary system, when time or space delays appear in switchings, the inequality $|\xi| \leq \Delta$ is satisfied and, consequently, the asymptotic stability is preserved, and the phase portrait has the form shown in Fig. 4.14*a*. For the standard VSS, when there is a time delay in switchings, the phase portrait coincides with the portrait in Fig. 4.14*a*, but, under a space delay, it differs from it and has the form shown in Fig. 4.14*b*. Consequently,

- a binary control system with CO-relay feedback is robust relative to these perturbations,

whereas the classical VSS is not robust. We have thus shown that in the absence of a wave model of O-perturbation a, i.e., $|a| \leq a^0$, and for any O-perturbation $b \geq b_{\min} > 0$, i.e., $b \in B$, the synthesized binary control algorithm with CO-relay feedback

- is as yet the only one known algorithm that robustly solves the stabilization problem for an uncertain nonstationary object.

This is the first practical result of the CO-theory of stabilization.

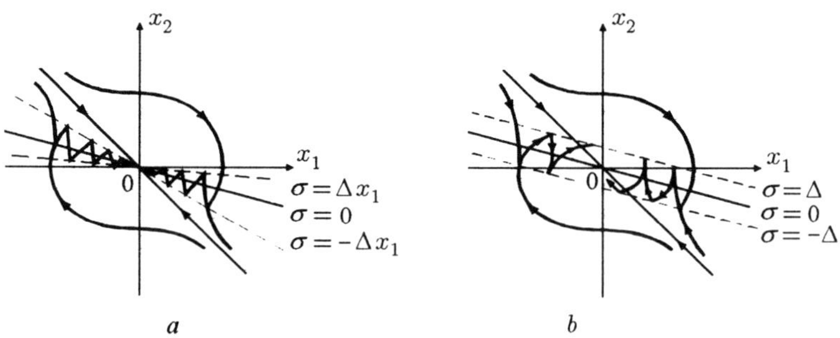

Figure 4.14.

One of the drawbacks of the algorithm that we have considered is the "pulsing" that follows the relay sliding mode in which practically all physical systems function. Another drawback is the laboriousness of exact digital realization of the sliding mode in real time since, in this case, we can only apply first-order approximation schemes, but then the "amplitude" of deviation from the line of sliding has the order of the discretization step h, i.e.,

$$|\sigma| \sim O(h).$$

Therefore it is important to answer the following question:

- are there continuous stabilization algorithms that provide a feedback system with the properties described above?

It is clear that in the general case the answer to this question is negative, but we can indicate such algorithms for special cases. Remaining in the framework of the CO-theory, we shall indicate some of them.

4.2.6. Linear CO-algorithms of stabilization

Consider a situation where in the process of stabilizing the object

$$\dot{\xi} = 2d\xi + b\mu + a$$

we can consider the parameters $a \in A$, $b \in B$ to be constant. For the unknown b, the method of asymptotic evaluation is inapplicable and we have to look for other techniques of stabilization. Techniques of this kind are developed in the classical control theory, and they presuppose the use of linear feedbacks with an integral component, the so-called PI- (proportionally integral) and PID- (proportional integro-differentiable) laws of control. We shall begin with proportion laws.

P-laws of control. In this case,

$$\mu = -k\xi, \quad k = \text{const},$$

the feedback control system is described by the equation

$$\dot{\xi} = -(bk - 2d)\xi + a,$$

and, when the condition

$$b^- k > 2d$$

is fulfilled, it is asymptotically stable with the equilibrium position at the point

$$\xi_\infty = \frac{a}{bk - 2d}.$$

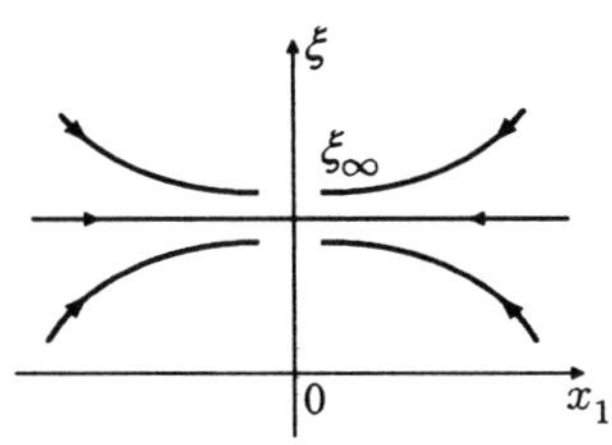

Figure 4.15.

The phase portrait of a feedback control system in coordinates (x_1, ξ) is shown in Fig. 4.15. The point ξ_∞ in the figure defines the statism of the stabilization system in coordinates (x_1, ξ). The phase portrait in the original coordinates (x_1, x_2) shown in Fig. 4.16, where the straight line $\sigma = \xi_\infty x_1$ is an

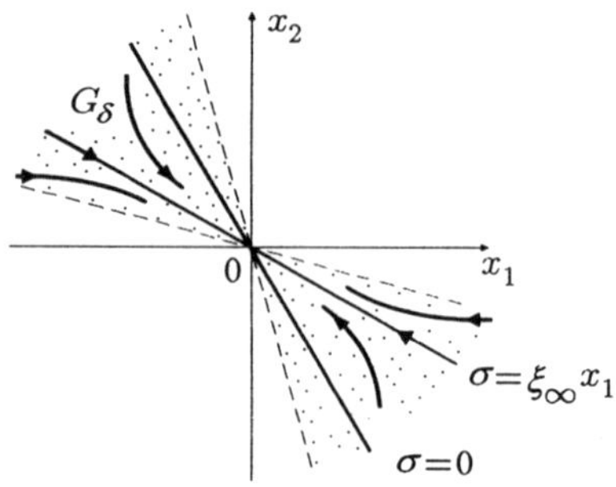

Figure 4.16.

asymptote defining the dominant motion, corresponds to Fig. 4.15. Recall that the analysis is carried out under the condition $|\xi| \leq \delta$, i.e., in the set

$$G_\sigma = \{(x_1, x_2) \mid |\sigma| \leq \delta |x_1|\},$$

and therefore the trajectories outside of G_δ are not yet shown in the figures.

Thus, if $|\xi_\infty| \leq \delta$, then the statism in the CO-loop of control does not violate the asymptotics of the control system in the initial space, but since the asymptote $\sigma = \xi_\infty x_1$ does not coincide with the required position of $\sigma = 0$, the asymptotic behavior (the degree of stability) becomes different, and this can be interpreted as a dynamical statism. To eliminate this dynamical statism, we can tend the coefficient k to infinity, $k \to \infty$, but this is equivalent to the introduction of a feedback with high gain in the CO-loop since $u = \mu x_1 = -k\xi x_1 = -k\sigma$, but then we shall obtain a nonrobust system, which is not the task of our investigation. Therefore we shall consider the possibilities provided by the PI-law of control in a CO-loop.

PI-laws of control. In this case we have

$$\mu = -k_2\xi - k_1 \int \xi \, dt, \tag{4.29}$$

and, using feedback for the object under investigation,

$$\dot{\xi} = 2d\xi + b\mu + a, \tag{4.30}$$

we obtain a second-order control system

$$\ddot{\xi} + (bk_2 - 2d)\dot{\xi} + bk_1\xi = 0. \tag{4.31}$$

We have obtained Eq. (4.31) as a result of the substitution of the differentiated relation (4.29) into the differentiated equation (4.30). We can clearly see from Eq. (4.31) that when conditions $b^- k_2 > 2d$, $k_1 > 0$ are fulfilled, the zero of the equation becomes asymptotically stable. Depending on the relationships between the parameters k_1, k_2, this stability may be oscillatory or aperiodic. In addition, the degree of stability may be assigned arbitrarily by a suitable choice of the parameters k_1, k_2.

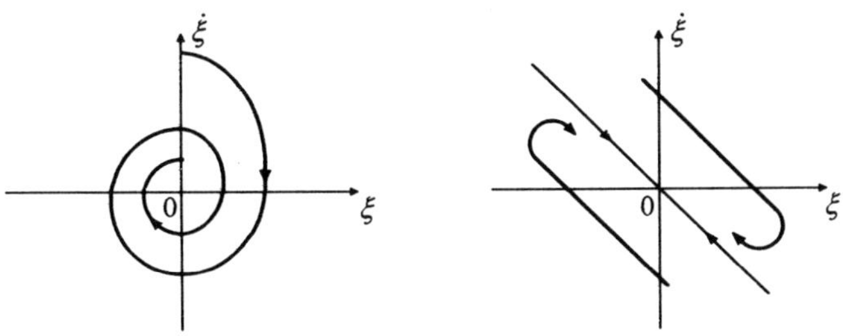

Figure 4.17.

Figure 4.17 shows, in coordinates $(\xi, \dot{\xi})$, the phase trajectories corresponding to the oscillatory and aperiodic transient processes in the feedback system $\ddot{\xi} + (bk_2 - 2d)\dot{\xi} + bk_1\xi = 0$. The corresponding projections of the phase trajectories onto the plane (x_1, x_2) are shown in Fig. 4.18.

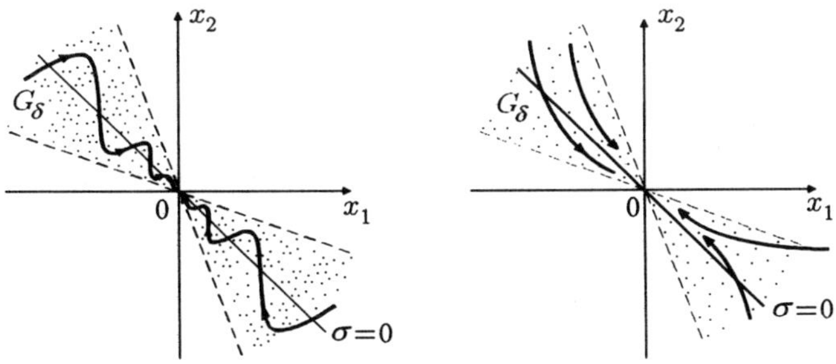

Figure 4.18.

The resulting algorithm of stabilization corresponding to the PI-law of the CO-feedback has the form

$$u = \mu x_1 = -k_2\xi x_1 - k_1 x_1 \int \xi \, dt = -k_2\sigma - k_1 x_1 \int \frac{\sigma}{x_1} \, dt. \tag{4.32}$$

Feedback (4.32) is associated (recall that this correspondence holds only in the limits of the set G_δ) with the block diagram shown in Fig. 4.19. It is easy to understand that the synthesized binary control system is robust relative to singular perturbations and, consequently, the main task of the synthesis is achieved, i.e.,

- the continuous control algorithm is described which ensures, for any constant $a \in A$, an asymptotic "motion" along the required trajectory $\sigma = \dot{x}_1 + dx_1 = 0$.

Consequently, the effect of small dependence of motion on the uncertainty factor is achieved not with the aid of a large gain factor or sliding mode but only by the application of an appropriate CO-feedback.

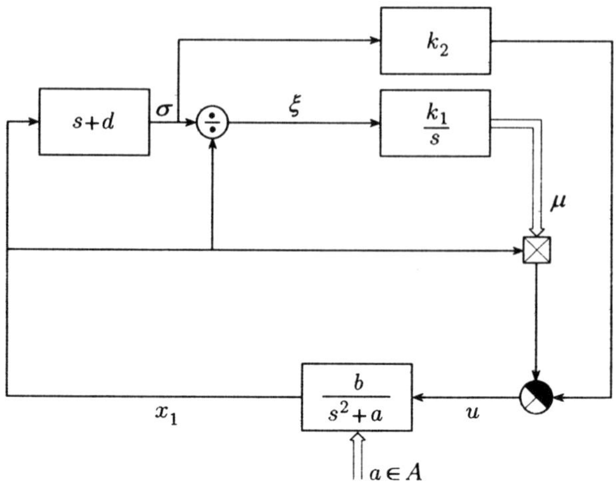

Figure 4.19.

4.2.7. Integro-relay CO-algorithm of stabilization

From the block diagram of the system shown in Fig. 4.19 we can see that in order to realize the PI-law in a CO-loop of feedback, we must use the division operation. However, we can get rid of this operation if we take a relay-integral component instead of an integral component. Let us study the consequences of using the law

$$\mu = -k_2\xi - k_1 \int \operatorname{sgn} \xi \, dt.$$

In this case, the behavior of the feedback system is defined by the equation

$$\ddot{\xi} + (bk_2 - 2d)\dot{\xi} + bk_1 \operatorname{sgn} \xi = 0. \tag{4.33}$$

For $b^- k_2 > 2d$, $k_1 > 0$ the phase portrait of Eq. (4.33) is formed by segments of the parabolic trajectories of the equations

$$\ddot{\xi} + (bk_2 - 2d)\dot{\xi} = \pm bk_1$$

"sewed" along the line of discontinuity $\xi = 0$ (Fig. 4.20*a*). Since damping occurs ($bk_2 - 2d > 0$), the motion along these trajectories asymptotically tends to zero. The "rate" of twisting is controlled by the parameter k_2. In the initial space (x_1, x_2) the corresponding transient processes are illustrated by Fig. 4.20*b*.

The block diagram of a binary system with this proportional relay-integral CO-feedback is given in Fig. 4.21, and we see that this controller is simpler in the realization of a CO-control with the PI-law. Note, in particular, that for $bk_2 = 2d$ the system in the O-space $(\xi, \dot{\xi})$ becomes oscillatory,

$$\ddot{\xi} + bk_1 \operatorname{sgn} \xi = 0.$$

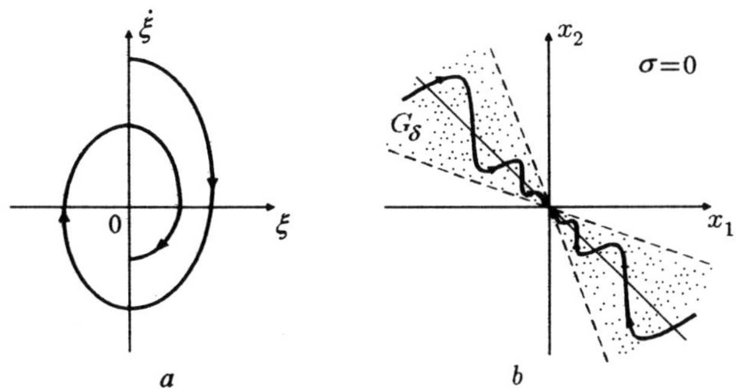

Figure 4.20.

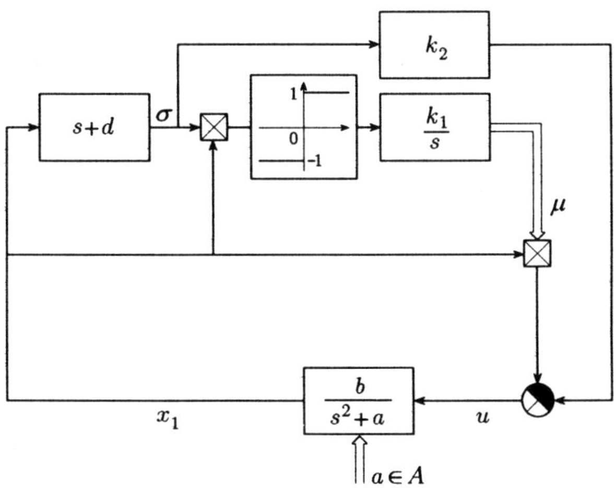

Figure 4.21.

And although the coordinate $\xi \not\to 0$, the asymptotic stability with respect to the principal variables is guaranteed if $|\xi(0)| \leq \delta$. The graphs corresponding to this case are given in Fig. 4.22.

It should be pointed out in conclusion that in the smooth binary systems that we have considered the error σ does not finally (as in a VSS) vanish, and therefore we can only speak about an asymptotic elimination of the dependence of the properties of the system on the uncontrolled factors. However, under more intricate laws of CO-control, we can finitely stabilize the error σ at zero, and, in this case, the system has a definite degree of smoothness.

A more detailed discussion of these problems is included in the following themes. Here we would like to emphasize that

- an asymptotic independence of motion in the control of an uncertain object can be achieved without the use of feedback with high gain or a sliding mode but only with the use of a nonlinear feedback of the CO-type; this is the second practical result of the CO-theory,

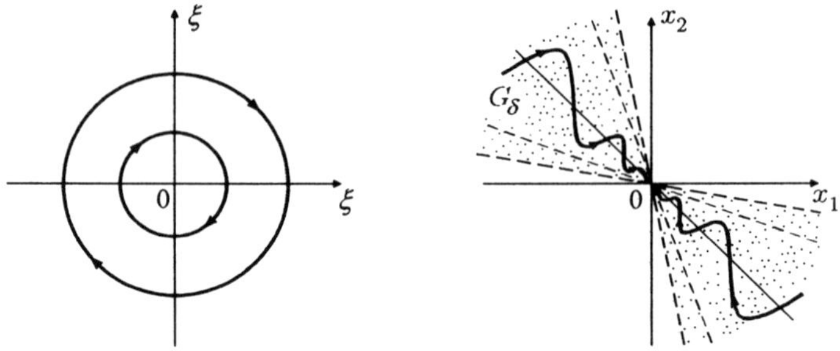

a *b*

Figure 4.22.

- the physical basis of the compensation for uncertainly consists in the use, at certain stages of motion, of a positive feedback which allows the control signal to "grow" to a value sufficient for the compensation for the uncertainty. This process is similar to that described in Chap. 2 where the solution of the problem of the reconstruction of the reference action with the aid of one integral is described.

Chapter 5

Higher-Degree Sliding Modes

The importance of the part played by the sliding mode in control problems under the conditions of uncertainty can be seen from what was said in the preceding chapters. It suffices to recall relay systems, variable structure systems, and systems with coordinate-operator feedback. The brief investigation carried out in the corresponding sections of the monograph has shown that although different sliding modes have certain similar features, they also have considerable distinctions which are revealed, in particular, under the action of uncertainty factors and disturbances. A more detailed analysis shows that there exist many kinds of sliding modes whose only common feature is the existence of discontinuities on the right-hand side of differential equations which describe the behavior of the corresponding dynamical systems. In what follows, we shall give a substantive outline of the division of the general theory of sliding modes which is connected with the degree of smoothness on the solutions of a system of the function which defines the surface of discontinuity. The degree of smoothness of a solution naturally puts the sliding modes in order, and a possibility appears of introducing a new concept, that of the degree of sliding.

5.1. Preliminaries from the Theory of Sliding Modes

We give here some facts from the theory of sliding modes which are necessary for understanding the further exposition.

Sliding modes appear in discontinuous dynamical systems, say, in control systems of the form

$$\dot{x} = f(x) + b(x)u,$$

where the control u (feedback) suffers discontinuities on the smooth manifold

$$M = \{x \in R^n \mid \sigma(x) = 0\},$$

namely,

$$u(x) = \begin{cases} u^+(x), & \sigma(x) > 0, \\ u^-(x), & \sigma(x) \leq 0, \quad u^+ \neq u^-. \end{cases}$$

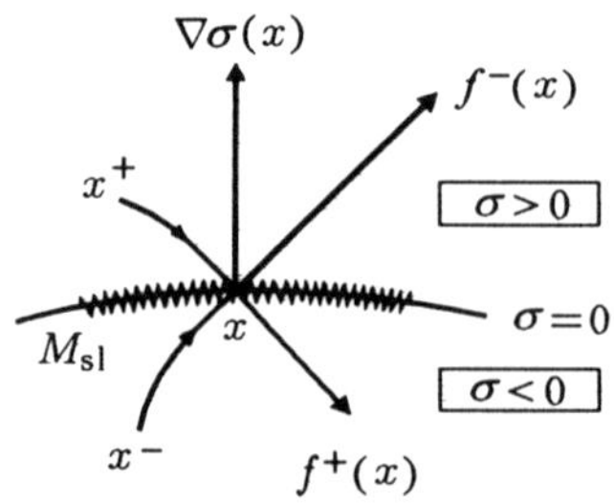

Figure 5.1.

A sliding mode occurs at the points of the manifold M which the phase trajectories $x^{\pm}(t)$ of the dynamical systems

$$\dot{x} = f^{\pm}(x) = f(x) + b(x)\, u^{\pm}(x)$$

approach from different sides of the manifold and intersect transversally (Fig. 5.1). The inequalities

$$\pm \left\langle \nabla\sigma(x), f^{\pm}(x) \right\rangle \big|_{\sigma(x)=0} < 0$$

are satisfied at these points. The union of all these points $x \in M$ forms a set $M_{\text{sl}} \subset M$ which is called in the sequel a domain of sliding (this domain is denoted M_{sl} and marked in Fig. 5.1).

5.1.1. Equations of sliding

In order to obtain equations which describe motion in a sliding mode (in the sequel they are called equations of sliding), we can use Filippov's procedure. Here is a necessary fragment of this procedure. We connect the endpoints of the phase vectors $f^{\pm}(x)$ by a straight line. Every vector $f_{\alpha}(x)$ which meets this straight line is given by a convex combination $f_{\alpha} = \alpha f^{+} + (1-\alpha) f^{-}$ of the phase vectors $f^{\pm}(x)$. One of these vectors is taken as the required one (Fig. 5.2). Namely, since $\sigma \equiv 0$ in the sliding mode, the required phase vector must satisfy the equation

$$\langle \nabla\sigma, f_{\alpha} \rangle = 0,$$

where $\nabla\sigma$ is the gradient of the function $\sigma(x)$ at the point x. Solving this equation for the parameter α, we obtain

$$\alpha^{*} = \frac{\langle \nabla\sigma, f^{-} \rangle}{\langle \nabla\sigma, (f^{-} - f^{+}) \rangle}.$$

The corresponding differential equation has the form

$$\begin{aligned} \dot{x} = f_{\alpha^{*}} &= f^{-} - \alpha^{*}(f^{-} - f^{+}) \\ &= f^{-} - \frac{(f^{-} - f^{+}) \langle \nabla\sigma, f^{-} \rangle}{\langle \nabla\sigma, (f^{-} - f^{+}) \rangle} = f_{\text{sl}}(x) \end{aligned}$$

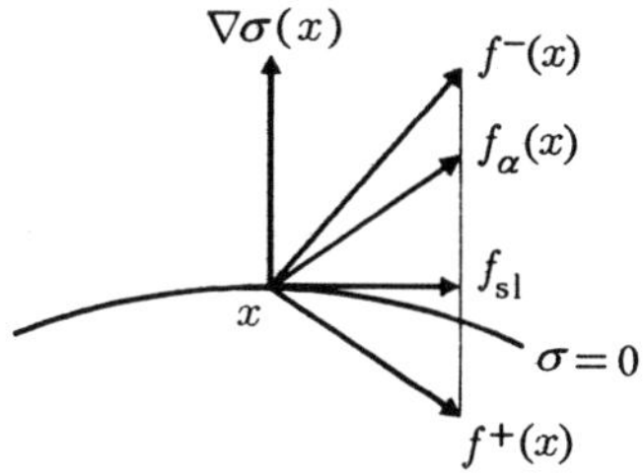

Figure 5.2.

and describes the motion over the level surfaces $\sigma(x) = \text{const}$. In order to obtain an equation of sliding, we must complete it by the relation

$$\sigma = 0, \quad x \in M_{\rm sl}.$$

The equations of sliding

$$\dot{x} = f_{\rm sl}(x), \quad \sigma(x) = 0, \quad x \in M_{\rm sl}$$

can be given a useful geometrical interpretation. On the manifold M we introduce into consideration an operator

$$P(x) = E - \frac{f^-(x) - f^+(x)\,\langle \nabla\sigma(x), (\cdot)\rangle}{\langle \nabla\sigma(x), (f^-(x) - f^+(x))\rangle}.$$

Since $P^2 = P$, this operator is an operator of local projection onto the manifold $\sigma = 0$ along the straight line (in parallel) defined by the vector $\Delta f = f^- - f^+$ (Fig. 5.3). Therefore the equations of sliding can be written as

$$\dot{x} = P_{\Delta f} \cdot f^-, \quad P_{\Delta f} = E - \frac{\Delta f\,\langle \nabla\sigma, (\cdot)\rangle}{\langle \nabla\sigma, \Delta f\rangle}.$$

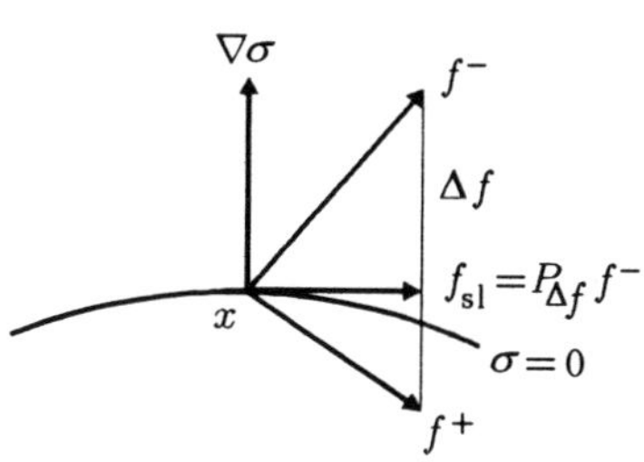

Figure 5.3.

For the equation $\dot{x} = f + bu$ in question, the vector

$$\Delta f = f + bu^- - f - bu^+ = b(u^- - u^+) = b\Delta u,$$

and, if $\Delta u \neq 0$, then the vectors Δf and b are collinear. Consequently, the relation

$$P_{\Delta f} = P_b = E - \frac{b\langle\nabla\sigma, (\cdot)\rangle}{\langle\nabla\sigma, b\rangle}$$

is valid, and since

$$P_b \cdot b = \left(E - \frac{b\langle\nabla\sigma, (\cdot)\rangle}{\langle\nabla\sigma, b\rangle}\right) b = b - \frac{b\langle\nabla\sigma, b\rangle}{\langle\nabla\sigma, b\rangle} = 0,$$

by the property of the projection operator we can write the equations of sliding in a more compact form

$$\dot{x} = P_b f, \quad \sigma = 0, \quad x \in M_{\text{sl}}.$$

5.1.2. On the invariance of an equation of sliding relative to disturbances which satisfy the matching condition

Recall that the external disturbance $\varphi(t, x)$ satisfies the matching condition if it acts in the control channel

$$\dot{x} = f + b(u + \varphi). \tag{5.1}$$

By virtue of the property indicated above,

$$P_b b = 0.$$

The equations of sliding over the manifold M for the disturbed (5.1) and undisturbed

$$\dot{x} = f + bu$$

objects coincide and are defined by the relations

$$\dot{x} = P_b f, \quad \sigma = 0, \quad x \in M_{\text{sl}} \subset M. \tag{5.2}$$

It is precisely this mathematical fact that explains why so much interest is paid in control theory and its applications to the idea of an integral manifold and its realization with the aid of discontinuous feedback and high gain feedback.

It should be pointed out here that Eqs. (5.2) also describe the motion in a system with high gain feedback

$$\dot{x} = f + bu\big|_{u=-k\sigma} = f - kb\sigma, \quad k = \text{const} > 0, \tag{5.3}$$

i.e., when the feedback factor tends to infinity ($k \to \infty$) if the condition

$$\langle\nabla\sigma, b\rangle > 0$$

is fulfilled.

In order to verify this fact, we can use the geometrical arguments given above or calculations presented below. By virtue of the motion equation we find that

$$\dot{\sigma} = \langle \nabla\sigma, f \rangle - k \langle \nabla, \sigma, b \rangle \sigma, \tag{5.4}$$

and, under the conditions indicated above, the representative point "instantaneously" attains the surface $\sigma(x) = 0$ and does not leave it. Hence, the relation $\dot{\sigma} = 0$ holds during the motion. Let us use now the following heuristic technique: we express the product $k\sigma$ from the equation $\dot{\sigma} = 0$ and substitute the result into Eq. (5.3), and then we obtain an equation

$$\dot{x} = f - b\frac{\langle \nabla\sigma, f \rangle}{\langle \nabla\sigma, b \rangle} = P_b f,$$

which, together with the relation $\sigma = 0$, defines the motion equation for a system with high gain feedback. We can see that it coincides with the equation of sliding of a discontinuous system obtained above.

In a certain sense, we can regard the standard discontinuous system and a system with high gain factor as two poles of realization of the same idea, namely, the idea of sliding on a smooth manifold. In one case, this sliding is infinitely smooth and in the other case it has a discontinuity of already the first derivative of the function which defines the surface of sliding. It turns out that between these two extreme systems there exist infinitely many systems that "slide" over the same surface but have different degrees of smoothness. In this chapter we speak about some of these intermediate systems.

5.1.3. Equations of real sliding

In practice, i.e., in actual situations, switchings of the discontinuous element occur not exactly on the manifold $M = \{x \mid \sigma(x) = 0\}$ but always in a certain neighborhood of the manifold,

$$O(M) = \{x \mid |\sigma(x)| \leq A(\Delta)\},$$

where $A(\Delta)$ is the "amplitude" of deviation of the trajectories of the discontinuous system $\dot{x} = f^{\pm}$ from the manifold M (Fig. 5.4). Here and in the sequel Δ is a parameter which characterizes nonideal switchings, for instance, a time or space delay, fast dynamics that cannot be simulated, etc. It is customary to call

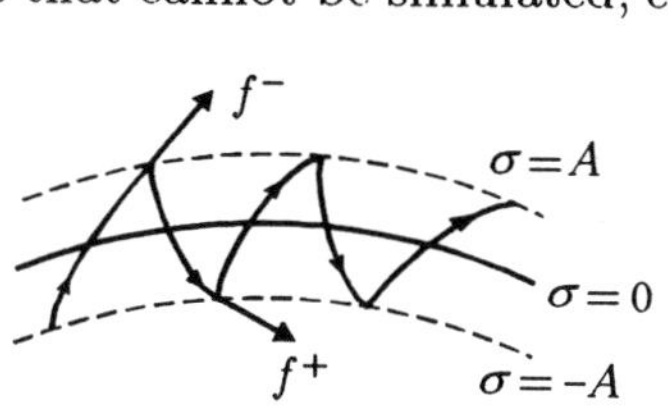

Figure 5.4.

this sliding mode a real sliding mode. When the amplitudes $A(\Delta)$ are small, it can also be characterized by the "frequency" of switchings $W(\Delta)$ for which the estimate

$$W(\Delta) \leq \frac{\pi\|f\|}{2A(\Delta)},$$

where $\|f\| = \min(\|f^-\|, \|f^+\|)$, holds true. When switchings are made ideal (i.e. as $\Delta \to 0$), the "amplitude" decreases, as a rule, and becomes zero, and the "frequency" increases to infinity, i.e.,

$$\lim_{\Delta\to 0} W(\Delta) = \infty, \quad \lim_{\Delta\to 0} A(\Delta) = 0.$$

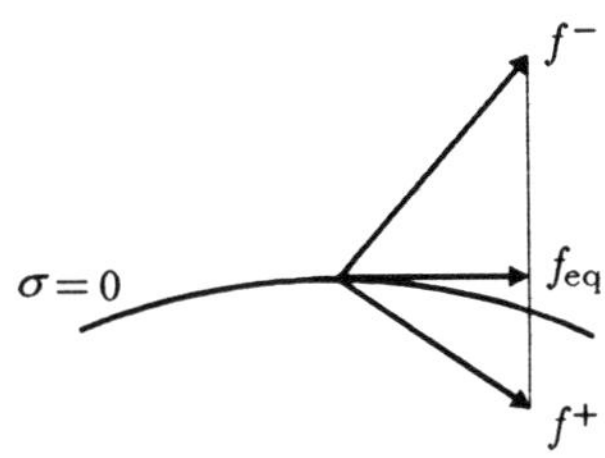

Figure 5.5.

Then the real sliding becomes ideal (Fig. 5.5), and it is very important to answer the question concerning the degree of this transition to an ideal sliding with respect to the parameter Δ. Before defining exactly what we have in mind, let us obtain an equation for real sliding. Recall that when the sliding is ideal ($\sigma = 0$), the required phase vector f_{eq} on the right-hand side of the motion equation (Fig. 5.5) can be found by solving the equation

$$\dot\sigma = \langle\nabla\sigma, f_{\text{eq}}\rangle = 0.$$

When the sliding is real, the inequality $|\sigma| \leq A(\Delta)$ is satisfied and the required vector f_{rs} on the right-hand side of the system of real sliding can be found from the nonhomogeneous equation

$$\dot\sigma = \langle\nabla\sigma, f_{\text{rs}}\rangle.$$

In particular, if the system has the form

$$\dot x = f + b(u + \varphi), \tag{5.5}$$

then we can write the last equation as

$$\dot\sigma = \langle\nabla\sigma, f\rangle + \langle\nabla\sigma, b\rangle\, u_{\text{rs}} + \langle\nabla\sigma, b\rangle\, \varphi.$$

If $\langle\nabla\sigma, b\rangle \neq 0$, then

$$u_{\text{rs}} = -\frac{\langle\nabla\sigma, f\rangle}{\langle\nabla\sigma, b\rangle} + \frac{\dot\sigma}{\langle\nabla\sigma, b\rangle} - \varphi,$$

and, substituting the obtained value into the motion equation (5.5), we find that

$$\dot{x} = f - b\frac{\langle \nabla\sigma, f\rangle}{\langle \nabla\sigma, b\rangle} + b\frac{\dot{\sigma}}{\langle \nabla\sigma, b\rangle}.$$

If we use the projection operator

$$P_b = E - b\frac{\langle \nabla\sigma, f\rangle}{\langle \nabla\sigma, b\rangle},$$

then the last equation assumes the form

$$\dot{x} = P_b f + b\frac{\dot{\sigma}}{\langle \nabla\sigma, b\rangle} \tag{5.6}$$

and, together with the constraint

$$|\sigma| \le A(\Delta), \tag{5.7}$$

defines the motion in the real sliding mode.

It is of interest to find out how close are the solutions of an ideal and a real sliding (we denote them by $x_{\text{id}}(t)$ and $x(t)$ respectively) that issue from the same initial point $x^0 \in M$ under the same disturbance $\varphi(t)$. The solution for the ideal sliding is described by the equations

$$\dot{x}_{\text{id}} = P_b f(x_{\text{id}}), \quad \sigma(x_{\text{id}}) = 0,$$

and the solution for the real sliding has the form

$$\dot{x} = P_b f(x) + \frac{b}{\langle \nabla\sigma, b\rangle}\dot{\sigma}, \quad |\sigma(x)| \le A(\Delta).$$

For simplicity and without loss of generality we assume that the vector

$$h = \frac{b}{\langle \nabla\sigma, b\rangle} = \text{const},$$

and then the operator $P_b = \text{const}$, and, subtracting the first equation from the second, we obtain the following equation for the error $\varepsilon = x - x_{\text{id}}$:

$$\dot{\varepsilon} = P_b\big[f(x_{\text{id}} + \varepsilon) - f(x_{\text{id}})\big] + h\dot{\sigma}.$$

According to the Lagrange theorem, we have a relation

$$f(x_{\text{id}} + \varepsilon) - f(x_{\text{id}}) = \frac{\partial f}{\partial x}(\Theta)\varepsilon$$

for a certain vector $\Theta \in [x_{\text{id}}, x_{\text{id}} + \varepsilon]$, and therefore, in the sequel, we deal with an equation of the form

$$\dot{\varepsilon} = N\varepsilon + h\dot{\sigma}, \quad N = P_b\frac{\partial f}{\partial x}(\Theta).$$

Since $\varepsilon(0) = 0$, the last differential equation is equivalent to the integral equation

$$\varepsilon(t) = \int_0^t N\varepsilon(\tau)\,d\tau + h\int_0^t \dot{\sigma}\,d\tau = \int_0^t N\varepsilon(\tau)\,d\tau + h\sigma(t).$$

Assuming that

$$||N|| \le D$$

on an interval $[0, T]$, we pass from the integral equation to a scalar inequality for the norm of the error

$$||\varepsilon(t)|| \le D\int_0^t ||\varepsilon(\tau)||d\tau + ||h||A(\Delta).$$

Applying the Gronwall–Bellman lemma, we find on the interval $[0, T]$ the desired estimate

$$||\varepsilon(t)|| = ||x(t) - x_{\text{id}}(t)|| \le ||h||e^{DT}A(\Delta).$$

Thus, since the trajectories of real sliding belong to the neighborhood of the manifold M,

$$O(M) = \{x \mid |\sigma| \le A(\Delta)\},$$

they are close, on a finite time interval, to the respective trajectories of ideal sliding with an accuracy to within the degree $A(\Delta)$, i.e., with an accuracy of real sliding. Then it follows that

- we should try to raise the degree of real sliding relative to the parameter Δ since then we can achieve a greater accuracy of approximation to ideal sliding when the switching element is not very ideal.

In other words, this is exactly the shortest way to achieving the robustness of a control system.

A similar problem arises in the discrete simulation of discontinuous systems when the role of the parameter of the nonideal switching is played by the step of discretization of the time scale

$$h = t_{j+1} - t_j,$$

and we observe discontinuities on the set of discrete time moments $\{t_j\}$. In this case, the neighborhood $O(M)$ of real sliding is characterized by the inequality

$$|\sigma(t_j)| \le A(h),$$

and, in order to raise the accuracy of simulation, we must diminish the discretization step h, but then the calculations become more complicated and lengthy. The advantages that we get when we diminish h increase with the growth of the order of smallness of the function $A(h)$ with respect to h, i.e., the higher the number r in the relation

$$A(h) \sim h^r,$$

the higher the accuracy of calculation for the specified integration step h.

The arguments that we have presented make the problem of finding the means of raising the degree of the function $A(\Delta)$ with respect to the small parameter Δ more actual since the degree of this function defines the "exactness" of real sliding.

5.1.4. Remarks concerning the degree of sliding

Let us find out on what the degree of sliding depends and what defines it.

Let us consider the motion in sliding mode of the discontinuous system

$$\dot{x} = f^{\pm}(x) = f(x) + b(x)\, u^{\pm}(x) \tag{5.8}$$

on the smooth manifold

$$M_0 = \{x \mid \sigma(x) = 0\}. \tag{5.9}$$

Relations (5.8), (5.9) yield

$$\dot{\sigma} = \langle \nabla\sigma, f^{\pm} \rangle = \langle \nabla\sigma, f \rangle + \langle \nabla\sigma, b \rangle\, u^{\pm}, \tag{5.10}$$

and if the vector fields $f^{\pm}(x)$ are transversal to M_0, then $\langle \nabla\sigma, b \rangle \big|_{M_0} \neq 0$ and a sliding mode is possible on the manifold M_0. In this case, $\sigma = 0$ and

$$\dot{\sigma}_{\text{eq}} = \langle \nabla\sigma, f \rangle + \langle \nabla\sigma, b \rangle\, u_{\text{eq}} = 0. \tag{5.11}$$

Subtracting (5.11) from (5.10) for $\sigma = 0$ and introducing the notation

$$u_1^{\pm} = u^{\pm} - u_{\text{eq}},$$

we obtain

$$\dot{\sigma}\big|_{\sigma=0} = \langle \nabla\sigma, b \rangle\, u_1^{\pm}.$$

Recall that we have already dealt with equations of this kind and analyzed their properties in the case of nonideal switchings.

Suppose, for definiteness, that the inequality $\langle \nabla\sigma, b \rangle > 0$ is satisfied, and then, for the control

$$u_1^{\pm} = -\operatorname{sgn}\sigma,$$

we have an equation of sliding

$$\dot{\sigma}\big|_{M_0} = -\langle \nabla\sigma, b \rangle \operatorname{sgn}\sigma,$$

which, in the case of ideal switchings, is related to Fig. 5.6*a* and in the case of nonideal switchings to Fig. 5.6*b*. We can see from these figures that the system under consideration has the first order of smallness with respect to the parameter of nonideal switchings Δ since in the case of real sliding we have

$$|\sigma| \leq \Delta, \quad |\dot{\sigma}| \leq \text{const}.$$

We assign the first order to this real sliding with respect to the parameter Δ. In the limit, as $\Delta \to 0$, we have relations

$$\sigma \equiv 0, \quad |\dot{\sigma}| \leq \text{const}$$

which define the standard mode of ideal sliding which *is characterized by the continuity of the variable σ and by the discontinuity of its derivative $\dot{\sigma}$*. It is also expedient to assign the first degree to this ideal sliding.

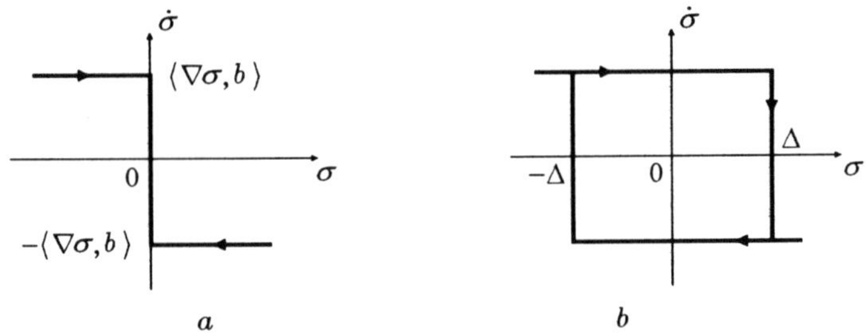

Figure 5.6.

It follows from what was said above that the transversality of intersection of the sliding manifold M_0 by the trajectories of systems $\dot{x} = f^{\pm}$ leads to a first-degree sliding (ideal and real). Therefore the sliding acquires a higher degree only when the vector fields $f^{\pm}$ touch the sliding manifold M_0, i.e., when we have an identity

$$\dot\sigma(x)\big|_{M_0} \equiv 0. \tag{5.12}$$

We denote the set of points x that satisfy (5.12) by M_1. When conditions (5.12) are fulfilled, we may have a sliding mode at the intersection $M_0 \cap M_1$ if the second derivative is discontinuous and alternating,

$$\ddot\sigma = \langle \nabla\dot\sigma, f^{\pm} \rangle = \langle \nabla\dot\sigma, f \rangle + \langle \nabla\dot\sigma, b \rangle \, u^{\pm},$$

i.e., $\langle \nabla\dot\sigma, b \rangle \neq 0$ on $M_0 \cap M_1$.

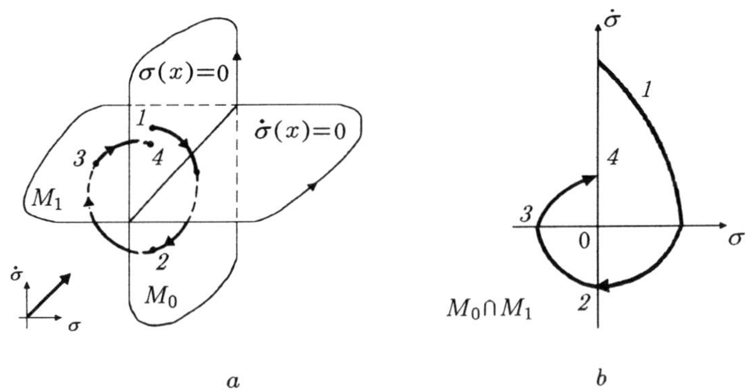

Figure 5.7.

Figure 5.7 illustrates what we have stated. Figure 5.7*b* shows the trajectories of this system when we view it along the intersection $M_0 \cap M_1$. Since the functions σ and $\dot\sigma$ are continuous in this case and the function $\ddot\sigma$ is discontinuous, we have relations $\sigma = 0$, $\dot\sigma = 0$, $|\ddot\sigma| \le$ const in an ideal sliding mode. Following the conception accepted above, we must assign the second degree to this sliding mode.

To determine the degree of the mode of real sliding with respect to the parameter Δ of nonideal switching, we assume that for a sufficiently small $\Delta > 0$ the real sliding mode is preserved in the neighborhood of the intersection $M_0 \cap M_1$, i.e., for certain constants $A_0(\Delta)$, $A_1(\Delta)$, $A_2^{\pm}(\Delta)$ we have inequalities

$$|\sigma| \leq A_0(\Delta), \quad |\dot{\sigma}| \leq A_1(\Delta), \quad A_2^-(\Delta) \leq |\ddot{\sigma}| \leq A_2^+(\Delta), \tag{5.13}$$

with

$$\lim_{\Delta\to 0} A_0(\Delta) = \lim_{\Delta\to 0} A_1(\Delta) = 0,$$
$$\lim_{\Delta\to 0} A_2^{\pm}(\Delta) > 0.$$

Let $\tau(\Delta)$ be an interval of uncertainty $\ddot{\sigma}$. It is clear that $\tau(\Delta) \to 0$ as $\Delta \to 0$. Then, according to the Lagrange theorem, we have

$$|\ddot{\sigma}(\Theta)| \leq \frac{2\max|\dot{\sigma}|}{\tau(\Delta)} \leq \frac{2A_1(\Delta)}{\tau(\Delta)} \tag{5.14}$$

for the point Θ from this interval.

Similarly, for a certain Θ' we have

$$|\dot{\sigma}(\Theta')| \leq \frac{2\max|\sigma|}{\tau(\Delta)} \leq \frac{2A_0(\Delta)}{\tau(\Delta)} \tag{5.15}$$

for the same interval.

Comparing (5.13), (5.14), and (5.15), we find that

$$A_1 \sim O(\tau), \quad A_0 \sim O(\tau^2),$$

and if $\tau \sim O(\Delta)$, and this is the case as a rule, then

$$A_1 \sim O(\Delta), \quad A_0 \sim O(\Delta^2).$$

This means that the real sliding mode under consideration is of the second degree with respect to the parameter Δ of nonideal switching.

In particular, if the parameter $\Delta = h$, where h is a discretization step of the time scale in the numerical simulation of a discontinuous system, then, for a real second-degree sliding, the corresponding second-degree accuracy with respect to h of the validity of the feedback $\sigma = 0$ is guaranteed and, consequently, a similar accuracy of approximation of the trajectory of real sliding to the corresponding trajectory of ideal sliding.

Figure 5.8 illustrates what was said above. The conception that we have described can be naturally generalized to an arbitrary degree of sliding. Namely, let $L_f(\cdot)$ be an operator of differentiation in the direction of the vector fields $f^{\pm}$, i.e., for the differentiable function $\varphi(x)$

$$L_f^l \varphi(x) = \langle \nabla\varphi, f^{\pm} \rangle, \quad f^{\pm} = f + bu^{\pm}.$$

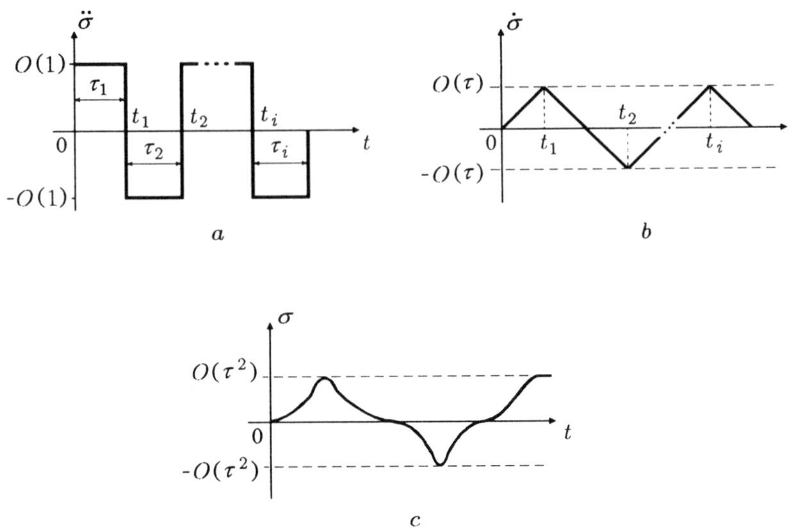

Figure 5.8.

Upon a multiple application of this operator, we have a relation

$$L_f^l \varphi = L_f(L_f^{l-1}\varphi), \quad l \text{ is an integer.}$$

Note that when this has sense, the following relationship between the time derivatives of the function $\sigma(x)$ and the system $\dot{x} = f(x)$ holds true:

$$\sigma^{(l)} = \frac{d^l \sigma}{dL^l} = L_f^l \sigma.$$

Suppose that in the successive differentiation of the function $\sigma(x)$ all its derivatives up to the order $r-1$ inclusive are continuous and the rth derivative $\sigma^{(r)} \equiv L_f^r \sigma$ is discontinuous and alternating. Then, if the intersection

$$M_r = M_0 \cap M_1 \cap \cdots \cap M_{r-1},$$

where

$$M_l = \left\{x \mid \sigma^{(l)}(x) = 0\right\}, \quad l = 0, 1, \ldots, r-1,$$

is nonempty, then it is an ideal sliding manifold of order r.

If, in addition, the inequalities

$$|\sigma^{(l)}| \le A_l(\Delta), \quad |\sigma^{(r)}| \le A_r(\Delta), \quad A_r(\Delta) \ge \text{const} > 0,$$

for $l = 0, 1, \ldots, r-1$, where Δ is a parameter characterizing the nonideal switchings, are valid in real sliding, then the accuracy of real sliding with respect to σ is of degree Δ^r since

$$|\sigma^{(l)}| \sim O(\Delta^{r-l}), \quad l = 0, 1, \ldots, r-1.$$

5.2. Algorithms of Second-Degree Sliding

Let us consider some algorithms of second-degree sliding.

First of all, let us define more exactly the mathematical model which we shall deal with. We shall again consider the smooth manifold

$$M_0 = \{x \mid \sigma(x) = 0\}.$$

We can study the behavior of the discontinuous system

$$\dot{x} = f + b\,u^{\pm}$$

in its neighborhood $O(M_0)$ using a scalar equation of the form

$$\dot{\sigma} = \langle \nabla\sigma, f\rangle + \langle \nabla\sigma, b\rangle\, u^{\pm}. \tag{5.16}$$

Let u_{eq} be a solution for $x \in O(M_0)$ of the equation

$$0 = \langle \nabla\sigma, f\rangle + \langle \nabla\sigma, b\rangle\, u_{\mathrm{eq}}, \tag{5.17}$$

and then, subtracting (5.17) from (5.16), we find that an equation

$$\dot{\sigma} = \langle \nabla\sigma, b\rangle\,(u^{\pm} - u_{\mathrm{eq}})$$

acts in the neighborhood $O(M_0)$.

If we denote the right-hand side of the last equation by u, then we can reduce, without loss of generality, the study of the motion in the neighborhood $O(M_0)$ to the standard equation

$$\dot{\sigma} = u. \tag{5.18}$$

When we analyze second-degree sliding modes, the peculiarity of this equation is that the function u is continuous and such that the phase trajectories of the original system touch the manifold at points of second-degree sliding, in other words, u vanishes on the intersection

$$M_0 \cap M_1,$$

where, as before,

$$M_1 = \{x \mid \dot{\sigma}(x) = 0\}.$$

The problem of stabilization of the scalar object (5.18) by continuous control can be reduced, by means of the changes of variables

$$\sigma_1 = \sigma, \quad \sigma_2 = u,$$

to a standard problem of stabilization at zero of a second-order object

$$\begin{gathered} \dot{\sigma}_1 = \sigma_2, \quad \dot{\sigma}_2 = \nu, \\ \sigma = \sigma_1 \end{gathered} \tag{5.19}$$

by the feedback $\nu(\sigma_1, \sigma_2)$ which may be discontinuous when the sliding is of the second degree. It stands to reason that the most preferable are feedbacks $\nu(\sigma_1, \sigma_2)$ which finitely stabilize object (5.19) at zero with respect to the output, but we will not neglect the asymptotically stabilizing feedbacks with respect to the state or to the output.

The construction described above admits a natural generalization. For instance, the problem of organizing an rth-degree sliding mode on the manifold M_0 is equivalent to the problem of finite stabilization with respect to the output at zero of the rth-order system

$$\begin{aligned} &\dot{\sigma}_i = \sigma_{i+1}, \quad i = 1, 2, \ldots, r-1, \\ &\dot{\sigma}_r = \nu, \quad \sigma = \sigma_1. \end{aligned} \tag{5.20}$$

Indeed, in this case, in the required mode we have relations

$$\sigma = \sigma_1 = \sigma_2 = \cdots = \sigma_r = 0$$

which mean that the motion is along the intersection

$$M_0 \cap M_1 \cap \ldots \cap M_r,$$

where, obviously, the sets M_i are defined by the relations

$$M_i = \{x \mid \sigma^{(i)}(x) = 0\}.$$

The degree of a sliding mode can be connected with the relative order of the original object with respect to control if the variable σ is assumed to be its output. Without loss of generality, we can easily demonstrate this by way of an example of a linear stationary object. Indeed, suppose that the relationship between the input u and the output σ of the object is defined by the rational transfer function $W(s) = \beta(s)/\alpha(s)$ (Fig. 5.9). Then, for a relative degree unity, i.e., when

$$\deg \alpha(s) - \deg \beta(s) = 1,$$

the derivative $\dot{\sigma}$ may suffer discontinuities if the control u is discontinuous and, consequently, a first-degree sliding mode is possible.

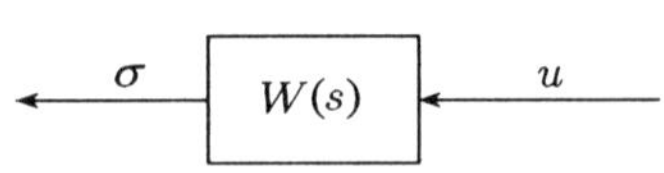

Figure 5.9.

Under the condition that

$$\deg \alpha(s) - \deg \beta(s) = r,$$

the functions $\sigma^{(l)}$, $l = 0, 1, \ldots, r-1$, do not explicitly depend on the control whereas the function $\sigma^{(r)}$ depends on the control, and, consequently, under the discontinuous control u only $\sigma^{(r)}$ may suffer discontinuities, and, hence, a sliding mode of a degree not lower than the rth is possible, in principle, in the system. Thus, it is the order of the function, which defines the surface of discontinuity, with respect to control (i.e., the number of the first Lie derivative which explicitly depends on the control) that defines the degree of the sliding mode if, of course, the mode is stable.

Let us consider specific examples of the algorithms of second-degree sliding in greater detail.

5.2.1. Asymptotic algorithms of the second-degree sliding

When solving the problem of asymptotic stabilization at zero of an object (Σ-system)

$$\dot{\sigma}_1 = \sigma_2, \quad \dot{\sigma}_2 = u,$$

we begin with investigating the potentiality of the linear feedback

$$u = -k_1\sigma_1 - k_2\sigma_2. \tag{5.21}$$

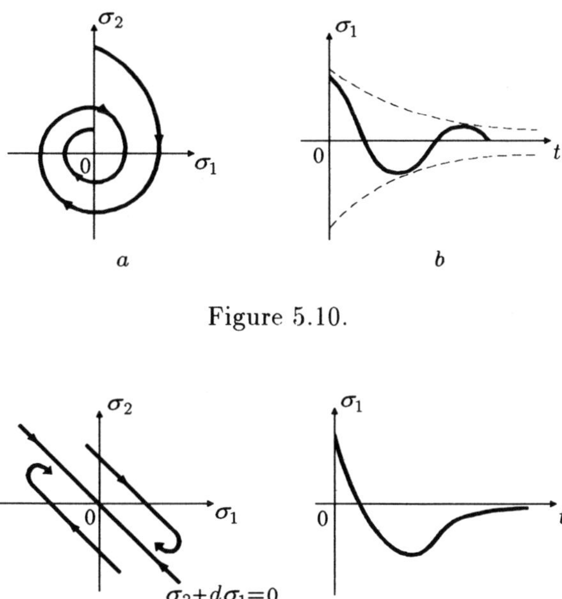

Figure 5.10.

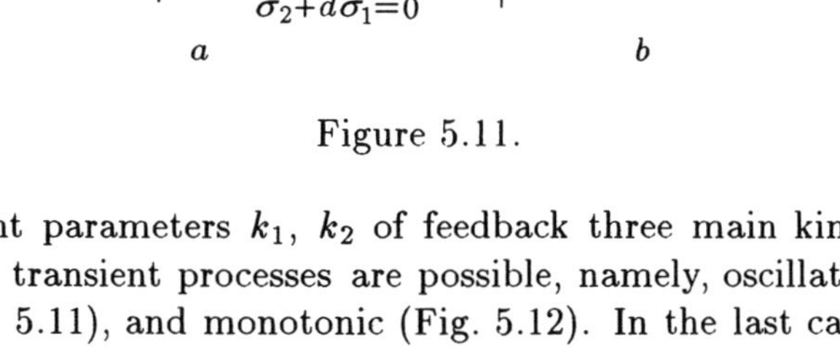

Figure 5.11.

For constant parameters k_1, k_2 of feedback three main kinds of exponentially damping transient processes are possible, namely, oscillatory (Fig. 5.10), aperiodic (Fig. 5.11), and monotonic (Fig. 5.12). In the last case, use is made

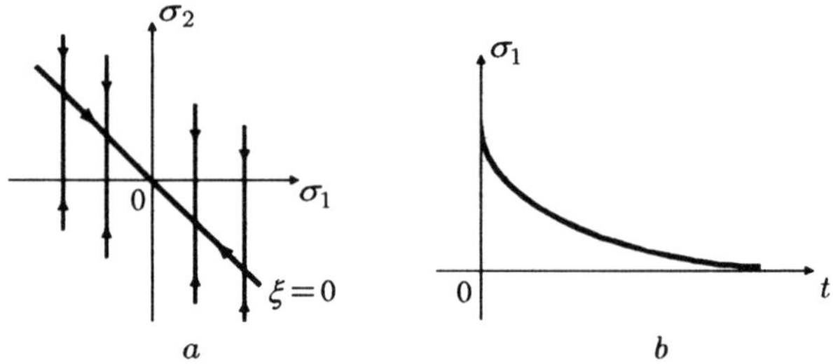

Figure 5.12.

of feedback with high gain introduced by tending the common factor k of the parameters k_1, k_2 to infinity, i.e., in this case, feedback (5.21) has the form

$$u = -k\xi = -k(\sigma_2 + d\sigma_1), \quad d = \text{const} > 0.$$

The situations in the initial phase space X of the discontinuous system

$$\dot{x} = f^{\pm} = f + bu^{\pm}$$

corresponding to Figs. 5.10–5.12 are illustrated by Figs. 5.13, 5.14. The peculiar features of these algorithms of sliding are caused by the characteristic properties of linear stabilization systems. In particular, in order to decrease the time of transient process (and this is desirable), we must increase damping, but this lowers the robustness of the system. In addition, when there is uncertainty in the description of the original system $\dot{x} = f^{\pm}$, we may have no information about the derivative σ_2, and its restoration by standard observation methods is impossible. Therefore we must use nonlinear stabilization algorithms.

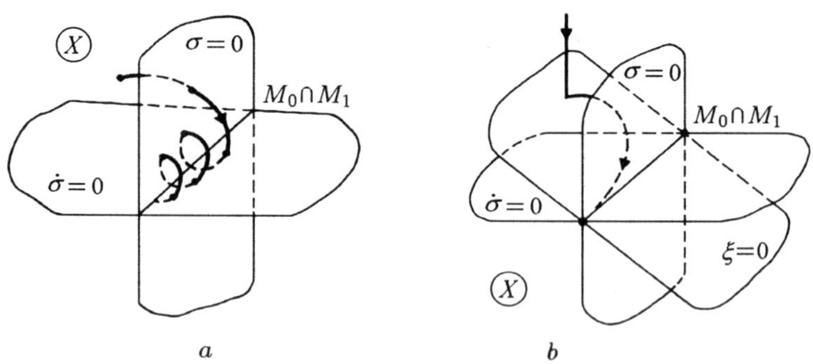

Figure 5.13.

Let us begin with a situation where the sliding manifold M_0, and only this manifold, is a manifold of discontinuity of feedback, for instance,

$$u = -k_2\sigma_2 - k_1 \operatorname{sgn} \sigma_1.$$

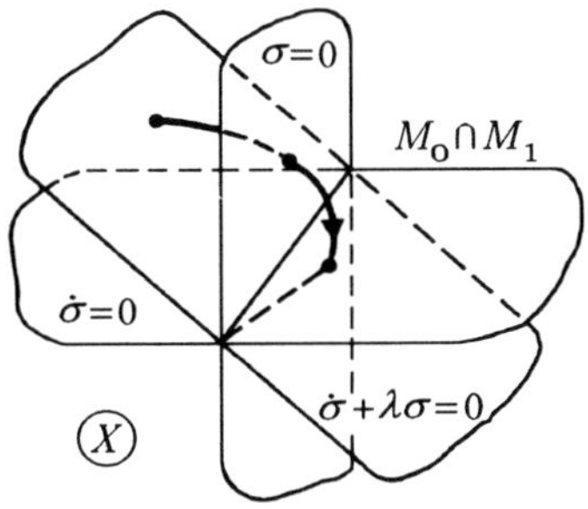

Figure 5.14.

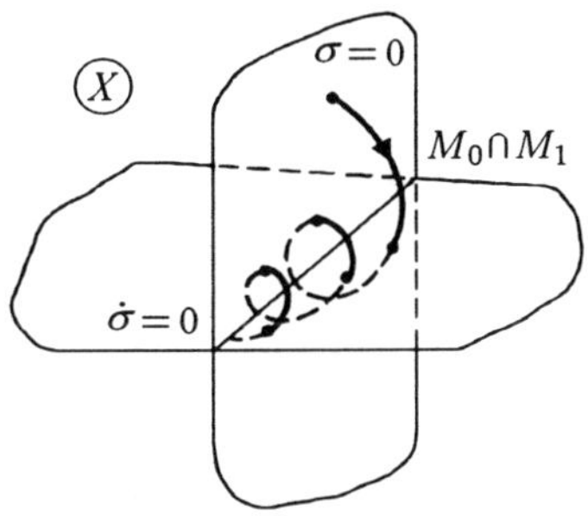

Figure 5.15.

In this case, the feedback system

$$\begin{aligned}\dot{\sigma}_1 &= \sigma_2,\\ \dot{\sigma}_2 &= -k_1 \operatorname{sgn} \sigma_1 - k_2 \sigma_2\end{aligned}$$

is asymptotically stable at zero, which fact can be verified by the test function

$$v = |\sigma_1| + \frac{1}{2k_1}\sigma_2^2$$

whose derivative has the form

$$\dot{v} = -\frac{k_2}{k_1}\sigma_2^2$$

and becomes identically zero only at the origin. Figures 5.15 and 5.16 give a qualitative idea of the character of motion in the neighborhood of the intersection $M_0 \cap M_1$ in this case.

5.2.2. Discontinuous asymptotic algorithms for the second-degree sliding

In the cases where control discontinuities are admitted not only on the sliding manifold M_0, the exponential stabilization of the Σ-system at zero is guaranteed

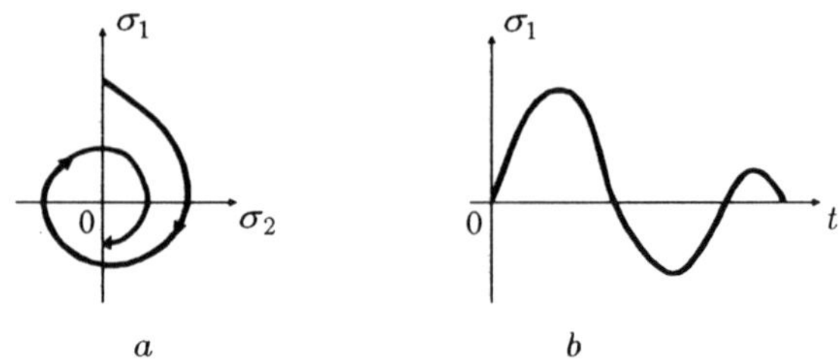

Figure 5.16.

by the following law of the VSS:

$$u = -k|\sigma_1| \operatorname{sgn} \xi,$$
$$\xi = \sigma_2 + d\sigma_1, \quad d = \text{const} > 0.$$

Indeed, equations of feedback have the form

$$\dot{\sigma}_1 = \sigma_2,$$
$$\dot{\sigma}_2 = -k|\sigma_1| \operatorname{sgn} \xi,$$

and, when the inequality

$$k > d^2$$

is satisfied, a sliding mode (Figs. 5.17, 5.18) appears on the straight line $\xi = 0$ in finite time (except for asymptotic trajectories). This is a first-degree sliding mode with respect to the variable ξ but a second-degree sliding mode with respect to the variable σ_1. Discontinuous sliding algorithms ensure a more robust solution of the stabilization problem than linear algorithms for the reasons that we have repeatedly indicated. It is of interest, however, to synthesize finite stabilization algorithms.

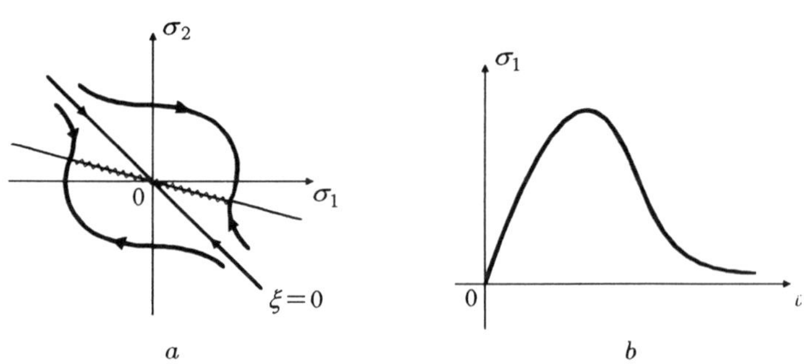

Figure 5.17.

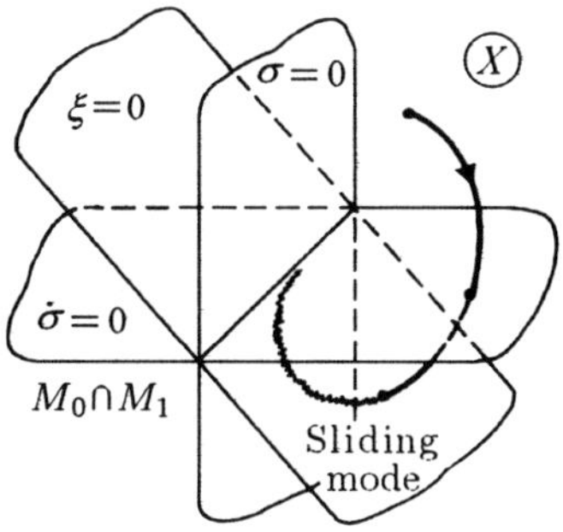

Figure 5.18.

5.2.3. Finite algorithms of second-degree sliding: linear feedback

Finite stabilization is only possible in the class of linear nonstationary or nonlinear (discontinuous) feedbacks.

First of all, it is natural to consider stabilizing potentialities of a linear feedback. Namely, for stabilizing at zero the Σ-system

$$\dot\sigma_1 = \sigma_2, \qquad \dot\sigma_2 = u$$

in a finite time $\eta > 0$ let us apply a linear feedback with variable parameters $u = -k\bar\sigma = -k_1\sigma_1 - k_2\sigma_2$, where, for convenience, we introduced vectors $k^\top = (k_1, k_2)$ and $\bar\sigma = (\sigma_1, \sigma_2)$; here $^\top$ is the transposition sign.

The general solution for the Σ-system is given by the Cauchy formula

$$\bar\sigma(t) = e^{tA}\left[\bar\sigma_0 + \int_0^t e^{-\tau A} bu\, d\tau\right],$$

where the matrix A and the vector b have the form

$$A = \begin{bmatrix} 0 & 1 \\ 0 & 0 \end{bmatrix}, \qquad b = \begin{bmatrix} 0 \\ 1 \end{bmatrix}.$$

We set

$$u(t) = -b^\top e^{-tA^\top} l,$$

where the vector l can be found from the relation $\bar\sigma(\eta) = 0$, i.e.,

$$\bar\sigma_0 = \left[\int_0^\eta e^{-\tau A} bb^\top e^{-\tau A^\top}\, d\tau\right] l = W(0,\eta)\, l.$$

Since the Σ-system is controllable, the controllability Gramian $W(0,\eta)$ is nondegenerate for any $\eta > 0$, and therefore $l = W^{-1}(0,\eta)\bar\sigma_0$, which allows us to find the control that solves the problem in the form of a "program"

$$u = -b^\top e^{-tA^\top} W^{-1}(0,\eta)\bar\sigma_0.$$

In order to find the corresponding feedback, we express $\bar{\sigma}_0$ from the relation

$$e^{-tA}\,\bar{\sigma}(t) = \bar{\sigma}_0 + W(0,t)\,W^{-1}(0,\eta)\,\bar{\sigma}_0$$

and substitute the result into the preceding relation for u. We obtain a relation

$$u = -\frac{b^\top e^{-tA^\top} W^{-1}(0,\eta)\,e^{-tA}}{1 - W(0,t)\,W^{-1}(0,\eta)}\,\bar{\sigma}$$

which defines the required feedback. Since

$$W(0,t)\,W^{-1}(0,\eta) \to 1$$

as $t \to \eta$, the feedback factor increases indefinitely in finite time, i.e.,

$$k(t) = -\frac{b^\top e^{-tA^\top} W^{-1}(0,\eta)\,e^{-tA}}{1 - W(0,t)\,W^{-1}(0,\eta)} \to \infty, \qquad t \to \eta.$$

The last procedure is, of course, unsuitable for applications. To express the parameters $k_1(t)$ and $k_2(t)$ explicitly in terms of the parameters of the Σ-system, we must use the relations

$$e^{tA} = \begin{bmatrix} 1 & t \\ 0 & 1 \end{bmatrix}, \quad W(0,t) = \begin{bmatrix} t^3/3 & -t^2/2 \\ t^2/2 & t \end{bmatrix},$$
$$W^{-1}(0,t) = \begin{bmatrix} 12/t^3 & 6/t^2 \\ 6/t^2 & 4/t \end{bmatrix}, \quad b^\top e^{-tA^\top} = [-t, 1].$$

5.2.4. Finite algorithms of second-degree sliding: relay feedback

Let us consider a nonlinear feedback which also ensures a finite stabilization but is bounded for any $\bar{\sigma}$. Let the scalar differentiable function $g(\sigma)$ be such that $g(0) = 0$, g' is bounded, a solution of the differential equation $\dot{\sigma} = g(\sigma)$ exists for any $\sigma(0)$, and this solution becomes zero in a finite time interval. For instance, the function $g(\sigma) = -d\operatorname{sgn}\sigma|\sigma|\rho$, $0.5 \le \rho < 1$, $d = \text{const} > 0$, possesses these properties.

Then the feedback which finitely stabilizes the Σ-system

$$\dot{\sigma}_1 = \sigma_2, \qquad \dot{\sigma}_2 = u$$

can be chosen to be a relay system, say, in the form

$$u = -k\operatorname{sgn}\xi(\sigma_1,\sigma_2),$$
$$\xi(\sigma_1,\sigma_2) = \sigma_2 - g(\sigma_1).$$

For a sufficiently large factor $k > 0$ the curve $\sigma_2 = g(\sigma_1)$ is a sliding curve, it is attained from any initial position in finite time, and this guarantees the finiteness of the stabilization time η (Fig. 5.19).

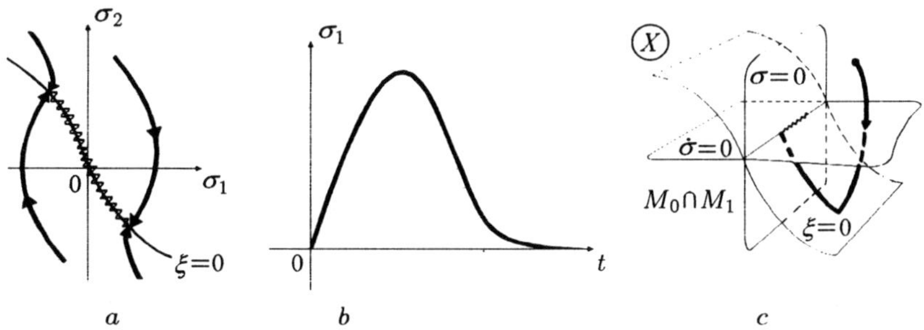

Figure 5.19.

5.2.5. Twisting algorithm

Let us investigate the stabilizing properties of a discontinuous feedback of the form

$$u = -k_1 \operatorname{sgn} \sigma_1 - k_2 \operatorname{sgn} \sigma_2,$$

where $k_1 > k_2$ are positive parameters. In this case, the feedback system is described by the equations

$$\begin{aligned} \dot\sigma_1 &= \sigma_2, \\ \dot\sigma_2 &= -k_1 \operatorname{sgn} \sigma_1 - k_2 \operatorname{sgn} \sigma_2. \end{aligned}$$

The trajectories corresponding to it, which consist of segments of the parabola, are shown in Fig. 5.20.

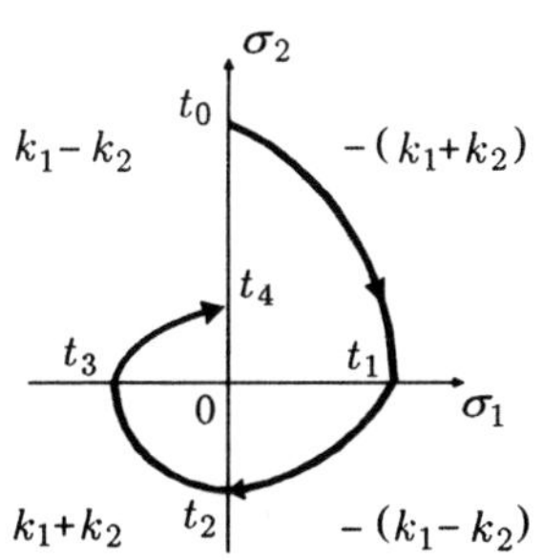

Figure 5.20.

Let t_0, t_1, t_2, t_3, and t_4 be successive moments of control switching and let $\tau_1 = t_1 - t_0$, $\tau_2 = t_2 - t_1$, etc., be time intervals between switchings. The relationships between the successive values of the variables σ_1, σ_2 at switching moments are evidently given by the following expressions: $\sigma_1 = \pm\frac{\sigma_2^2}{k_1+k_2}$ in the first and third quadrants and $\sigma_1 = \mp\frac{\sigma_2^2}{k_1-k_2}$ in the second and fourth quadrants. Therefore the relation

$$\sigma_2^2(t_2) = \frac{k_1 - k_2}{k_1 + k_2}\sigma_2^2(0) = \kappa\sigma_2^2(0)$$

holds between the squares of $\sigma_2(0)$ and $\sigma_2(t_2)$. Similarly, we have $\sigma_2^2(t_4) = \kappa\sigma_2^2(t_2) = \kappa^2\sigma_2^2(0)$.

Let i be the number of a full rotation of the phase vector about zero. Then the successive values of the variable σ_2 are related as

$$|\sigma_2(i)| = \kappa|\sigma_2(i-1)| = \cdots = \kappa^i|\sigma_2(0)|, \quad i = 1, 2, \ldots,$$

whence follows the convergence of the transient processes of the Σ-system to zero, provided that $\kappa < 1$. Let us prove that this convergence is finite.

Let T_i be the time of the ith rotation. It is obvious that

$$T_i = \tau_1^i + \tau_2^i + \tau_3^i + \tau_4^i,$$

and we can easily obtain expressions

$$\tau_1^i = \frac{|\sigma_2(t_0^i)|}{k_1 + k_2}, \quad \tau_2^i = \frac{|\sigma_2(t_2^i)|}{k_1 - k_2}, \quad \tau_3^i = \frac{|\sigma_2(t_2^i)|}{k_1 + k_2}, \quad \tau_4^i = \frac{|\sigma_2(t_4^i)|}{k_1 - k_2}$$

for the intervals τ_j^i $(j = 1, 2, 3, 4)$. Using these relations, we can easily make sure that the estimate

$$\begin{aligned} T_i &\le \frac{1}{k_1 - k_2}\left(|\sigma_2(t_0^i)| + 2|\sigma_2(t_2^i)| + |\sigma_2(t_4^i)|\right) \\ &= \frac{1}{k_1 - k_2}\left(1 + 2\sqrt{\kappa} + \kappa\right)|\sigma_2(t_0^i)| = \frac{(1+\sqrt{\kappa})^2}{k_1 - k_2}|\sigma_2(i)| \end{aligned}$$

is valid. Since the upper estimate

$$T = \sum_{i=1}^{\infty} T_i \le \frac{(1+\sqrt{\kappa})^2}{k_1 - k_2}\sum_{i=1}^{\infty}|\sigma_2(i)| = \frac{(1+\sqrt{\kappa})^2|\sigma_2(0)|}{k_1 - k_2}\sum_{i=1}^{\infty}\kappa^{i-1}$$

is valid for the time of the transient process and the series $\sum_{i=0}^{\infty}\kappa^i$ converges for $\kappa < 1$, this means that the time T of the transient process is limited. In other words, any phase trajectory is "twisted" into zero in a finite time interval, and this explains the name of the algorithm. Figure 5.21 illustrates this behavior.

To conclude this section, we want to point out that the relay algorithm and the twisting algorithm can also be applied in discrete measurements, but then they turn into second-degree real sliding algorithms. Namely, let t_i be moments of measurement of the function $\sigma(t)$ with a constant step $h = t_{i+1} - t_i$. Then the second-degree real sliding relay algorithm is defined by the expression

$$u(t) = g - k\,\mathrm{sgn}[\delta_i\sigma_1 - g(\sigma_1(t_i))], \quad t \in [t_i, t_{i+1}],$$

where $\delta_i\sigma_1 = \left[\sigma_1(t_i) - \sigma_1(t_{i-1})\right]/[t_i - t_{i-1}]$ and the algorithm for discrete twisting is defined by the expression

$$u(t) = -k_1\,\mathrm{sgn}\,\sigma_1(t_i) - k_2\,\mathrm{sgn}[\delta_i\sigma_1], \quad t \in [t_i, t_{i+1}].$$

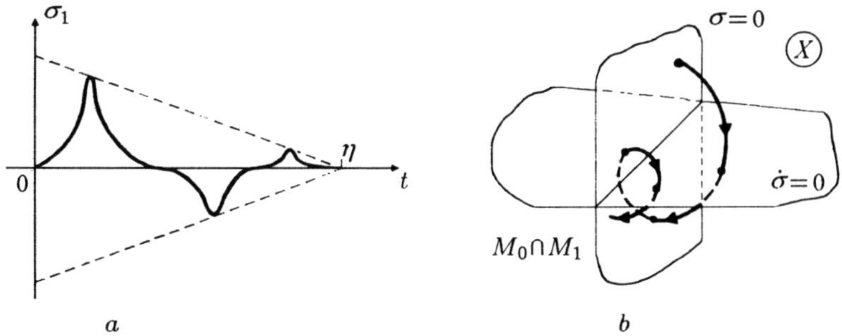

Figure 5.21.

5.3. Output Finite Stabilization

Let us again consider a problem of finite stabilization at zero of the Σ-system

$$\dot{\sigma}_1 = \sigma_2, \quad \dot{\sigma}_2 = u$$

defined by the equation for output

$$\sigma = \sigma_1.$$

It is convenient to reduce the equations of the Σ-system to one second-order equation $\ddot{\sigma} = u$. We take a feedback of the form

$$u = -k_1 \operatorname{sgn} \sigma - \frac{k_2}{|\sigma|^{1/2}} \dot{\sigma}, \quad k_1, k_2 = \text{const} > 0, \tag{5.22}$$

whose peculiarity is that the damping factor $k_2/|\sigma|^{1/2}$ in the equation for the feedback system

$$\ddot{\sigma} + \frac{k_2}{|\sigma|^{1/2}} \dot{\sigma} + k_1 \operatorname{sgn} \sigma = 0 \tag{5.23}$$

increases indefinitely when we approach the sliding manifold

$$M_0 = \{x \mid \sigma(x) = 0\}.$$

It is precisely this "physical" effect that we hope to use for the finite stabilization of (5.23).

Let us now analyze the closed-loop control system defined by Eq. (5.23). Note, first of all, that as $\sigma \to -\sigma$, the form of the equation is preserved and, hence, its phase trajectories are preserved, and therefore it suffices to analyze its solution for $\sigma > 0$ since the trajectories in the quadrants $\sigma\dot{\sigma} > 0$ $(\sigma\dot{\sigma} < 0)$ are similar.

Qualitatively, the phase trajectories of the equation

$$\ddot{\sigma} + \frac{k_2}{|\sigma|^{1/2}} \dot{\sigma} + k_1 \operatorname{sgn} \sigma = 0$$

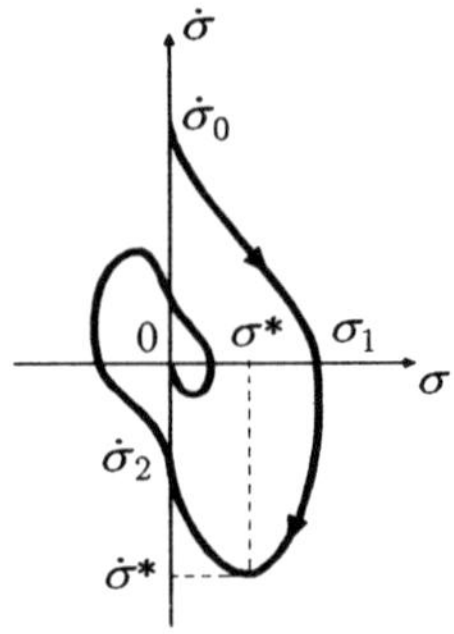

Figure 5.22.

are shown in Fig. 5.22 in which $\dot\sigma_0$ is the initial value of the velocity, $\dot\sigma_2$ is the point of the second intersection of the phase trajectory with the axis $\sigma = 0$ (i.e., the second intersection of the manifold M_0), $\dot\sigma_* = \dot\sigma(t_*)$ is the minimal value of the velocity on the interval of motion $[t_1, t_2]$, where t_1 is the moment when the velocity $\dot\sigma$ becomes zero (i.e., the moment at which the representative point is most distant from the manifold M_0), and t_2 is the moment at which the representative point gets into M_0. For $\sigma > 0$ we deal with the equation

$$\ddot\sigma + \frac{k_2}{\sqrt{\sigma}}\dot\sigma + k_1 = 0$$

which can be easily reduced to the form

$$\frac{d}{dt}\left[\dot\sigma + k_1 t + 2k_2\sigma^{1/2}\right] = 0,$$

and then, since the relations $\dot\sigma(0) = \dot\sigma_0$, $\sigma(0) = 0$ are valid for $t = 0$, the integration results in the relation

$$\dot\sigma + k_1 t + 2k_2\sigma^{1/2} = \dot\sigma_0. \tag{5.24}$$

For $t = t_1$ we find from (5.24) (taking into account that $\dot\sigma(t_1) = 0$) that

$$k_1 t_1 + 2k_2\sigma_1^{1/2} = \dot\sigma_0.$$

Since $t_1 > 0$, the last relation yields the first of the inequalities that we shall need, namely,

$$\sigma_1^{1/2} < \frac{\dot\sigma_0}{2k_2}. \tag{5.25}$$

We substitute $t = t_*$ into (5.23). Then, taking into account that $\ddot\sigma(t_*) = 0$, we get a relation

$$k_2\dot\sigma_* + k_1\sigma_*^{1/2} = 0, \quad \text{or} \quad \sigma_*^{1/2} = -\frac{k_2}{k_1}\dot\sigma_*.$$

However, it is obvious (Fig. 5.22) that $\sigma_* < \sigma_1$, $\dot\sigma_* < \dot\sigma_2$, and therefore the last relation yields a chain of inequalities

$$\sigma_1^{1/2} > \sigma_*^{1/2} = \frac{k_2}{k_1}(-\dot\sigma_*) > \frac{k_2}{k_1}(-\dot\sigma_2)$$

from which we can get the second of the required relations, namely,

$$-\dot\sigma_2 < \frac{k_1}{k_2}\sigma_1^{1/2}. \tag{5.26}$$

Substituting (5.25) into (5.26), we obtain the following inequality between the velocities of two successive intersections of the sliding manifold M_0:

$$\frac{|\dot\sigma_2|}{\dot\sigma_0} < \gamma = \frac{k_1}{2k_2^2}.$$

By virtue of the symmetry of the equations of the feedback system that we have pointed out, a similar inequality holds for any two successive moments of intersection of the manifold M_0, i.e., the inequality

$$\left|\frac{\dot\sigma_{i+1}}{\dot\sigma_i}\right| \le \gamma$$

holds true. Hence we have an estimate of the form $|\dot\sigma_i| \le \gamma^i |\dot\sigma_0|$, and if the condition

$$\gamma = \frac{k_1}{2k_2^2} < 1$$

is fulfilled, then the stabilizability is proved since, in this case,

$$|\dot\sigma_i| \to 0, \quad t \to \infty,$$

and it follows from an inequality of type (5.25) that

$$|\sigma_i| \to 0, \quad t \to \infty$$

as well, where σ_i are the maxima of deviations from the sliding manifold M_0.

Let us prove that the time of the transient process is finite. To this end, we substitute $t = t_2$ into Eq. (5.24) and get a relation

$$\dot\sigma_2 + k_1 t_2 = \dot\sigma_0,$$

and, consequently, the following estimate holds for the time interval $T_1 = t_2 - 0$ between two intersections of the manifold M_0:

$$T_1 = t_2 = \frac{\dot\sigma_0 - \dot\sigma_2}{k_1} < \dot\sigma_0 \frac{1+\gamma}{k_1}.$$

It is natural that a similar estimate is valid for every ith interval, i.e.,

$$T_i < |\dot\sigma_{i-1}|\frac{1+\gamma}{k_1}.$$

However, the time T of the transient process is finite since

$$|\dot{\sigma}_i| < \gamma |\dot{\sigma}_{i-1}|, \quad 0 < \gamma < 1,$$

in the inequality

$$T = \sum_{i=1}^{\infty} T_i < \frac{1+\gamma}{k_1} \sum_{i=1}^{\infty} |\dot{\sigma}_{i-1}|$$

as we proved earlier, and, hence, the series $\sum_{i=1}^{\infty} |\dot{\sigma}_i|$ is bounded.

Thus, we have a finite stabilization of the Σ-system by feedback of form (5.22). The investigation that we have carried out substantiates Fig. 5.23 which shows a finite transient process which terminates by the time moment η.

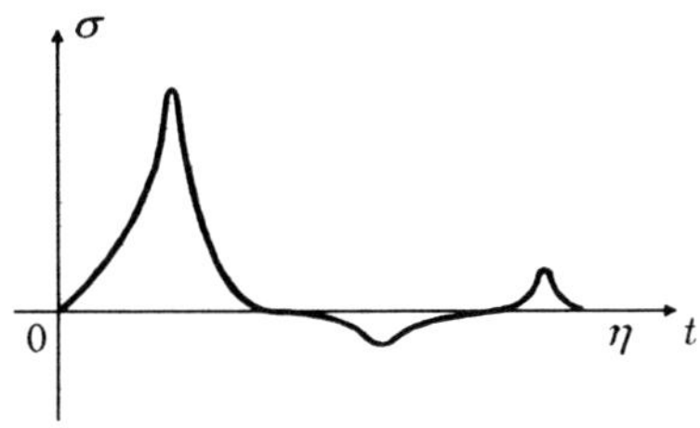

Figure 5.23.

To complete this section, we shall indicate the reason why the feedback

$$u = -k_1\, d \operatorname{sgn} \sigma - k_2 \frac{\dot{\sigma}}{|\sigma|^{1/2}}$$

is called a feedback with respect to output. The matter is that the behavior of the Σ-system which was analyzed above is the same as that of the system which is studied below and which is described by the integro-differential equation

$$\dot{\sigma}(t) = -k_1 \int_0^t \operatorname{sgn} \sigma \, d\tau - k_2 \int_0^t |\sigma|^{1/2} \operatorname{sgn} \sigma \, d\tau + \dot{\sigma}_0,$$

but to realize the control which appears on the right-hand side of the last equation, we need information only about the output σ.

It stands to reason that the conclusions are the same in the case where we use feedback of the form

$$u = -k_1 \operatorname{sgn} \sigma - k_2 \frac{\dot{\sigma}}{|\sigma|^{\rho}}, \quad 0 < \rho < 1.$$

Note finally that all the results presented above can be generalized, without any essential exceptions, to an arbitrary degree of sliding. However, it should be pointed out that when we control an uncertain system, we may have no information about the variables σ_i, $i \geq 1$, and their evaluation by means of standard observations may turn to be impossible. This is a serious and as yet unsolved problem.

Chapter 6

Theory of Operator Feedback

In this chapter the properties of stabilization systems from Chap. 4 are improved by the use of an additional operator type nonlinear feedback. Recall that "by legend" the operator feedback must "bridge the gap" between the dynamical properties of the reference device and the control system with coordinate and coordinate-operator feedbacks. This process of making the properties similar must be carried out without using any information concerning the uncertainty factors but only by establishing an additional nonlinear feedback between the variables of the system. The use of this feedback must result in the rise of the quality of transient processes and the weakening of their dependence on the uncertainty factors.

6.1. The Purpose of Operator Feedback

In the theory of coordinate-operator feedback (Chap. 4) we considered different algorithms of stabilization of the simplest uncertain object

$$\dot{\xi} = 2d\xi + b\mu + a \tag{6.1}$$

with unknown parameters belonging to known sets

$$a \in A, \quad b \in B.$$

The proportional–integral–relay stabilization law

$$\mu = -k_2\xi - k_1 \int_0^t \operatorname{sgn} \xi \, dt \tag{6.2}$$

turned out to be the most efficient law simple in realization. Here k_1, k_2 are fixed parameters of feedback. The block diagram of the feedback system (6.1), (6.2) is given in Fig. 6.1 (the meaning of arrows in the integrator in law (6.2) will be elucidated later). The motion equations of the feedback system in the range of the linear zone of the integrator (Fig. 6.1) and for constant parameters

a and b in the coordinate-operator space follow from (6.1), (6.2) and have the form

$$\begin{aligned} \ddot{\xi} + (bk_2 - 2d)\dot{\xi} + bk_1 \operatorname{sgn} \xi &= 0, \\ \dot{x}_1 &= -dx_1 + \xi x_1, \end{aligned} \tag{6.3}$$

whence it is clear that the larger the factor k_2 (the higher the gain of the linear feedback) the higher the rate of damping of the operator variable ξ. This is precisely the purpose of the stabilization problem being solved. However, the expression

$$u = \mu x_1 = -k_2\sigma - k_1 x_1 \int \operatorname{sgn}(\sigma x_1)\, dt$$

is valid for the real control. We can see from this expression that with the growth of the parameter k_2 the action with respect to the variable σ increases considerably and, consequently, the feedback system becomes less robust.

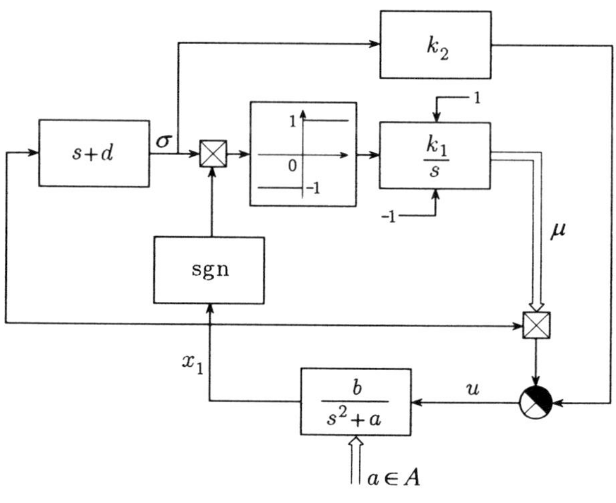

Figure 6.1.

Now if the parameter k_1 grows, then the transient process with respect to the operator variable becomes oscillatory. Therefore it is important to find means which

- will ensure sufficient damping of transient processes in system (6.3) without increasing the factor k_2 and, maybe, make it possible to do without the linear component in feedback at all ($k_2 = 0$),

- will make it possible to eliminate the oscillation of the operator variable ξ, which is desirable for the reasons indicated above.

Moreover, if the parameter b varies, then the stabilization law (6.2) inevitably leads to statism with respect to the operator variable ξ, and this, as we know, is equivalent to the dynamical statism with respect to the principal variable x_1.

We want to point out, without detailed comments, that law (6.2) admits variations of the parameter a at a limited rate, i.e., when

$$|\dot{a}| < \text{const} < bk_1.$$

In this connection, the question arises as to the attainability, in principle, of the purpose of control formulated above and the means of its achievement.

Of course, by analyzing Eq. (6.1)

$$\dot{\xi} = 2d\xi + b\mu + a,$$

we can try to guess the corresponding algorithm of stabilization of the operator variable ξ at zero, which is possible. However, we deem it more preferable to try to use the recommendations of the general theory exposed in Chap. 3, namely, to use the operator feedback (O-feedback, for brevity) for solving the posed problem. This operator feedback should be realized according to the block diagram shown in Fig. 6.2. This is quite natural since, from the viewpoint of physics, the purpose of O-feedback consists in bringing closer the dynamical properties of the reference device S_e and of the object P_e. It is obvious that when these properties become similar it is "easier" to control the object, and therefore the gain factors in the CO-loop may be diminished.

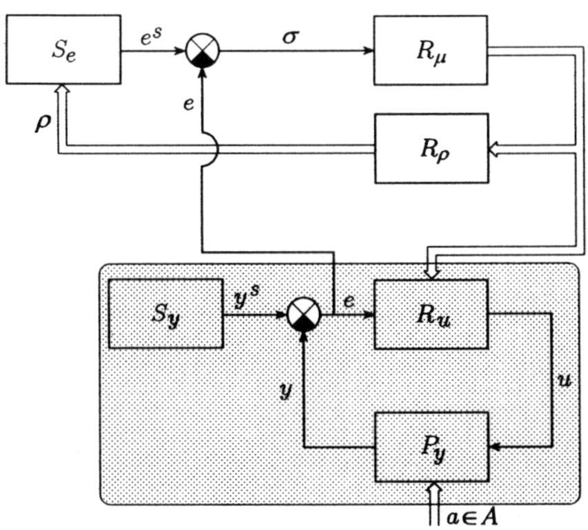

Figure 6.2.

If, in the preceding block diagram, we use the symbol $\tilde{P}_e$ to denote an object of an operator type with output μ and input ρ, then we can reduce the problem of synthesis of the operator R_ρ of O-feedback, at least as concerns its substance, to the standard problem of synthesis of feedback illustrated by Fig. 6.3. It stands to reason that, in this case, the important peculiarities of control of such an object will be revealed, the most significant being the nonlinearity of its motion

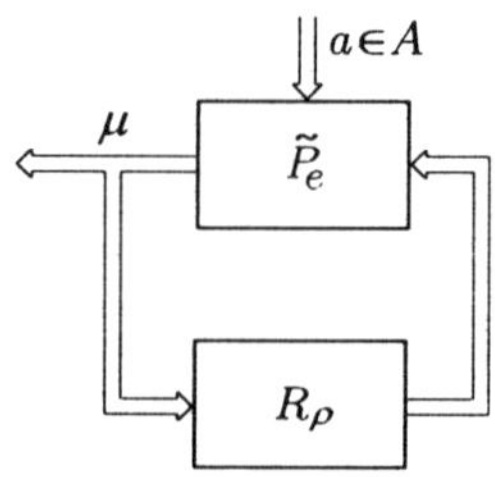

Figure 6.3.

equations. Therefore we shall begin with obtaining the motion equations of the system under consideration since the parameters d of the reference device S_e in the block diagram in Fig. 6.2 are variable now and depend on ρ, i.e., $d_\rho = d(\rho)$, and if $\rho = \rho(t)$, then $d_\rho = d_\rho(t)$.

6.2. Motion Equations in the Coordinate-Operator Space

Recall that in the original (x_1, x_2)-coordinate space the object being stabilized is described by the equations

$$\dot{x}_1 = x_2, \quad \dot{x}_2 = ax_1 + bu.$$

Following the general procedure of synthesis of binary control, we introduce a CO-error

$$\sigma_\rho = x_2 + d_\rho x_1 = \dot{x}_1 + d_\rho x_1,$$

which we shall denote by ρ, distinguishing it in this way from the previous error $\sigma = x_2 + dx_1$. We can now define, in the domain

$$G_\delta = \{(x_1, x_2) \mid |\sigma| \le \delta |x_1|, \ \delta = \text{const}\},$$

a new operator variable

$$\xi = \sigma_\rho / x_1. \tag{6.4}$$

To pass to the description in the coordinate-operator space (CO-space) (x_1, ξ), we shall need an equation of variation of the operator variable ξ. Differentiating expression (6.4) and using relations $\dot{x}_2 = ax_1 + bu$, $x_2 = \sigma_\rho - d_\rho x_1$, we successively obtain

$$\begin{aligned}\dot{\xi} &= \frac{\dot{\sigma}_\rho}{x_1} - \frac{\sigma_\rho}{x_1}\frac{\dot{x}_1}{x_1} = \frac{\dot{x}_2 + d_\rho x_2 + \dot{d}_\rho x_1}{x_1} - \frac{\sigma_\rho}{x_1}\frac{x_2}{x_1} \\ &= \frac{ax_1 + bu + d_\rho\sigma_\rho - d_\rho^2 x_1 + \dot{d}_\rho x_1}{x_1} - \frac{\sigma_\rho}{x_1}\frac{\sigma_\rho - d_\rho x_1}{x_1} \\ &= d_\rho^2\xi + b\frac{u}{x_1} + (a - d_\rho^2) + \dot{d}_\rho - \xi^2.\end{aligned}$$

If, as before, we use the simple binary operation $u = \mu x_1$ in the main loop and consider the stabilization in the small when we can neglect the quadratic terms, then we can write the required equations in the form

$$\begin{aligned} \dot{x}_1 &= -d_\rho x_1 + \xi x_1, \\ \dot{\xi} &= 2d_\rho \xi + b\mu + a - d_\rho^2 + \dot{d}_\rho. \end{aligned} \tag{6.5}$$

These equations differ from the equations

$$\begin{aligned} \dot{x}_1 &= -dx_1 + \xi x_1, \\ \dot{\xi} &= 2d\xi + b\mu + a - d^2 \end{aligned} \tag{6.6}$$

that we used before by the dependence of the parameter d of the reference device S_e on the operator variable ρ and by the presence of a derivative $\dot{d}_\rho \neq 0$ on the right-hand side of (6.5) since we now have $d_\rho = d_\rho(t)$. Since the O-law (i.e., the ρ function) can be formed without using the CO-law (i.e., the function μ), we can interpret the term $\dot{d}_\rho$ as an additional control applied for stabilizing the O-variable ξ at zero.

However, it is more convenient, in practice, to synthesize not the derivative $\dot{d}_\rho$ but the operator R_ρ of O-feedback (Fig. 6.2) for the chosen functional dependence $d(\rho)$ (for instance, without loss of generality, in the form $d_\rho = d + \rho$) and, in separate cases, the operator R_μ of CO-feedback.

In order to simplify further the equation of the $\tilde{P}_e$-object (Fig. 6.3) and obtain it in the final form. we shall assume that upon all variations the parameter ρ is small as compared to d. This assumption is fully justified, moreover, we must strive to guarantee it since in the desired stabilization mode, where $\xi = 0$, the principal variable x_1 changes, by (6.5), in accordance with the equation

$$\dot{x}_1 = -d_\rho x_1 = -(d + \rho)x_1.$$

Then the quality of the transient process is close to the standard one, defined by the equation $\dot{x}_1 = -dx_1$, if the order relationship between the parameters d and ρ, mentioned above, is satisfied. However, if we accept this assumption, then we can make an approximate substitution

$$d_\rho = (d + \rho)^2 \cong d^2 + 2d\rho.$$

Finally, if, as before, we first introduce, for convenience, the notation

$$a^* = a - d^2,$$

and then, for simplicity, omit the "star" (i.e., make a substitution $a^* \to a$), then the final form of the motion equation of the object investigated in the O-theory will be

$$\dot{x}_1 = -(d + \rho)x_1 + \xi x_1, \tag{6.7}$$

$$\dot{\xi} = 2(d + \rho)\xi + b\mu + a - 2d\rho + \dot{\rho}, \quad a \in A, \; b \in B. \tag{6.8}$$

Now, for convenience, we denote object (6.7), (6.8) by P_ρ. If we consider only the second equation of the P_ρ-object (6.8), then we denote it by $\tilde{P}_e$ as we agreed above.

Equations (6.7), (6.8) show the nonlinear, to be more precise, bilinear, character of the motion equations (this circumstance is essential and was repeatedly pointed out above), but the property which is significant and which ensures the success of the whole action is the affinity of the motion equations with respect to control (i.e., with respect to the pair (μ, ρ)). This allows us to apply the methods of the classical stabilization theory for solving the stabilizing problem.

6.3. Statical Operator Feedback

Considering the equation of the $\tilde{P}_e$-object

$$\dot{\xi} = 2(d+\rho)\xi + b\mu + a - 2d\rho + \dot{\rho}, \qquad a \in A, \quad b \in B,$$

it is easy to see that the feedback $\rho = R_\rho \mu$ may be statical or dynamical depending on the form of the operator R_ρ. However, only in the second case, since there is a derivative $\dot{\rho}$ on the right-hand side of (6.8), can the CO-feedback $\mu = R_\mu \sigma_\rho$ be discontinuous.

It is natural to begin the investigation with the analysis of potentialities of the negative statical O-feedback when

$$\rho = -q\mu, \tag{6.9}$$

where $q > 0$ is the gain of the O-feedback. As a result, we have motion equations

$$\begin{aligned} \dot{x}_1 &= -(d - q\mu)x_1 + \xi x_1, \\ \dot{\xi} &= 2(d - q\mu)\xi + \tilde{b}\mu + a - q\dot{\mu}, \\ \tilde{b} &= b + 2qd. \end{aligned} \tag{6.10}$$

After choosing the O-feedback operator $R_\rho = -q$, we are free to choose the CO-feedback operator R_μ. Applying different operators R_μ, we get versions of a binary control system with O-feedback. It is natural to begin the investigation with the simplest variants of choice of the operator R_μ.

6.3.1. Statical operator and coordinate-operator feedbacks

When we use the statical CO-feedback

$$\mu = -k\xi, \quad k = \text{const} > 0,$$

the second equation of motion in (6.10) assumes the form

$$\dot{\xi} = 2(d + qk\xi)\xi - \tilde{b}k\xi + qk\dot{\xi} + a.$$

After collecting terms and separating the principal part, we have a linear equation

$$\dot{\xi} = -\frac{(\tilde{b}k - 2d)}{1 - qk}\xi + \frac{a}{1 - qk}, \qquad \tilde{b} = b + 2qd,$$

if, of course, $1 > qk$. When the inequality

$$(b^- + 2qd)k > 2d$$

is satisfied, the equation is asymptotically stable and the limit (i.e., steady-state) value of the variable $\xi(t)$ is defined by the expression

$$\xi_\infty = \frac{a}{(b + 2qd)k - 2d} = \frac{a}{bk - 2d(1 - qk)}. \tag{6.11}$$

The error ξ_∞ defines the statical O-error which generates dynamical statism in the main loop.

Since in the limit the principal derivative satisfies the equation

$$\dot{x}_1 = -(d + qk\xi_\infty)x_1 + \xi_\infty x_1 = -[d + (qk - 1)\xi_\infty]x_1,$$

the value of the indicated dynamical statism is given by the expression

$$\xi^*_\infty = \frac{a\,(1 - qk)}{bk - 2d(1 - qk)}.$$

It should be pointed out that the statical error in system (6.10) without the O-feedback (i.e., for $q = 0$) is defined by the expression

$$\xi_\infty = \frac{a}{bk - 2d},$$

comparing which to (6.11) we find out that in a binary system with O-feedback the given value of dynamical statism is attained for a smaller value of the parameter k than in a binary system without O-feedback. This immediately follows from the fact that

$$\frac{1}{bk - 2d} > \frac{1 - qk}{bk - 2d(1 - qk)}.$$

Indeed, after clearing the fractions, we first find that

$$(bk - 2d) + 2dqk > (bk - 2d) - qk(bk - 2d),$$

and, collecting terms, get a trivial inequality

$$0 > -qbk^2.$$

Moreover, since the dynamical statism

$$\xi^*_\infty(q) = \frac{a\,(1 - qk)}{bk - 2d(1 - qk)}$$

decreases monotonically with the growth of the parameter q, the value of the dynamical statism $\xi^*_\infty(q)$ can be made arbitrarily close to zero as $qk \to 1$, although zero is unattainable since, for $q = 0$, all these equations are invalid. Consequently,

- the use of statical O-feedback is justifiable even in the trivial case where only statical nonlinear feedbacks are employed.

To interpret the last inference appropriately, it is useful to obtain an expression for the control law with respect to the principal variable, i.e., a law expressed in terms of the original variables x_1, x_2. Making an inverse substitution $\mu = -k\xi$ and using the relation $\rho = -q\mu$, we successively obtain

$$\begin{aligned} u &= \mu x_1 = -k\xi x_1 = -k\sigma_\rho = -k[x_2 + (d+\rho)x_1] \\ &= -k[x_2 + (d - q\mu)x_1] = -k(x_2 + dx_1) + qk\mu x_1 = -k\sigma + qku. \end{aligned}$$

Solving the last relation for the control u, we find that

$$u = \frac{k}{1-qk}\sigma, \quad \sigma = \dot{x}_1 + dx_1.$$

In other words, the nonlinear binary control system that we have constructed on the set G_δ is equivalent to the linear system with gain factor

$$k_{\text{eq}} = \frac{k}{1-qk},$$

and, as $qk \to 1$, we have an effect of high gain although finite gain factors are used in all loops of the control system. This effect is the direct result of the use of nonlinearity and positive feedback.

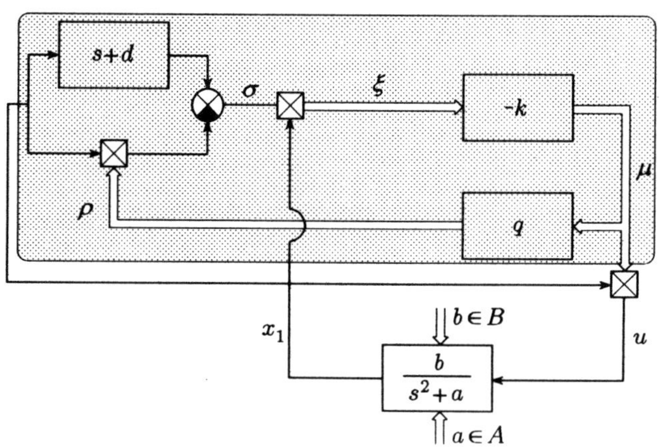

Figure 6.4.

The block diagram of the system that we have considered is given in Fig. 6.4. Recall that this diagram is efficient only under the constraint $|\sigma| \le \delta|x_1|$. The phase portrait of this control system in coordinates (x_1, x_2) within the set G_δ is given in Fig. 6.5 in which there are no oscillations about the line $\sigma = 0$. One of the drawbacks of this diagram is the presence of the inner statical loop (it is shown in Fig. 6.4) of feedback in the controller. This makes the system nonrobust since the effect produced by uncertain dynamics may lead to undesirable consequences, namely, oscillations or instability. Therefore we must think how to raise the robustness of a system with statical O-feedback.

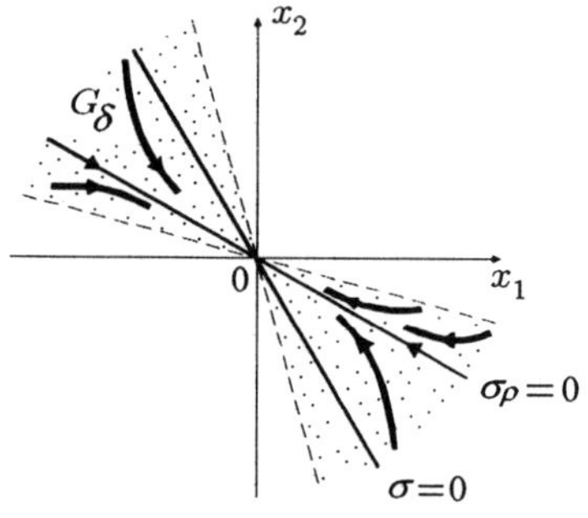

Figure 6.5.

6.3.2. Statical operator and dynamical coordinate-operator feedbacks

Let us study the possibility of stabilizing the $\tilde{P}_e$-object

$$\dot{\xi} = 2(d+\rho)\xi + b\mu + a - 2d\rho + \dot{\rho},$$
$$a \in A, \quad b \in B,$$

with the aid of the statical O-feedback

$$\rho = -q\mu, \quad q = \text{const},$$

and different types of dynamical CO-feedback

$$\mu = R_\mu \sigma,$$

where R_μ is a differential, integral, or some other type of operator of dynamical transformation.

Removing ρ, we get an equation for the object being stabilized which depends only on the CO-law of control in the following form:

$$\dot{\xi} = 2(d - q\mu)\xi + \tilde{b}\mu + a - q\dot{\mu}, \quad \tilde{b} = b + 2qd. \tag{6.12}$$

We shall begin with the case where use is made of a coordinate-operator feedback of a different type.

6.3.3. Inertial coordinate-operator feedback

In this case, we assume the parameter $b \in B$ to be known and therefore can take the CO-law in the form

$$q\dot{\mu} - \tilde{b}\mu = k\xi, \quad k = \text{const} > 0, \tag{6.13}$$

for instance. When the inequalities $q < 0$, $\tilde{b} = b + 2qd > 0$ are satisfied, it is natural to call this CO-law an inertial law since the transfer function from

ξ to μ has the form of an inertial link (see Fig. 6.6*a*, or, for a more detailed form, Fig. 6.6*b*). Substituting (6.13) into (6.8), we obtain an equation for the feedback system in the form

$$\begin{aligned} \dot{\xi} &= 2(d - q\mu)\xi + a - k\xi, \\ q\dot{\mu} - \tilde{b}\mu &= k\xi. \end{aligned} \tag{6.14}$$

In the equilibrium position (μ_∞, ξ_∞) of this system the relations

$$\begin{aligned} 2(d - q\mu_\infty)\xi_\infty + a &= k\xi_\infty, \\ \tilde{b}\mu_\infty &= -k\xi_\infty \end{aligned}$$

are satisfied solving which with an accuracy to within the values of order $1/k$, we obtain

$$\xi_\infty \cong \frac{a}{k - 2d}, \quad \mu_\infty \cong -\frac{k}{\tilde{b}} \frac{a}{k - 2d}.$$

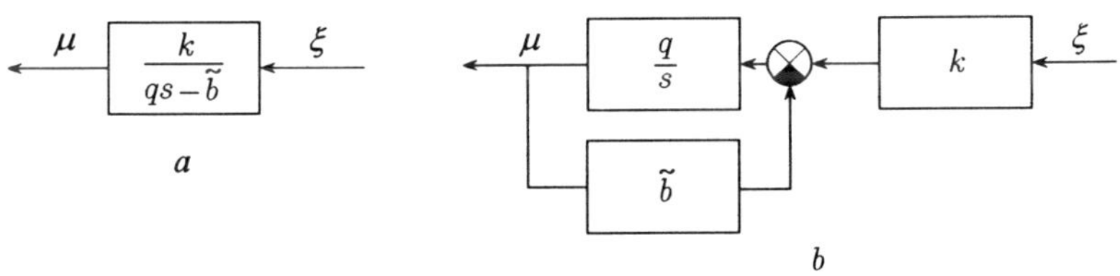

Figure 6.6.

In order to investigate the stability of this equilibrium position in the small, we write the motion equations of system (6.14) in deviations

$$\xi_1 = \xi - \xi_\infty, \quad \xi_2 = \mu - \mu_\infty,$$

preserving only terms of the first order of smallness. After elementary transformations we obtain equations

$$\dot{\xi}_1 = -(k - d)\xi_1, \quad \dot{\xi}_2 = \frac{\tilde{b}}{q}\xi_2 + \frac{k}{q}\xi_1,$$

analyzing which we see that the point (μ_∞, ξ_∞) is asymptotically stable if

$$k > 2d, \quad \frac{\tilde{b}}{q} < 0.$$

In order to define the limit motion of the principal coordinate x_1, we shall use the first equation of system (6.10). We obtain

$$\dot{x}_1 = -(d - q\mu_\infty)x_1 + \xi_\infty x_1 = -(d - q\mu_\infty - \xi_\infty)x_1.$$

Consequently, the number

$$\xi_\infty^* = -(q\mu_\infty + \xi_\infty) = -\frac{a}{\tilde{b}}\,\frac{(\tilde{b} - qk)}{k - 2d}$$

defines, in this case, the magnitude of statism. It is impossible to eliminate this statism by tending the parameter to infinity, $k \to \infty$, since

$$\lim_{k\to\infty} \xi_\infty^* = \frac{aq}{\tilde{b}} = \frac{aq}{b + 2qd}.$$

However, diminishing the parameter q, we can make it arbitrarily small although we cannot eliminate it completely in this way since $q < 0$.

The block diagram of the stabilization system that we have investigated is shown in Fig. 6.7. Recall that it is valid only under the constraint $|\sigma| \le \delta|x_1|$. The figure showing the projections of the phase trajectories of this system onto the set G_δ in the plane (x_1, x_2) is similar to Fig. 6.5, and therefore we do not give it here.

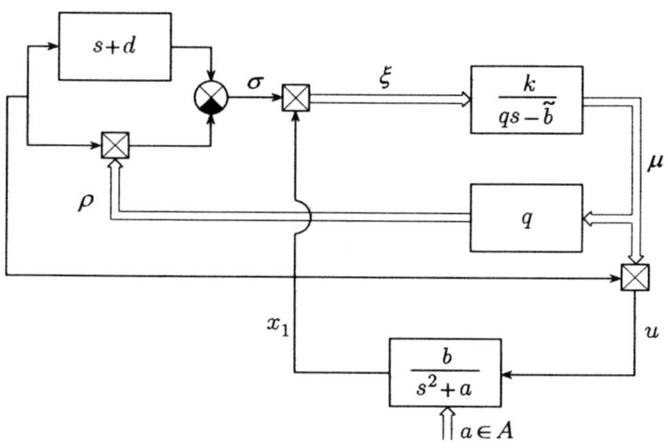

Figure 6.7.

Since the robustness of this system is doubtless thanks to the presence of dynamics in the O-loop, we can state that

- in the binary system that we have considered the stabilization problem can be solved by means of continuous control with arbitrary accuracy and there are no oscillations of the phase vector about the line $\sigma = 0$.

However, to realize this system, we must know the parameter $b \in B$, and the realization itself is complicated since it requires the division operation. In order to overcome these drawbacks, let us consider a different type of CO-feedback.

6.3.4. Inertial-relay coordinate-operator feedback

Let us again assume that the parameter $b \in B$ is known and define the CO-feedback by an equation of the form

$$q\dot{\mu} - \tilde{b}\mu = k \operatorname{sgn} \xi, \quad k = \text{const} > 0. \tag{6.15}$$

Using this feedback for the $\tilde{P}_e$-object

$$\dot{\xi} = 2(d - q\mu)\xi + \tilde{b}\mu + a - q\dot{\mu},$$

we obtain a feedback stabilization system with the motion equation

$$\dot{\xi} = 2(d - q\mu)\xi - k \operatorname{sgn} \xi + a. \tag{6.16}$$

The advantage of the CO-law (6.15) is that for $q < 0$ the variable μ is automatically bounded in absolute value, and this is useful, of course, since $u = \mu x_1$ and the variable μ in the principal loop of the control acts as a gain factor which, as we know, must be bounded for the reason of robustness. Note that this effect was absent in other systems with the CO-law that we have considered. In those systems this constraint had to be specially ensured. We

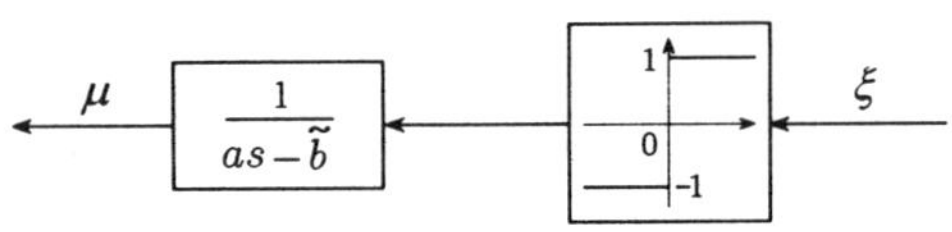

Figure 6.8.

shall speak about all this in greater detail later on and will now give a block diagram that explains the name of the law (Fig. 6.8) Since the variable μ is bounded, it is obvious, from the analysis of Eq. (6.16)

$$\dot{\xi} = 2(d - q\mu)\xi - k \operatorname{sgn} \xi + a,$$

that for a sufficiently large value of the parameter k, namely, $k > a$, a sliding mode appears at the point $\xi = 0$. To determine the motion equation in the sliding mode, we shall first find from the equations $\xi = 0$, $\dot{\xi} = 0$ an equivalent value of the discontinuous element

$$k \operatorname{sgn}_{\text{eq}} \xi = a,$$

then substitute the obtained value into Eq. (6.15) defining the variation of the variable μ, and obtain a motion equation

$$q\dot{\mu} - \tilde{b}\mu = a. \tag{6.17}$$

This equation is stable for $q < 0$ and has an equilibrium position

$$\mu_\infty = -a/\tilde{b}$$

for a fixed parameter a.

We can now derive a motion equation for a system in sliding mode. Since for $\xi = 0$ the equation

$$\dot{x}_1 = -(d - q\mu)x_1 \tag{6.18}$$

acts for the principal variable x_1, in the sliding mode this equation must be supplemented with Eq. (6.17). However, if the parameter q is sufficiently small, then we may assume that $\mu(t) \cong \mu_\infty$ and Eq. (6.18) is reduced (by the substitution $\tilde{b} = b + 2qd$) to a simpler equation

$$\dot{x}_1 = -(d - q\mu_\infty)x_1 = -\big(d + qa/\tilde{b}\big)x_1.$$

From the last equation it follows, in particular, that dynamical statism is defined by the relation

$$\xi^*_\infty = q\frac{a}{b + 2qd}$$

and can be made arbitrarily small by diminishing the parameter q. The statism cannot be completely eliminated since the value $q = 0$ is inadmissible. Therefore the projection of the phase trajectories of this system onto the set G_δ in the plane (x_1, x_2) is like that shown in Fig. 6.5.

To explain the physical effect exploited in the stabilization system under consideration, it is useful to write the algorithm of stabilization in the original variables. Since $\xi = 0$ but $\sigma_\rho = \xi x_1$ in the sliding mode, it follows that $\sigma_\rho = 0$ as well. However, $u = \mu x_1$, and, consequently,

$$\sigma_\rho = \dot{x}_1 + dx_1 - q\mu x_1 = \sigma - qu = 0.$$

Thus, we have a relation $u = \sigma/q$ which means that in the sliding mode the nonlinear feedback under investigation is equivalent to a linear feedback with a gain factor $1/q$ which can be made arbitrarily large as $q \to -0$, and this ensures the solution of the stabilization problem.

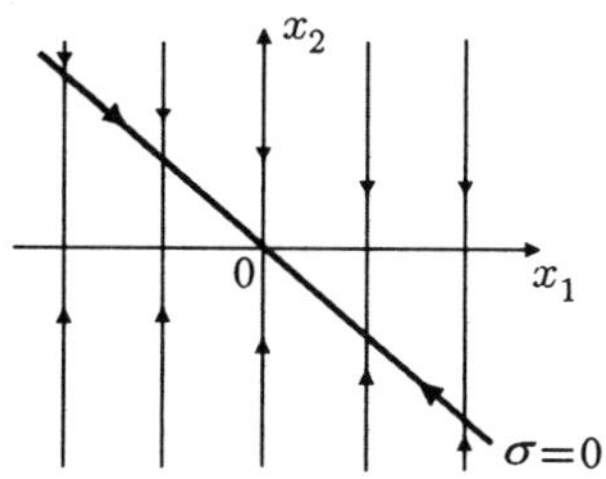

Figure 6.9.

Indeed, in this case, in the original variables, the system is described by the equations

$$\dot{x}_1 = x_2$$

$$\dot{x}_2 = ax_1 + \frac{1}{q}(x_2 + dx_1)$$

and, as $q \to 0$, it is associated with the phase portrait shown in Fig. 6.9. Thus,

- the proposed feedback solves the posed stabilization problem as exactly as we need, but with an unremovable dynamical statism.

In this case, the oscillations of the operator variable ξ are completely eliminated ($\xi = 0$) and the gain factor k decreases as compared to the preceding law.

It is also easy to realize that this system is robust. Indeed, suppose that there is a delay $\tau > 0$ in switchings, i.e., instead of the law (6.15) we deal with a CO-law of the form

$$q\dot{\mu} - \tilde{b}\mu = k \operatorname{sgn}_\tau \xi, \quad k = \text{const} > 0. \tag{6.19}$$

In the coordinates $(q\dot{\mu} - \tilde{b}\mu, \xi)$, Eqs. (6.15) and (6.19) are associated with the graphs in Fig. 6.10, whence we immediately obtain an estimate $|\xi| \leq \Delta(\tau)$, and this means that the system is robust since $\Delta(\tau) \to 0$ as $\tau \to 0$.

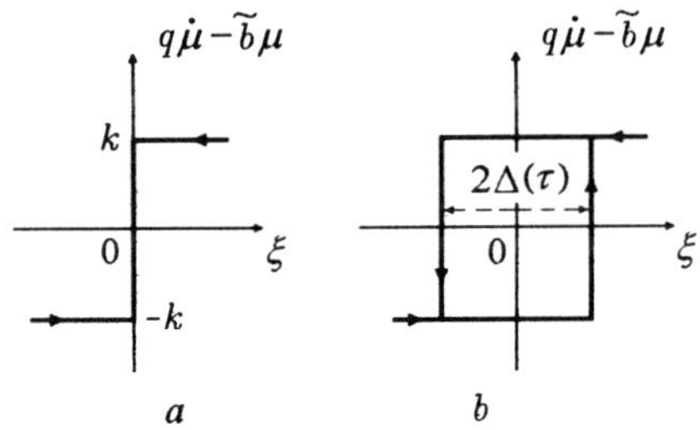

Figure 6.10.

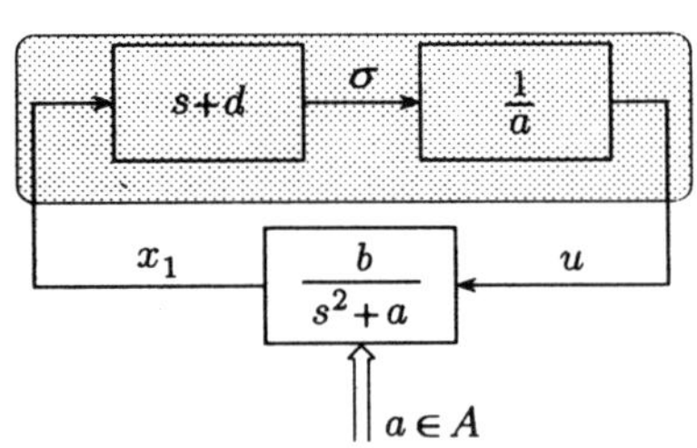

Figure 6.11.

The block diagram of the synthesized binary system is shown in Fig. 6.11, and, for the sake of comparison, Fig. 6.12 shows a system which is equivalent to it in the case of appearance of a sliding mode. Comparing Fig. 6.7 and Fig. 6.11, we see that the latter system has the same potentialities but can be more easily realized. However, one can think of its further simplification and, if possible, make it unnecessary to use the information about the parameter $b \in B$.

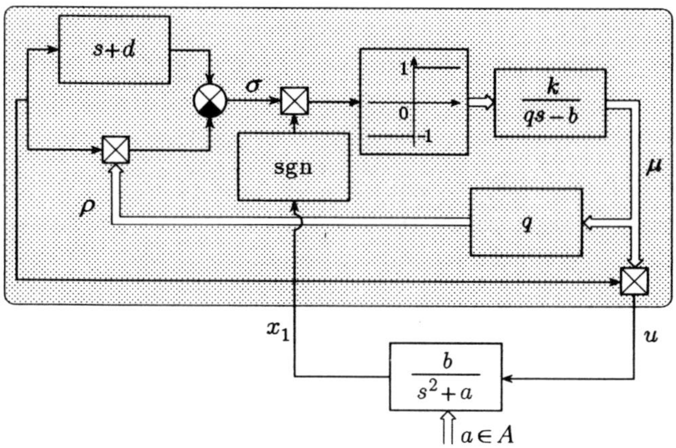

Figure 6.12.

6.3.5. Inertial-relay coordinate-operator feedback with an unknown parameter in the control

In this section, to stabilize the $\tilde{P}_e$-object (6.8),

$$\dot{\xi} = 2(d - q\mu)\xi + \tilde{b}\mu + a - q\dot{\mu}, \quad \tilde{b} = b + 2qd$$
$$a \in A, \quad b \in B,$$

we can use an inertial-relay CO-feedback of the form

$$\begin{aligned} q\dot{\mu} - \tilde{b}^{-}\mu &= k \operatorname{sgn} \xi, \quad k = \text{const} > 0, \\ \tilde{b}^{-} &= b^{-} + 2qd, \end{aligned} \tag{6.20}$$

which can also be used when the value of the parameter b is not known. Note that arbitrary variations of the unknown parameters a, b in time are admissible. The structure of control (6.20) coincides with that of control in Fig. 6.8 when $\tilde{b}$ is replaced by $\tilde{b}^{-}$.

Substituting (6.20) into (6.8), we get an equation for a feedback system in the form

$$\dot{\xi} = 2(d - q\mu)\xi + (\tilde{b} - \tilde{b}^{-})\mu - k \operatorname{sgn} \xi + a. \tag{6.21}$$

Since in this case as well the variable μ is uniformly bounded (for $q < 0$), it is obvious that there exists a number a^0 such that a sliding mode appears at the point $\xi = 0$ for $k > a^0$.

In the sliding mode we can assume the relations

$$\xi = 0, \quad \dot{\xi} = 0$$

to be satisfied, and then the equivalent value of the discontinuous signal can be defined by Eq. (6.21) as

$$k \operatorname{sgn}_{\text{eq}} \xi = a + (\tilde{b} - \tilde{b}^{-})\mu.$$

Substituting this value into the equation for control (6.20), we find the motion equation of the operator variable μ in the sliding mode

$$q\dot{\mu} - \tilde{b}\mu = a, \tag{6.22}$$

which exactly coincides with the equation of sliding for the variable μ from the preceding section. Therefore all the references that we made above are valid and, in particular, the conclusion that when the transient process of Eq. (6.22) terminates, the equation of sliding with respect to the principal variable is given by the expression

$$\dot{x}_1 = -(d + qa/\tilde{b})x_1.$$

Consequently,

- in this case as well we have an unremovable (although arbitrarily small as $q \to -0$) dynamical statism.

All the remarks concerning the robustness of the control system and the physical meaning of the effect of compensation for the perturbations a and b are also preserved.

The projections of phase trajectories onto the set G_δ in the plane (x_1, x_2) are shown in Fig. 6.5, and, to find the structure of the system, it is sufficient to replace the parameter $\tilde{b}$ by $\tilde{b}^-$ in the block diagram in Fig. 6.11 and add a double arrow $\Downarrow b \in B$ that acts on the object P and reflects the effect produced by the uncertainty factor.

The advantage of control (6.20) is that it acts efficiently for any unknown parameters $a \in A$, $b \in B$.

6.3.6. Integral-relay coordinate-operator feedback

The maximal possible simplification of a control system is attained when we use the integral-relay law

$$q\dot{\mu} = k \operatorname{sgn} \xi. \tag{6.23}$$

In this case, the feedback control system is described by (6.23) and by the equation

$$\dot{\xi} = 2(d - q\mu)\xi + \tilde{b}\mu + a - k \operatorname{sgn} \xi, \quad \tilde{b} = b + 2qd. \tag{6.24}$$

When a sliding mode appears, we can consider the relations

$$\xi = 0, \quad \dot{\xi} = 0$$

to be satisfied in the sense that we need, and therefore the equivalent value of the discontinuous signal is defined by the relation

$$k \operatorname{sgn}_{\text{eq}} \xi = \tilde{b}\mu + a.$$

Substituting it into the equation of control (6.23), we find the exact and asymptotic equations of sliding for the variables μ and x_1 in the form of expressions

$$q\dot{\mu} - \tilde{b}\mu = a, \quad \dot{x}_1 = -(d - q\mu)x_1;$$
$$\mu_\infty = -a/\tilde{b}, \quad \dot{x}_1 = -(d + qa/\tilde{b})x_1$$

that we already know. Therefore all inferences made in Secs 6.3.4. and 6.3.5. are valid.

The only difficulty in carrying out the reasoning is that the fact of appearance of the sliding mode at the point $\xi = 0$ cannot be easily revealed from the equations of feedback system (6.23), (6.24). Since we are interested in the analysis in the small, this problem can be investigated with the aid of a Lyapunov function of the form

$$v = \frac{1}{2}\xi^2. \tag{6.25}$$

Indeed, assuming, for simplicity, that the parameters $a, b = \text{const}$, we find, after the differentiation of function (6.25), by virtue of the motion equations (6.23), (6.24), that

$$\begin{aligned} \dot{v} = \dot{\xi}\xi &= \xi[2(d - q\mu)\xi + \tilde{b}\mu + a - \alpha \operatorname{sgn} \xi] \\ &= 2(d - q\mu)\xi^2 - |\xi|\,[d - (\tilde{b}\mu + a)\operatorname{sgn}\xi]. \end{aligned} \tag{6.26}$$

It should also be pointed out that in the equilibrium position

$$\xi = 0, \quad \tilde{b}\mu + a = 0,$$

and therefore the term $-d|\xi|$ dominates in its neighborhood in (6.26). Consequently, there exists a number $\gamma > 0$ such that

$$\dot{v} \le -\gamma|\xi| = -\gamma\sqrt{2v} = -\gamma^*\sqrt{v}.$$

Solving the last differential inequality, we find an estimate

$$\sqrt{v(t)} - \sqrt{v(0)} \le -\gamma^* t,$$

from which it follows that in a certain neighborhood of the equilibrium position the sliding mode always appears in finite time, and this is what we had to prove.

The block diagram of the synthesized binary control system is shown in Fig. 6.13. It appears to be the most simple of all the diagrams considered before, with all the stabilizing properties being preserved.

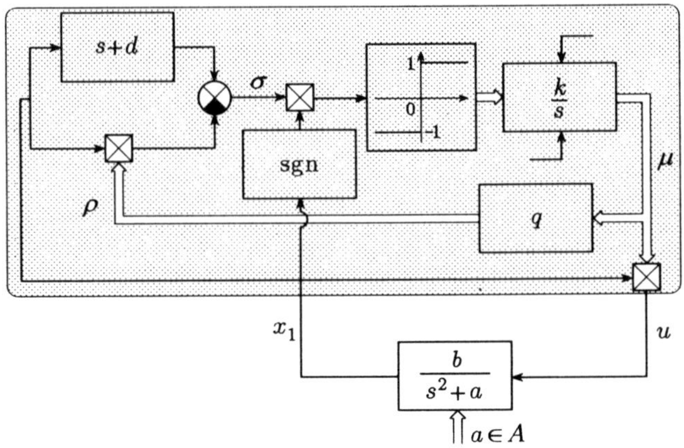

Figure 6.13.

The investigation that we have carried out shows that using the O-feedback, we can considerably raise the quality of transient processes of a binary control system, in particular,

- eliminate the oscillation of the phase vector in the set G_δ,
- raise the robustness of the control system by diminishing the gain factors in the main channel and, moreover, obviate the necessity of using the damping component $\mu = -k_2\xi$ in CO-laws,
- lower the dimension of the feedback system by two orders, say, from the third order to the first one (in conjunction with a dynamical discontinuous CO-feedback),
- arbitrarily exactly attain (to the prejudice of robustness, to tell the truth) the required quality of transient processes.

In this case, the statical O-feedback is inevitably followed by statism (dynamical statism in principal variables), this, however, being natural for any statical feedback. We do not consider here the problem of using dynamical O-feedback. In the next chapter we use the OC-feedback for eliminating this statism.

Chapter 7

Theory of Operator-Coordinate Feedback

This chapter is devoted to the qualitative investigation of binary automatic control systems with different types of feedback. Potentially, nonlinear dynamical stabilization systems of this type must possess the most perfect properties which are close to those ideal properties that are observed in control systems with high gain feedback. However, in contrast to the latter, a binary stabilization system is robust in the class of regular and singular perturbations. Very likely, in this chapter we demonstrate, to the full extent, the use of the nonlinearity effect in problems of stabilization of an essentially uncertain object. The final motion equations of a stabilization system are nonlinear in principle, sufficiently complicated, and possess globally stable solutions which depend but slightly (and do not depend at all, in certain cases) on uncertainty factors and can be described by simple differential equations. Nonlinear equations of this kind are a direct result of application of the theory worked out in this monograph.

Nothing is known today as concerns the possibility of obtaining these equations by any other methods.

7.1. Dynamical Statism and Operator-Coordinate Feedback

As was established in the preceding chapters, the stabilization at zero of an uncertain object

$$\dot{x}_1 = x_2,$$
$$\dot{x}_2 = ax_1 + bu, \quad a \in A, \ b \in B,$$

with the aid of three types of feedback (Fig. 7.1) can be achieved for any, including variable, parameters of the object $a \in A$, $b \in B$.

In this case, in the set

$$G_\delta = \{x \mid |\sigma| \leq \delta |x_1|\}$$

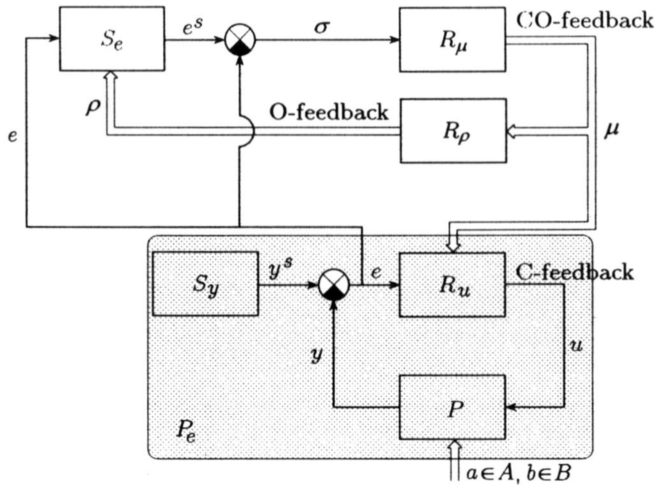

Figure 7.1.

the limit motion is described by the equation

$$\dot{x}_1 = -(d + \rho_\infty)x_1,$$

in which the number ρ_∞ defines the dynamical statism

$$\rho_\infty = -qa/\tilde{b}, \quad \tilde{b} = b + 2qd,$$

which deviates this limit motion from the standard motion defined by the equation $\dot{x}_1 = -dx_1$ (Fig. 7.2)

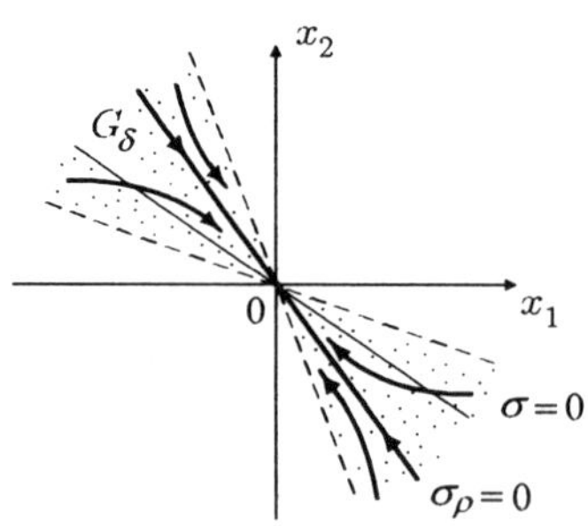

Figure 7.2.

The statism can be eliminated by tending k to infinity, $k \to \infty$, but this lowers the robustness of the control system. Conversely, we can tend the parameter q to zero, $q \to 0$, which cannot be excluded, but in the limit (i.e., for $q = 0$) the O-feedback does not operate and then, to achieve the similar quality of the transient process, we must increase the gain factors or consent to the oscillations of the phase vector about the manifold $\sigma = 0$.

It should be pointed out that the tending of the parameter q to zero, $q \to 0$, as it follows, for instance, from the equation of an integral-relay CO-control

$$\dot{\mu} = \frac{k}{q} \operatorname{sgn} \xi,$$

means, in essence, an introduction of feedback with high gain in the CO-loop of control since in this case $k/q \to \infty$, and this, as we know, leads to the loss of robustness and, although we can consider the values of hidden parameters in the controller to be smaller than those in the object and, consequently, the critical value of the gain factor of this loop, with whose increase stability is lost, can be considered to be higher, nevertheless, in principle, we must look for new means for an exact solution of the stabilization problem under consideration.

It is also important to note that in systems with statical O-feedback and discontinuous CO-feedback the control in the main loop is described by the expression $u = (k/q)\sigma$ when sliding mode appears, and, although it may happen that $k/q > k_{\mathrm{cr}}$, instability does not occur. The justification of this fact is almost obvious and is not given here.

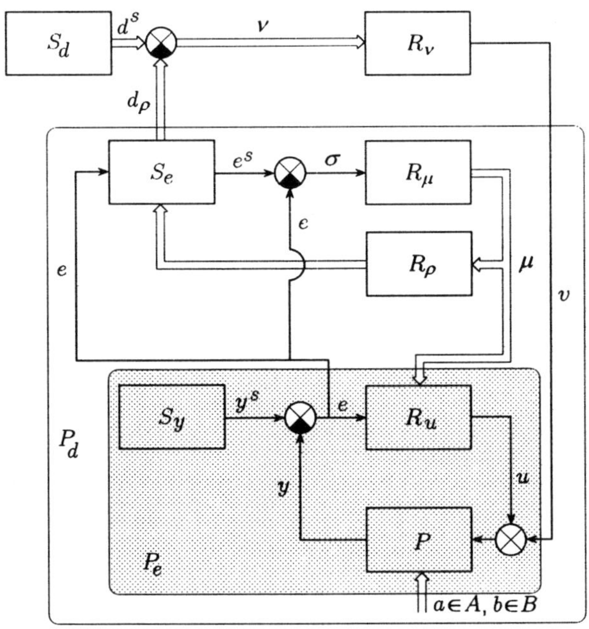

Figure 7.3.

In order to eliminate dynamical statism, we shall use an operator-coordinate feedback, or, briefly, OC-feedback, enveloping the control object as is shown in Fig. 7.3. In this block diagram the operators of the O-reference device S_d and of the OC-controller R_ν must be specified or synthesized. Recall that $d_\rho = d(\rho)$. The marked part of the block diagram in Fig. 7.3 is a typical OC-object, and

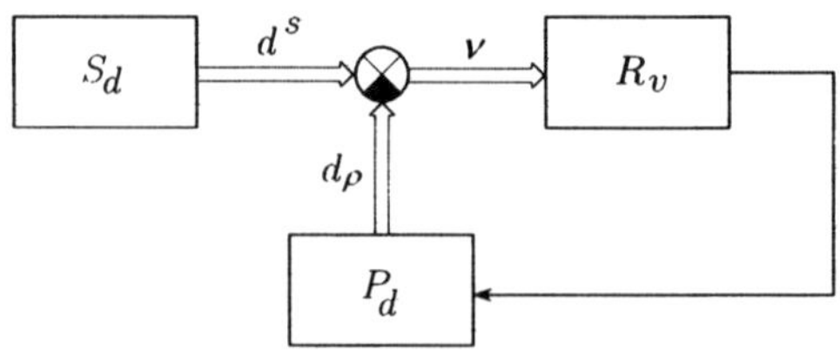

Figure 7.4.

therefore the problem under consideration is given in an enlarged form in Fig. 7.4. Since this diagram is standard for the feedback theory, the methods of solving the problem may also be standard.

7.2. Motion Equations for an Operator-Coordinate Object

The original equations of an object in a coordinate space have now the form

$$\dot{x}_1 = x_2, \quad a \in A$$
$$\dot{x}_2 = ax_1 + b(u+v), \quad \sigma = dx_1 + x_2, \; b \in B.$$

The presence of OC-feedback is given here by the component v in the control. Introducing, with due account of the action of O-feedback, an error

$$\sigma_\rho = x_2 + d_\rho x_1$$

and an operator variable $\xi = \sigma_\rho / x_1$ in the set

$$G_\delta = \left\{ \begin{pmatrix} x_1 \\ x_2 \end{pmatrix} \;\middle|\; |\sigma| \leq \delta |x_1| \right\},$$

and acting as usual (see Chaps. 4, 6), we find an equation for the variation of the operator variable in the set G_δ, which, in the first approximation, has the form

$$\dot{\xi} = 2d_\rho \xi - d_\rho^2 + \dot{\rho} + a + b\frac{u+v}{x_1}. \tag{7.1}$$

Let us assume, as usual, that the binary operation in the C-loop is statical,

$$u = \beta_u(\mu, x_1) = \mu x_1,$$

and, similarly, choose a binary operation in the OC-loop,

$$v = \beta_v(\eta, x_1) = \eta x_1,$$

where, as we accepted above, μ and η are operator variables. Upon this choice, Eq. (7.1) assumes the form

$$\dot{\xi} = 2d_\rho\xi - d_\rho^2 + \dot{\rho} + a + b(\mu + \eta), \tag{7.2}$$

where μ and η can be regarded as controls.

As we did before, we assume that the O-feedback is statical:

$$\rho = -q\mu, \quad q = \text{const},$$

and the action on the parameter of the S_e-reference device is adaptive:

$$d_\rho = d + \rho.$$

Then, for the integral-relay CO-controller

$$q\dot{\mu} = k_1 \operatorname{sgn} \xi, \quad k_1 = \text{const},$$

the equation for the variation of the operator error ξ (7.2) assumes the form

$$\dot{\xi} = 2(d - q\mu)\xi + \tilde{b}\mu - k_1 \operatorname{sgn} \xi + a + b\eta, \quad \tilde{b} = b + 2qd. \tag{7.3}$$

Equation (7.3) completed by the equation for the variation of the principal variable

$$\dot{x}_1 = -(d + \rho)x_1 + \xi x_1 \tag{7.4}$$

and the equation of the OC-controller

$$\eta = R_\eta \nu, \tag{7.5}$$

where the OC-error of control

$$\nu = d_\rho - d = \rho \tag{7.6}$$

and R_η is the operator of OC-controller, describe in the set G_δ a feedback control system with four types of feedback. Since

$$\dot{\nu} = \dot{\rho} = -k_1 \operatorname{sgn} \xi, \tag{7.7}$$

we must appropriately choose the operator R_η which would guarantee stabilization of the error ν at zero. Let us consider some versions.

7.3. Statical OC-Controller

Suppose that the operator R_η in (7.5) is a simple linear operator, i.e., $\eta = k_2\nu$. Then we must test for stability the following system of equations which we denote by Σ^c for convenience:

$$\dot{\xi} = 2(d - q\mu)\xi + \tilde{b}\mu - k_1 \operatorname{sgn}\xi + k_2 b\nu + a, \tag{7.8}$$
$$\dot{\nu} = -k_1 \operatorname{sgn}\xi. \tag{7.9}$$

Since $\nu = \rho$, $\rho = -q\mu$, we can introduce a notation

$$b^* = (1 - qk_2)b + 2qd$$

and rewrite this system in a more convenient form

$$\dot{\xi} = 2(d - q\mu)\xi + b^*\mu - k_1 \operatorname{sgn}\xi + a,$$
$$q\dot{\mu} = -k_1 \operatorname{sgn}\xi.$$

Equations similar to Eq. (7.8) for the Σ^c-system were investigated in detail in Chaps. 4 and 6, and therefore we can assert that when the inequality $k_1 > a^0$ is satisfied, a sliding mode appears at the point $\xi = 0$. Since, in addition, $\dot{\xi} = 0$ in a certain sense, the equivalent value of the discontinuous signal is defined by the expression

$$k_1 \operatorname{sgn}_{\text{eq}} \xi = b^*\mu + a.$$

Substituting this expression into Eq. (7.9) of the Σ^c-system, we find the equation for variation of the O-error ν in the form $\dot{\nu} = -b^*\mu - a$. Since $\nu = \rho = -q\mu$, the final form of the equation for the error ν is

$$\dot{\nu} = \frac{b^*}{q}\nu - a. \tag{7.10}$$

Since $b^*/q < 0$, Eq. (7.10) is asymptotically stable and has an equilibrium position at the point

$$\nu_\infty = \frac{qa}{b^*}, \quad b^* = (1 - qk_2)b - 2qd.$$

Consequently,

- the statical OC-feedback does not eliminate dynamical statism but only decreases its magnitude as compared to the control systems considered before.

Indeed, in a system with a statical O-feedback, for instance, the value of statism is defined by the expression

$$\rho_\infty = qa/\tilde{b}, \quad \tilde{b} = b - 2qd,$$

but, since $b^* > \tilde{b}$, we obviously have the desired $\nu_\infty < \rho_\infty$.

By tending k_2 to infinity, we can eliminate the statism ν_∞, but this will impair the robustness of the system. Indeed, in this case, the second component of the control is linearly connected with the principal control

$$v = \eta x_1 = k_2 \nu x_1 = -q k_2 \mu x_1 = -q k_2 u$$

and the complete control $u' = u + v$ is also proportional to the principal control

$$u' = (1 - q k_2) u. \tag{7.11}$$

And this precisely means that the action of the statical OC-controller is equivalent to an increase in the gain factors of the CO-controller $(1 - qk_2)$ times, and this, as was indicated above, lowers the robustness of the control system. Relation (7.11) also elucidates the result of the use of a statical OC-controller. This system is illustrated by Fig. 7.5 which shows the projections of the phase trajectories of the system onto the set G_δ when a sliding mode appears.

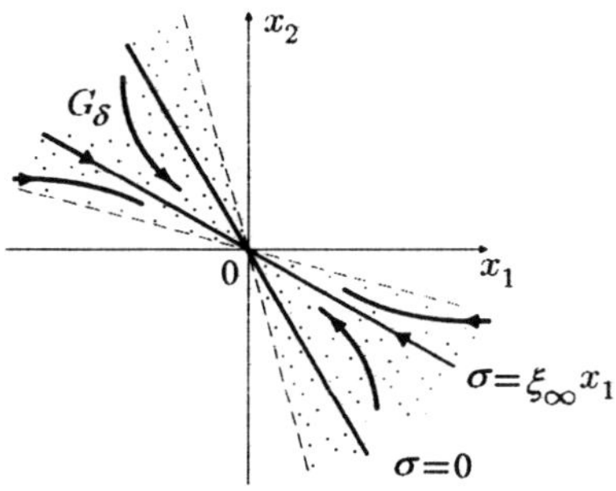

Figure 7.5.

7.4. Integral OC-Controller

We know from the classical theory of control that statism can be eliminated by the use of an integral feedback, and therefore we shall consider an OC-controller of the form

$$\dot{\eta} = -k_2 \nu, \quad k_2 = \text{const}\,. \tag{7.12}$$

Equation (7.12), in conjunction with the equations of CO- and O-controllers

$$q\dot{\mu} = k_1 \operatorname{sgn} \xi, \quad k_1 = \text{const}, \tag{7.13}$$

$$\rho = -q\mu, \quad q = \text{const}, \tag{7.14}$$

the equations of binary elements

$$u' = u + v = \mu x_1 + \eta x_1 = (\mu + \eta) x_1, \tag{7.15}$$

and also the motion equations for the object in the CO-space (x_1, ξ)

$$\dot{x}_1 = -(d+\rho)x_1 + \xi x_1, \tag{7.16}$$

$$\dot{\xi} = 2(d+\rho)\xi + b\mu + 2d\rho + \dot{\rho} + b\eta + a, \quad a \in A,\ b \in B, \tag{7.17}$$

defines the behavior of a feedback control system with four types of feedback on the set G_δ.

As before, we set

$$\nu = d_\rho - d = \rho, \tag{7.18}$$

$$\xi x_1 = \sigma, \quad \sigma = \dot{x}_1 + d x_1, \quad \sigma_\rho = \dot{x}_1 + d_\rho x_1. \tag{7.19}$$

The block diagram of the binary system being investigated is given in Fig. 7.6 (where the synthesized second-order nonlinear dynamical controller is shaded and the full order of the feedback system is equal to four). Now the problem is to choose parameters of the system such that the stabilization of the variable x_1 at zero would be guaranteed at the asymptotically vanishing dependence of the transient process on the uncertain (constant, for simplicity) parameters $a \in A$, $b \in B$.

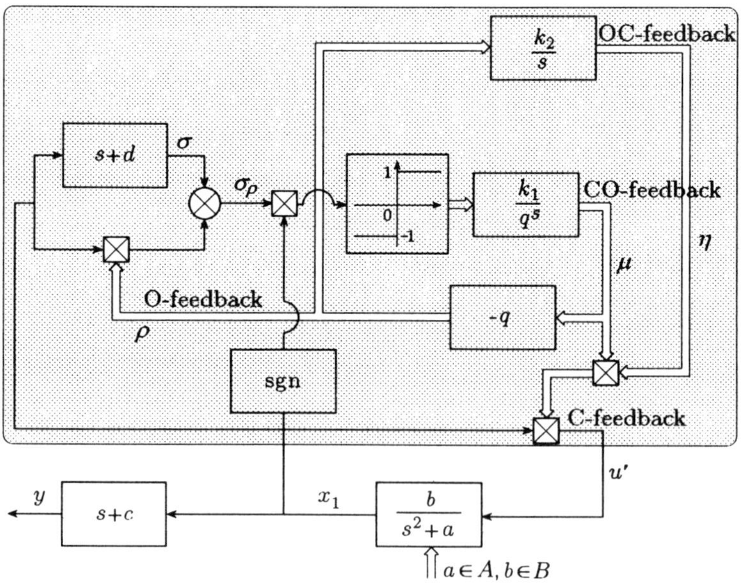

Figure 7.6.

To analyze the behavior of the synthesized binary system, we shall consider the equations for the variation of the operator variables (ξ, μ, η). Substituting relations (7.17), (7.14) into (7.12) and (7.13), (7.14) into (7.17), we get a collection of differential equations describing the so-called Σ_i^c-system

$$\begin{aligned}&\dot{\xi} = 2(d - q\mu)\xi + \tilde{b}\mu - k_1 \operatorname{sgn}\xi + b\eta + a, \quad \tilde{b} = b - 2qd,\\ &q\dot{\mu} = k_1 \operatorname{sgn}\xi, \quad \dot{\eta} = qk_2\mu.\end{aligned} \tag{7.20}$$

It is obvious that the equilibrium position of the Σ_i^c-system is at the point $(\xi_\infty, \mu_\infty, \eta_\infty) = (0, 0, -a/b)$.

Let us make sure that a sliding mode arises in a small neighborhood of the equilibrium position at a point $\xi = 0$ in a finite time. To this end, it suffices to multiply termwise the first equation for the Σ_i^c-system by ξ and obtain an expression

$$\xi\dot{\xi} = -k_1|\xi| + 2(d - q\mu)\xi^2 + \tilde{b}\mu\xi + b\eta\xi + a\xi,$$

analyzing which we see that when the condition

$$k_1 > a^0$$

is fulfilled, there exist constants α, $\beta > 0$ such that in the neighborhood

$$|\xi| \leq \beta$$

the inequality

$$\xi\dot{\xi} < -\alpha|\xi|$$

is satisfied which is equivalent to the differential inequality

$$\dot{\xi} < -\alpha \operatorname{sgn}\xi.$$

These inequalities imply that a sliding mode arises at the point $\xi = 0$ in a finite time and that the principal variable x_1 converges to zero in G_δ since (7.16) yields an equation

$$\dot{x}_1 = -(d + \rho)x_1.$$

This is not enough, however, and therefore we shall continue our investigation.

In the sliding mode, as usual, we determine from the equalities $\xi = \dot{\xi} = 0$ an equivalent value of the discontinuous signal in the form

$$k_1 \operatorname{sgn}_{\text{eq}}\xi = \tilde{b}\mu + b\eta + a.$$

Substituting the obtained equivalent control into Eq. (7.20) of the Σ_i^c-system, we get the following collection of differential equations for the $\tilde{\Sigma}_i^c$-system:

$$\begin{aligned}q\dot{\mu} &= \tilde{b}\mu + b\eta + a,\\ \dot{\eta} &= qk_2\mu.\end{aligned} \tag{7.21}$$

The equilibrium position of the $\tilde{\Sigma}_i^c$-system is at the point $(0, -a/b)$, which, naturally, coincides with the point $(\mu_\infty, \eta_\infty)$. The stability of the equilibrium position is defined by the asymptotic properties of the dynamical system

$$\dot{\mu} = \frac{\tilde{b}}{q}\mu + \frac{b}{q}\eta, \quad \dot{\eta} = qk_2\mu. \tag{7.22}$$

System (7.22) is asymptotically stable if and only if its characteristic polynomial

$$\varphi(s) = \det(sE - A) = \det \begin{bmatrix} s - \tilde{b}/q & -b/q \\ -qk_2 & s \end{bmatrix} = s^2 - \frac{\tilde{b}}{q}s - k_2 b$$

is Hurwitzian. This is the case when the inequalities $\tilde{b}/q < 0$, $k_2 < 0$ are satisfied. Note that the stability of system (7.22) increases as $q \to 0$, i.e., the degree of stability can be chosen arbitrarily without any change in the gain factors of the principal C-loop of control. Moreover, (7.14) and (7.18) yield a relation $\nu = -q\mu$, and therefore $\nu \to 0$ as $\mu \to 0$. Thus, the integral OC-controller solves the posed problem concerning the asymptotic elimination of the dynamical statism at an arbitrary rate.

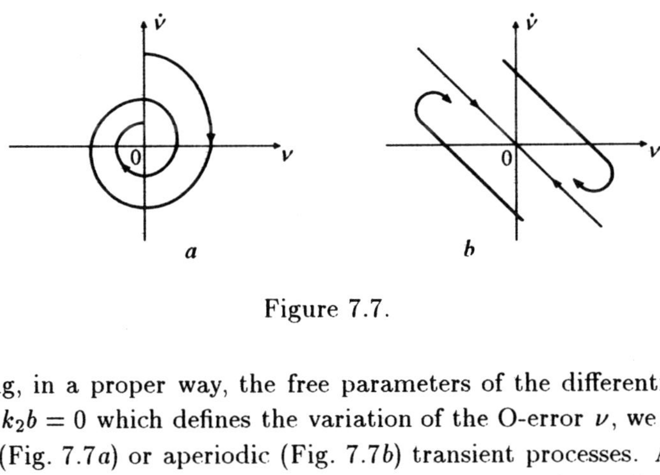

Figure 7.7.

Choosing, in a proper way, the free parameters of the differential equation $\ddot{\nu} - \dot{\nu}\tilde{b}/q + k_2 b = 0$ which defines the variation of the O-error ν, we can achieve oscillatory (Fig. 7.7*a*) or aperiodic (Fig. 7.7*b*) transient processes. Accordingly, the character of transient processes is defined in the original coordinate (x_1, x_2)-space (Fig. 7.8).

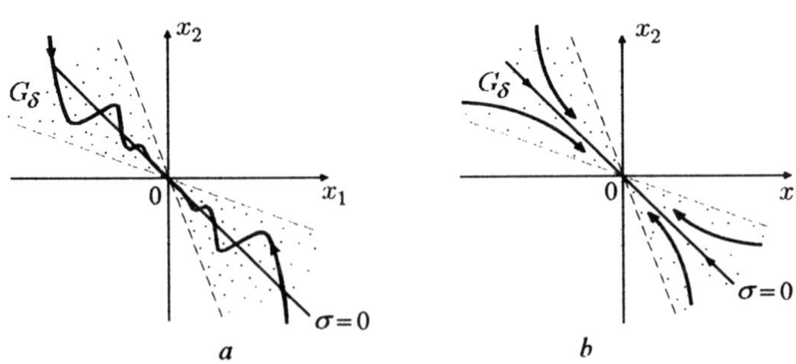

Figure 7.8.

7.5. The Main Properties and Specific Features of Binary Stabilization Systems with Different Types of Feedback

Consider the main properties and peculiarities of binary stabilization systems with different types of feedback.

The difference from the identification approach. In the binary system that we have considered there is not even an indirect effect of identification. The absence of identification follows from the relation

$$\eta(t) \to \eta_\infty = -\frac{a}{b}$$

and distinguishes, in principle, the proposed scheme of control from the adaptation approach.

The effect on which the compensation for uncertainty is based. According to relation (7.15), the control signal

$$u' = u + v = (\mu + \eta)x_1 = \left(1 + \frac{\eta}{\mu}\right)\mu x_1. \tag{7.23}$$

On the other hand, in the sliding mode $\xi = 0$, and this is equivalent, according to formula (7.19), to the relation

$$\sigma_\rho = \dot{x}_1 + dx_1 + \rho x_1 = 0. \tag{7.24}$$

However, $\rho = -q\mu$ (see (7.14)), and therefore we get from (7.24) a relation

$$\mu x_1 = \frac{\dot{x}_1 + dx_1}{q} = \frac{\sigma}{q}.$$

Substituting it into (7.23), we obtain

$$u' = \left(1 + \frac{\eta}{\mu}\right)\frac{\sigma}{q}.$$

According to what we proved above,

$$\eta \to \eta_\infty = -a/b, \quad \mu \to 0,$$

and, hence, $\eta/\mu \to \infty$ as well. In other words,

- an integral OC-controller ensures the effect of an infinite increase in the gain factor of the control error $\sigma = \dot{x}_1 + dx_1$, i.e.,

$$\left(1 + \frac{\eta}{\mu}\right)\frac{1}{q} \to \infty, \quad t \to \infty.$$

This fact elucidates the effect of compensation for the uncertainty factors in terms customarily used in the control theory.

On the robustness of a binary system. By virtue of the use of finite gain factors, the synthesized binary system is robust since the unmodeling dynamics or functional uncertainties lead only to a deviation of the variable ξ from zero, i.e., $|\xi| \leq \Delta$, $\Delta = \text{const} < \delta$, and this, obviously, does not change the asymptotic properties of the system in the original variables (x_1, x_2).

On the stabilization of an object with variable parameters. The variation of the parameter $b \in B$ in time does not change, in principle, the picture described above since, in this case, the motion equations remain the same. However, it should be pointed out that when proving stability, it is preferable to use the second Lyapunov method rather than operator methods and the Laplace transformation, as we did before. Now if the parameter $a \in A$ changes, then all motion equations are preserved but their asymptotic behavior changes.

Thus, for instance, the Σ_i^c-system (7.21)

$$\begin{aligned} q\dot{\mu} &= \tilde{b}\mu + b\eta + a, \\ \dot{\eta} &= qk_2\mu, \end{aligned}$$

which was asymptotically stable, becomes dissipative, i.e., in a finite time its solution attains the invariant ball

$$\mu^2 + (b\eta + a)^2 \leq \text{const}\,.$$

Since $\nu = \rho = -q\mu$, this also implies dissipativity with respect to ν, i.e., $|\nu| \leq \text{const}$. This means that the elimination of dynamical statism is not guaranteed. But this is what we had to expect since the integral law annihilates only constant perturbations, but to annihilate the arbitrary wave perturbation $Ka \equiv 0$ in the control, we must employ the operator K^{-1} which is the inverse of the annihilating operator K.

On the degree of a feedback control system. From the block diagram of a feedback system shown in Fig. 7.6 we can see that the initial degree of the system is four. After the appearance of a sliding mode in the O-controller, it becomes lower by unity and is described by the equations

$$\begin{aligned} q\dot{\mu} &= \tilde{b}\mu + b\eta + a, \\ \dot{\eta} &= qk_2\mu \\ \dot{x}_1 &= -(d - q\mu)x_1. \end{aligned}$$

If we set the degree of stability of the $\tilde{\Sigma}_i^c$-system much higher than the number d, and this can always be done by a proper choice of the control parameters, then we can consider its motions to be fast relative to the main motion

$$\dot{x}_1 = -dx_1, \tag{7.25}$$

and, consequently, in actual fact, the degree of the feedback system is equal to unity. In other words, in the final analysis, a complicated uncertain nonlinear system behaves as the scalar system (7.25).

It should be emphasized, however, that this dynamical lowering of the degree is valid only in asymptotics, and if we accept a discontinuous OC-feedback, then we can achieve the same effect finitely, i.e., in finite time.

7.6. Discontinuous OC-Feedback

It is clear from the preceding consideration that an integral OC-controller eliminates dynamical statism only in asymptotics and does not guarantee this elimination for the variable parameters $a \in A$, $b \in B$. Therefore we consider in this section the following problem:

- to propose methods of finite elimination of dynamical statism which would also be efficient for the variable parameters of the object.

We know from the classical control theory that when there is no information about the character of change of perturbation, we can use a high gain factor and discontinuous (relay or VSS) feedback in order to achieve this aim.

By raising the gain factor, we lower the robustness of the control system, and therefore we shall investigate the potentialities of a discontinuous OC-feedback for solving the formulated problem.

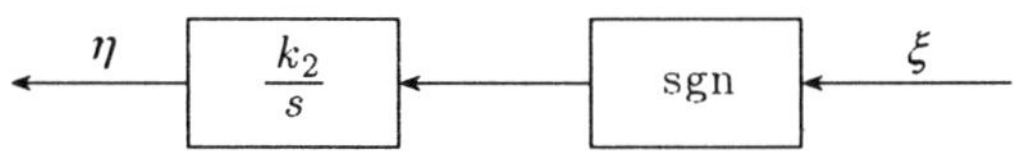

Figure 7.9.

7.6.1. Integral-relay OC-controller

In this case, the equation for the OC-controller has the form

$$\dot{\eta} = -k_2 \operatorname{sgn} \nu, \tag{7.26}$$

which is shown by the block diagram in Fig. 7.9 that elucidates the name of the controller. Equation (7.26), the equation for the variation of the O-variable ξ

$$\dot{\xi} = 2d_\rho \xi + \tilde{b}\mu - k_1 \operatorname{sgn} \xi + b\eta + a, \tag{7.27}$$

and the equation of the CO-controller

$$q\dot{\mu} = k_1 \operatorname{sgn} \xi \tag{7.28}$$

define, with due account of the relation

$$\nu = -q\mu, \tag{7.29}$$

the feedback system of differential equations under investigation. If we complete Eqs. (7.26)–(7.29) by the equation for the variation of the principal variable

$$\dot{x}_1 = -(d + \rho)x_1 + \xi x_1, \tag{7.30}$$

in which use is made of the variables

$$\xi x_1 = \sigma, \quad \sigma = \dot{x}_1 + dx_1, \quad \sigma_\rho = \sigma + \rho x_1 = \dot{x}_1 + d_\rho x_1,$$

then we obtain motion equations for the feedback system of the binary equation for Σ_R^c on the set G_δ. The block diagram of the binary system of control for Σ_R^c is given in Fig. 7.10. It should be pointed out that this system includes two discontinuous elements in the CO- and OC-controllers. Therefore a sliding mode may appear not only in the O-loop as before but also in the OC-loop. It is precisely this plan of compensation for the perturbation that we accept for the further analysis.

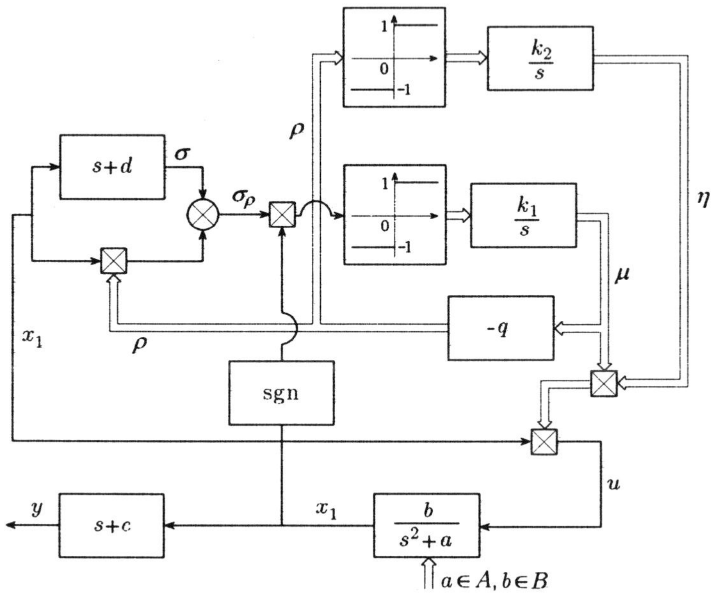

Figure 7.10.

Considering Eqs. (7.26)–(7.29), we see that the equilibrium position of the system in question is at the point

$$\xi_\infty = 0, \quad \mu_\infty = 0, \quad b\eta_\infty = -a.$$

Therefore, analyzing the equation

$$\dot{\xi} = 2d_\rho \xi + \tilde{b}\mu - k_1 \operatorname{sgn} \xi + b\eta + a, \quad \tilde{b} = b - 2qd, \tag{7.31}$$

we make sure, as in all similar cases considered above, that when the condition $k_1 > a^0$ is fulfilled, there exists at the point $\xi = 0$ a sliding mode with the following equivalent value of the discontinuous signal:

$$k_1 \operatorname{sgn}_{\text{eq}} \xi = \tilde{b}\mu + b\eta + a.$$

Substituting the obtained equivalent value into the equation for the CO-controller (7.28), we obtain a $\tilde{\Sigma}_R^c$-system of equations which must be subjected to a further analysis:

$$\begin{aligned} q\dot{\mu} &= \tilde{b}\mu + b\eta + a, \\ \dot{\eta} &= -k_2 \operatorname{sgn} \mu. \end{aligned} \tag{7.32}$$

To analyze this $\tilde{\Sigma}_R^c$-system, it is convenient to set $b =$ const and make a change of variable according to the relation

$$e = b\eta + a,$$

and then its motion equations will assume the form

$$\begin{aligned} q\dot{\mu} &= \tilde{b}\mu + e, \\ \dot{e} &= -k_2 \operatorname{sgn} \mu + \dot{a}, \end{aligned}$$

and the equilibrium position will shift to zero:

$$(\mu = 0, e = 0).$$

If we assume that

$$-k_2 b > \sup_{t \geq 0} |\dot{a}|,$$

then the qualitative behavior of the $\tilde{\Sigma}_R^c$-system and the system described by the equations

$$\begin{aligned} q\dot{\mu} &= \tilde{b}\mu + e, \\ \dot{e} &= -k_2 b \operatorname{sgn} \mu, \end{aligned} \tag{7.33}$$

is the same, and therefore we deal with simpler equations (7.33) assuming, in advance, that precisely these equations define the $\tilde{\Sigma}_R^c$-system.

Recall that $\tilde{\Sigma}_R^c$ is a relay system. We have already considered systems of this kind in detail, and therefore we shall simply use the results of the analysis that we carried out in the first part of the monograph (see Chap. 2). Sewing along the axis $\mu = 0$ the phase trajectories that correspond to the solutions of the systems

$$\begin{aligned} q\dot{\mu} &= \tilde{b}\mu + e, \quad & \dot{e} &= -k_2 b, \\ q\dot{\mu} &= \tilde{b}\mu + e, \quad & \dot{e} &= +k_2 b, \end{aligned}$$

we obtain a phase portrait of the $\tilde{\Sigma}_R^c$-system (Fig. 7.11). From Fig. 7.11c we can see that the trajectories tend to "twist" to zero. To analyze the asymptotic behavior of this motion, we shall use the second Lyapunov method. Note that it suffices to prove that $\mu \to 0$, since $\nu = -q\mu$ and, consequently, $\nu \to 0$ as well.

Let us take the test function

$$v = |\mu| + \frac{e^2}{2k_2 bq}.$$

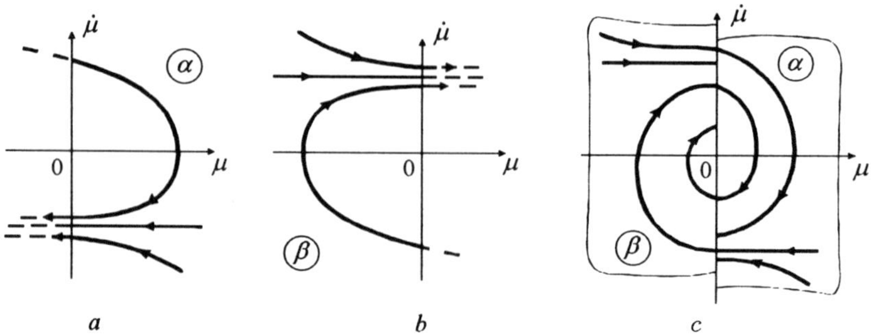

Figure 7.11.

By virtue of the $\tilde{\Sigma}^c_R$-system (7.33), its derivative has the form

$$\dot{v} = \dot{\mu}\operatorname{sgn}\mu + \frac{e}{k_2 q b}\dot{e} = \left(\frac{\tilde{b}}{q}\mu + \frac{e}{q}\right)\operatorname{sgn}\mu - \frac{e}{q}\operatorname{sgn}\mu = \frac{\tilde{b}}{q}|\mu|.$$

Since $|\mu| = v - e^2/2k_2bq$, we successively have

$$\dot{v} = \frac{\tilde{b}}{q}v - \frac{\tilde{b}}{2k_2bq^2}e^2 \le \frac{\tilde{b}}{q}v.$$

This inequality implies the exponential estimate

$$v(t) \le v(0)\exp\left(\frac{\tilde{b}}{q}t\right).$$

Recall that $q < 0$, and, consequently,

- an integral-relay OC-controller can compensate for the unknown and the variable perturbation $a \in A$, but it does it only asymptotically.

More intricate controllers are required to achieve the finite compensation.

7.6.2. Second-degree sliding modes in an OC-loop

We shall consider here the case of a finite stabilization of the error ν of an OC-loop which naturally occurs when we use a discontinuous OC-feedback of the form

$$\dot{\eta} = -k_2\operatorname{sgn}\left(\dot{\nu} + \sqrt{|q\nu|}\operatorname{sgn}\nu\right). \tag{7.34}$$

We omit the reasoning establishing the finiteness of appearance of sliding mode at the point $\xi = 0$ since it is similar to that carried out above and pass to the analysis of sliding equations in the (μ, e)-space or, to put it otherwise, the equations of the $\Sigma_{\mathrm{sl}}^{\mathrm{c}}$-system. For $b = \mathrm{const}$ we have equations

$$\begin{aligned} q\dot{\mu} &= \tilde{b}\mu + e, \\ \dot{e} &= -k_2 b \operatorname{sgn}\left(\dot{\mu} + \sqrt{|\mu|}\operatorname{sgn}\mu\right). \end{aligned} \tag{7.35}$$

When deriving these equations, we took into account that $\nu = -q\mu$.

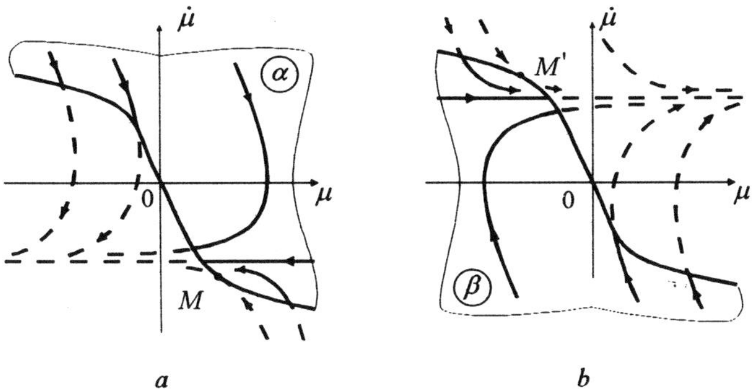

Figure 7.12.

Equations similar to those of the $\Sigma_{\mathrm{sl}}^{\mathrm{c}}$-system were studied in optimal-relay systems, similar equations were considered in Chap. 5 in the process of analyzing higher-order sliding modes. The sewing of the phase trajectories of the Σ_a- (Fig. 7.12a) and Σ_b- (Fig. 7.12b) systems in the space $(\mu, \dot{\mu})$ along the line of discontinuity

$$\xi = \dot{\mu} + \sqrt{|\mu|}\operatorname{sgn}\mu = 0 \tag{7.36}$$

gives the original phase portrait of the $\Sigma_{\mathrm{sl}}^{\mathrm{c}}$-system (Fig. 7.13). These figures show that there exists a segment MM' of the "parabola" $\xi = \dot{\mu} + \sqrt{|\mu|}\operatorname{sgn}\mu = 0$

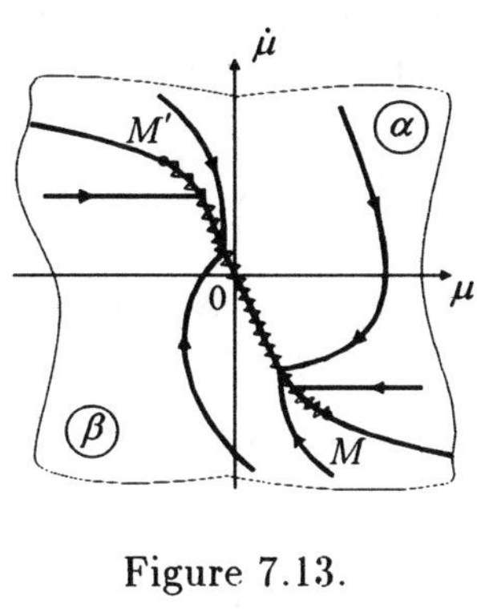

Figure 7.13.

in whose neighborhood the phase trajectories are directed towards one another, i.e., a sliding mode with the motion equation $\dot{\mu} = -\sqrt{|\mu|}\,\mathrm{sgn}\,\mu$ appears on this segment.

The exact analytic solution of this equation has a simple form $\sqrt{\mu(t)} = \sqrt{\mu(0)} - t/2$ which implies the finiteness of the attainment of zero at $t = 2\sqrt{\mu(0)}$.

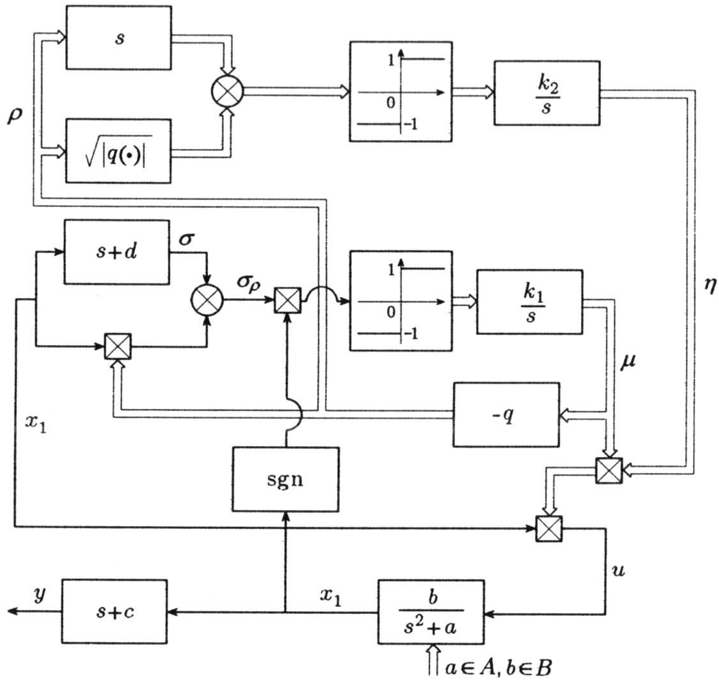

Figure 7.14.

The block diagram of the binary system under investigation is shown in Fig. 7.14 and is rather complicated. However, the finite stabilization of the OC-error ν can be ensured by a simpler OC-controller which realizes a switching mode. Let us consider this controller in greater detail.

The equation of the OC-controller has the form

$$\dot{\eta} = -k_2(\nu, \dot{\nu})\,\mathrm{sgn}\,\nu = \begin{cases} -k_2'\,\mathrm{sgn}\,\nu, & \nu\dot{\nu} \geq 0, \\ -k_2''\,\mathrm{sgn}\,\nu, & \nu\dot{\nu} < 0, \end{cases} \tag{7.37}$$

where k_2' and k_2'' are positive constants. We again omit the standard reasoning and pass for $b = \text{const}$ to the analysis of the $\tilde{\Sigma}^{c}_{\mathrm{sl}}$-system

$$\begin{aligned} q\dot{\mu} &= \tilde{b}\mu + e, \\ \dot{e} &= -k_2(\mu, \dot{\mu})b. \end{aligned} \tag{7.38}$$

The phase portrait of the $\tilde{\Sigma}^{c}_{\mathrm{sl}}$-system (7.38) can be obtained by "sewing" the phase portraits of the Σ^a- and Σ^b-systems along the coordinate axes, each

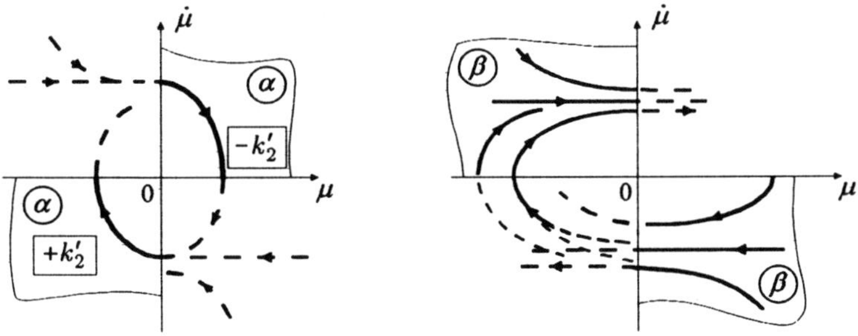

Figure 7.15.

of these systems acting in "its own" part of the plane (Fig. 7.15), This phase portrait is shown in Fig. 7.16. We established above that for $k_2'/k_2'' > 1$ the "twisting" of the trajectories to zero occurred in a finite time, and this proves the finiteness of stabilization of the error ν.

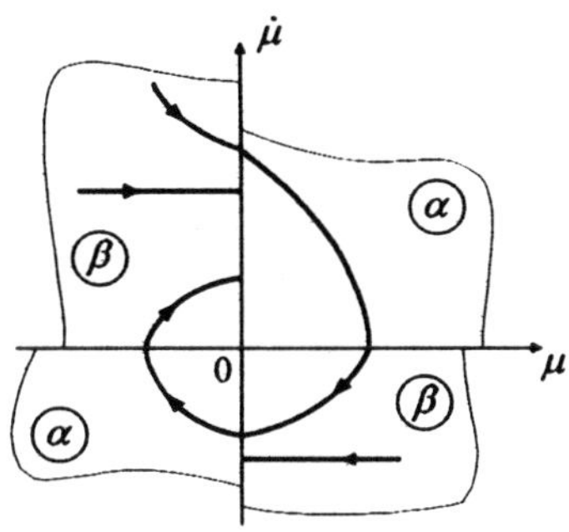

Figure 7.16.

Note now that if we define two numbers α and β by the relations

$$\beta + \alpha = k_2', \quad \beta - \alpha = k_2'',$$

then we can write the equation for the OC-controller used above in the form

$$\dot{\eta} = -\alpha \operatorname{sgn} \dot{\nu} - \beta \operatorname{sgn} \nu.$$

This allows us to represent the block diagram in the form shown in Fig. 7.17. Comparing it to the diagram in Fig. 7.14, we see that for the same quality of control we have achieved a considerable simplification of the structure of the nonlinear controller. In the OC-controllers that we have considered which ensure a finite elimination of the dynamical statism, we assumed that use should be made of the derivative $\dot{\nu}$ or, what is the same in this case, the derivative $\dot{\mu}$. By the hypothesis

$$\dot{\mu} = -k_1 \operatorname{sgn} \xi$$

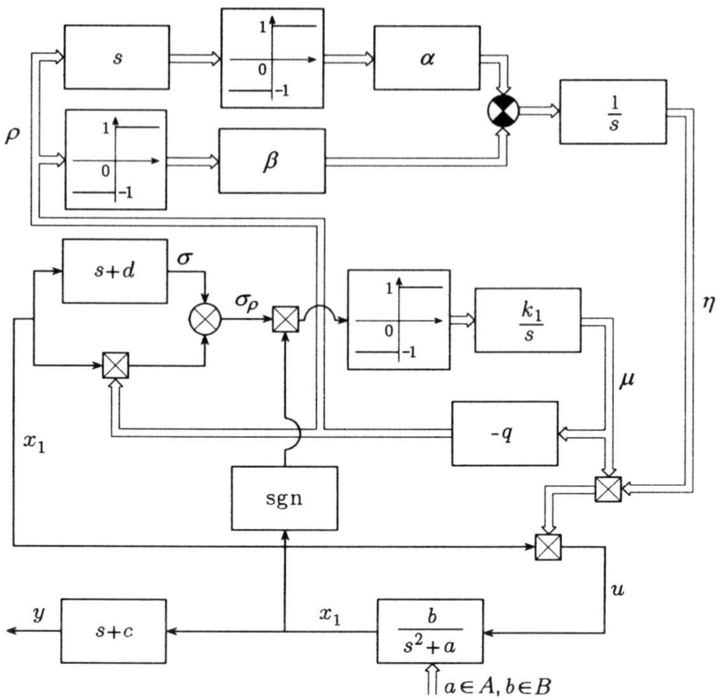

Figure 7.17.

and is discontinuous, and therefore we spoke, in essence, about using not $\dot{\mu}$ but a certain coinciding with it but differentiable function $\dot{\mu}_{\rm eq}$. To obtain the function $\dot{\mu}_{\rm eq}$, it suffices to use some averaging operation, say, in the form of a sliding mean (Fig. 7.18*a*) or an inertial link with sufficiently small time constant $\tau > 0$ (Fig. 7.18*b*). However, if we use in the OC-loop a second-degree sliding mode (described in detail in Chap. 5) which does not use information about the derivative $\dot{\mu}_{\rm eq}$, then we can do without solving the related problems. We omit the details.

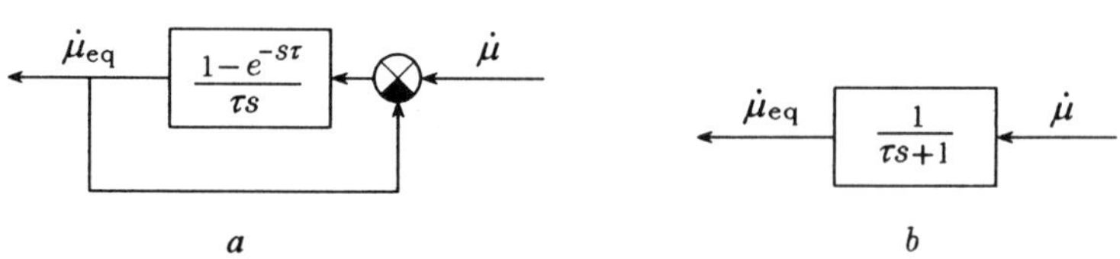

Figure 7.18.

The investigation that we have carried out allows us to make the following essential inference:

- when using the properly organized discontinuous OC-feedback in conjunction with three other types of feedback, we can achieve a finite independence of motion in a parametrically uncertain system without a sliding mode in the main loop of controller for the finite transfer factors in each loop of feedback as well.

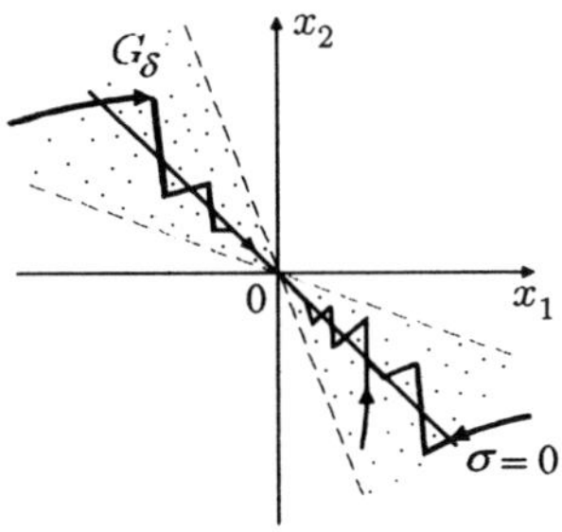

Figure 7.19.

Figure 7.19 shows typical projections of phase trajectories of the binary stabilization system on the set G_δ synthesized in this way. A question naturally arises concerning the physical basis of this effect of compensation for the perturbation. It is also important to work out definite rules of action of feedbacks outside of the set G_δ, and then we shall be able to speak about solving the problem of global stabilization. We shall deal with all this in the next chapter.

Chapter 8

Constraints, Physical Foundations of the Compensation for Disturbances, and Stabilization of Forced Motion in Binary Systems

Up till now, when synthesizing stabilization systems, we studied the motion only on the set $G_\delta = \{x \mid |\sigma| \le \delta |x_1|\}$ (Fig. 8.1) and considered the number $\delta > 0$ to be so small that we were able to use only the first approximation equations when analyzing the systems. In this connection, the following *first* question is relevant.

- What is the right way to modify the algorithm of binary control so that it could be used under any initial conditions? It stands to reason that all its stabilizing qualities must be preserved.

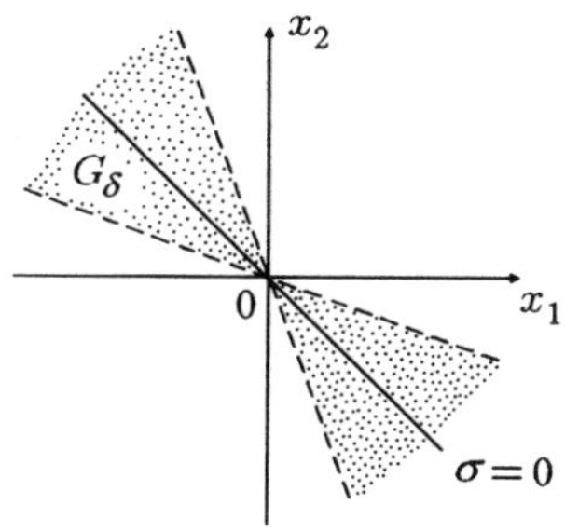

Figure 8.1.

A significant thesis, which was one of the main theses underlying the development of the binary control theory, is that concerning the *boundedness of the gain factor and continuity of control signals*, first of all, in the principal C-loop of control. Therefore the following *second* question arises.

- How can these fundamental requirements be reflected in the binary control algorithm and what secondary effects will appear as a result?

The following, *third*, question which will be answered in the sequel is

- what is the physical foundation of the effect compensating for the uncertainty in binary systems?

We know that in linear systems this foundation is an unlimited increase in the feedback gain factor and in continuous systems (VSS, relay systems) it is a sliding mode. However, there may be no sliding mode in binary systems and gain factors may be finite, and therefore the third question that we have formulated arises. Finally, the *fourth* question should be considered.

- What are the changes that must be introduced into a binary feedback in order to extend its stabilizing properties to problems of control of a forced motion when there is no detailed information about the coordinate disturbance?

We mean here the problem of stabilization of the object

$$\begin{aligned} \dot{x}_1 &= x_2, \\ \dot{x}_2 &= ax_1 + bu + f, \quad a \in A, \ b \in B, \ f \in F. \end{aligned}$$

8.1. Constraints Imposed on the Operator Variable

We begin the consideration from the second question. In synthesized binary systems the control signal is defined by the expression $u = \mu x_1$ or $u' = (\mu + \eta) x_1$. It is obvious that μ or $(\mu + \eta)$ can be interpreted as the gain factor of C-controller with respect to the variable x_1. Therefore, for the reasons of robustness, the O-variables μ and η must be bounded, i.e.,

$$|\mu(t)| \leq \text{const}, \quad |\eta(t)| \leq \text{const}$$

for all $t > 0$.

In practice, it is convenient to deal with constraints of the form

$$|\mu| \leq 1, \quad |\eta| \leq 1,$$

but actually $\mu(t)$ and $\eta(t)$ are output variables of a certain dynamical system (of the CO-controller and OC-controller respectively), and the problem of boundedness of such a variable is not at all trivial but requires a thorough analysis. By the way, a seemingly obvious method of restriction with the use of a saturator does not serve this purpose.

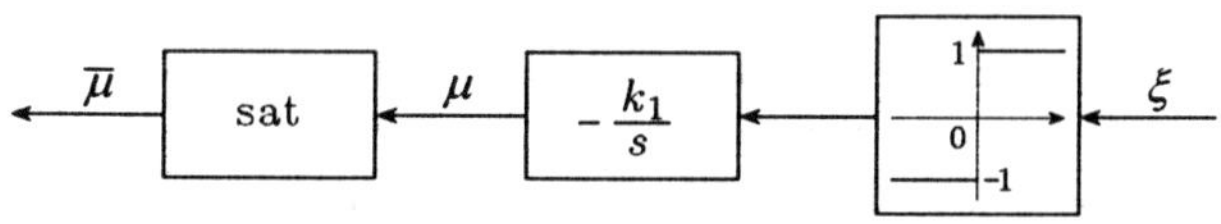

Figure 8.2.

Indeed, a possible block diagram corresponding to this proposition is shown in Fig. 8.2. The equations of the system represented in Fig. 8.2 have the form

$$\dot{\mu} = -k_1 \operatorname{sgn} \xi, \quad \bar{\mu} = \operatorname{sat}(\mu) = \begin{cases} \operatorname{sgn} \mu, & |\mu| > 1, \\ \mu, & |\mu| \leq 1. \end{cases} \tag{8.1}$$

Now the equation for the change of the O-variable ξ must include the variable $\bar{\mu}$ and not μ, as before, since $u = \bar{\mu}x$, and therefore

$$\dot{\xi} = 2d\xi + b\bar{\mu} + a. \tag{8.2}$$

As a result of the differentiation of $\bar{\mu}$, we can see from (8.1) that $\dot{\bar{\mu}} = \dot{\mu}$ only for $|\mu| < 1$, and $\dot{\bar{\mu}} = 0$ in the other cases. This changes qualitatively the problem of stabilization of the object (8.2) by the control $\bar{\mu}$ and has negative consequences since for $|\mu| > 1$ this object is uncontrollable. Therefore we must look for more "intricate" solutions of the problem concerning the restriction of the operator variable μ.

Let us consider some techniques of restricting the O-variable μ which preserve the controllability of the object $\dot{\xi} = 2d\xi + b\mu + a$.

Technique I. We set

$$\dot{\mu} = \begin{cases} -\omega\mu, & |\mu| > 1, \\ -k \operatorname{sgn} \xi, & |\mu| \leq 1, \end{cases} \tag{8.3}$$

where ω, k are positive constants. If $\xi > 0$ and $|\mu| < 1$, then $\dot{\mu} = -k$ and μ decreases linearly up to the value $\mu = -1$; on passing through $\mu = -1$, the equation for the variation of μ changes to $\dot{\mu} = -\omega\mu$, and the variable μ increases and cuts transversally the straight line $\mu = -1$ (Fig. 8.3). Thus, a sliding mode

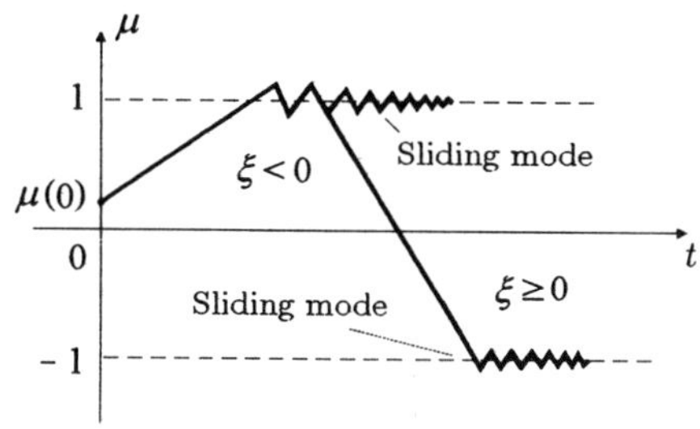

Figure 8.3.

appears on the lines $|\mu| = 1$, and this guarantees the fulfilment of constraint $|\mu| \leq 1$, perhaps, from some time moment.

The advantage of this technique is that the controllability of the object is not lost when the restriction is achieved since upon the change of sign of ξ the variable μ immediately changes the sign too, and this is what we had to achieve. Note that we can write the law (8.3) as one relation if we introduce a function

$$\varphi(\mu) = (1 - |\operatorname{sat}(\mu)|)/(1 - |\mu|), \tag{8.4}$$

(its graph is given in Fig. 8.4) and a binary operation $\beta(\varphi, \xi) = \varphi(\mu) \operatorname{sgn} \xi$.

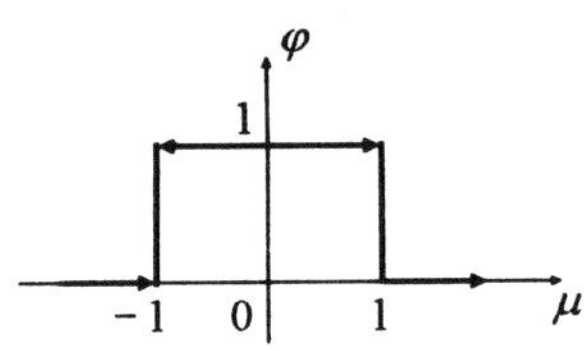

Figure 8.4.

In the new notation the law (8.3) is equivalent to the more simple law

$$\dot{\mu} = -k\beta(\varphi, \xi). \tag{8.5}$$

Technique II. Note that the function φ from (8.4) can be represented by the difference

$$\varphi(\mu) = [\operatorname{sg}(1 + \mu) - \operatorname{sg}(\mu - 1)]/2, \tag{8.6}$$

where $\operatorname{sg}(\xi)$ is a unity jump function (its graph is given in Fig. 8.5*a*). Then, as a result of the subtraction of the graphs in Figs. 8.5*b* and 8.5*c*, we get the required graph shown in Fig. 8.4. Substituting (8.6) into (8.5), we obtain

$$\dot{\mu} = -k \operatorname{sg}(1 + \mu) \operatorname{sgn} \xi + k \operatorname{sg}(\mu - 1) \operatorname{sgn} \xi, \quad k = \text{const} > 0. \tag{8.7}$$

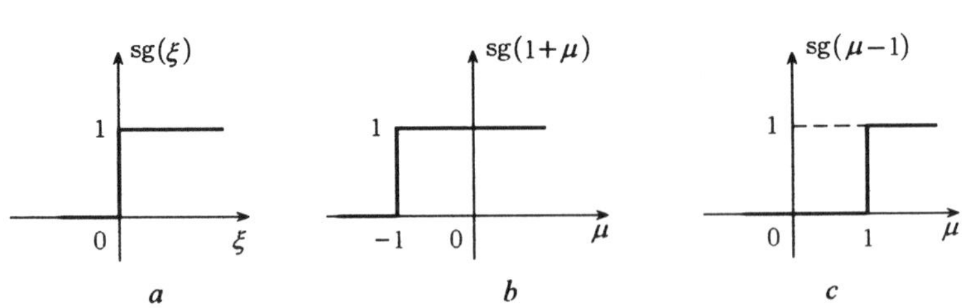

Figure 8.5.

Taking into account the sign multiplication, we can realize relation (8.7) without a binary operation but only by using two channels of signal propagation (Fig. 8.6). This scheme operates as follows. If $\xi > 0$, then $-\operatorname{sgn} \xi = -1$, and

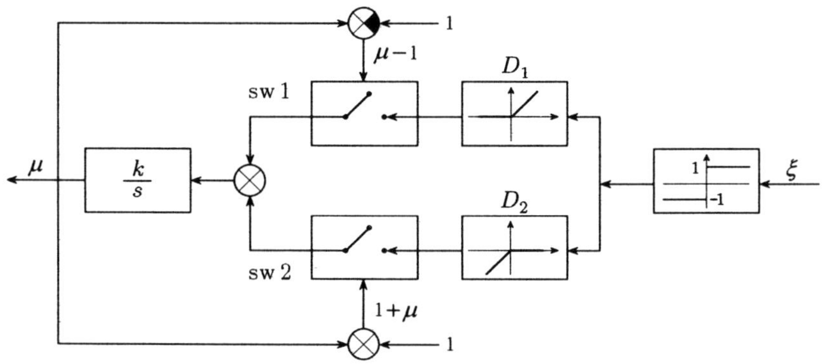

Figure 8.6.

out of the two diodes D_1 and D_2, only D_2 transmits the signal. If, in this case, $\mu + 1 > 0$, then the switch sw2 is closed and $\dot{\mu} = -k$, and this is what was required. This goes on until μ attains the value $\mu = -1$, when sw2 opens, and then $\mu = -1$ when $\xi > 0$, or until ξ changes sign. Then D_1 and not D_2 transmits the signal, i.e., the upper circuit operates, and if $1-\mu > 0$, then the switch sw1 is closed and $\dot{\mu} = k$, and this is what must happen. As soon as μ becomes equal to unity, $\mu = 1$, unless ξ changes sign, sw1 will open and fix $\mu = 1$. The advantage of this scheme is that no sliding mode arises in it on $|\mu| = 1$.

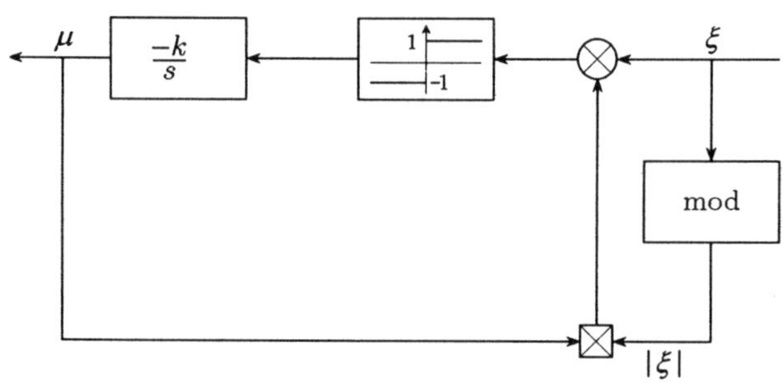

Figure 8.7.

Technique III. Perhaps the most compact notation for the operator-constraint is given by the following relation with the simplest binary operation:

$$\dot{\mu} = -k \operatorname{sgn}\big(\xi + \mu|\xi|\big), \quad k = \text{const} > 0. \tag{8.8}$$

This expression is associated with the block diagram in Fig. 8.7. Here is the principle of operation of this device. As long as $|\mu| < 1$, the condition $\operatorname{sgn}(\xi + \mu|\xi|) = \operatorname{sgn}\xi$ is fulfilled. As long as $|\mu| > 1$, the condition $\operatorname{sgn}(\xi + \mu|\xi|) = \operatorname{sgn}\mu$ is fulfilled. Thus, $|\mu| < 1$ and $\dot{\mu} = -k \operatorname{sgn}\xi$ in the lower zone and $\dot{\mu} = -k \operatorname{sgn}\mu$ outside of this zone. The latter relation ensures the fulfilment of the constraint $|\mu| \leq 1$ in the sliding mode on the boundaries of $|\mu| = 1$ and the immediate

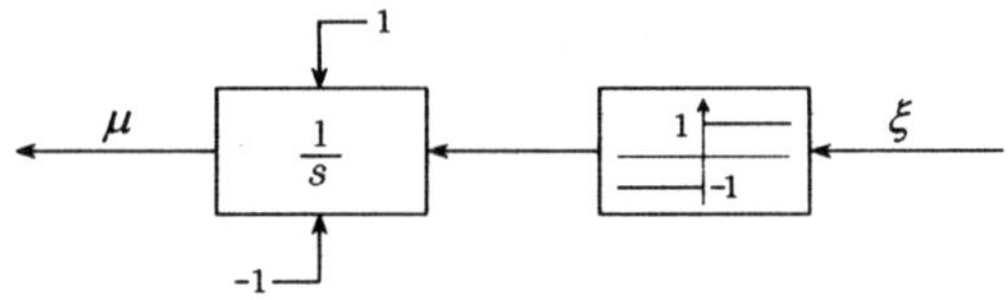

Figure 8.8.

getting off the boundaries of $|\mu| = 1$ when ξ changes sign, and this is precisely what is needed for controllability.

Thus, the operator with a controlled constraint on the output described above is a complicated nonlinear dynamical element. The unified designation of elements of this type is given in Fig. 8.8. Note that when the inertial-relay CO-controller $q\dot{\mu} = b\mu - k\,\mathrm{sgn}\,\xi$ is used, the constraint $|\mu| \leq$ const is automatically fulfilled. However, if we use an ordinary inertial element $q\dot{\mu} = b\mu - k\xi$, then a constraint on the output is necessary.

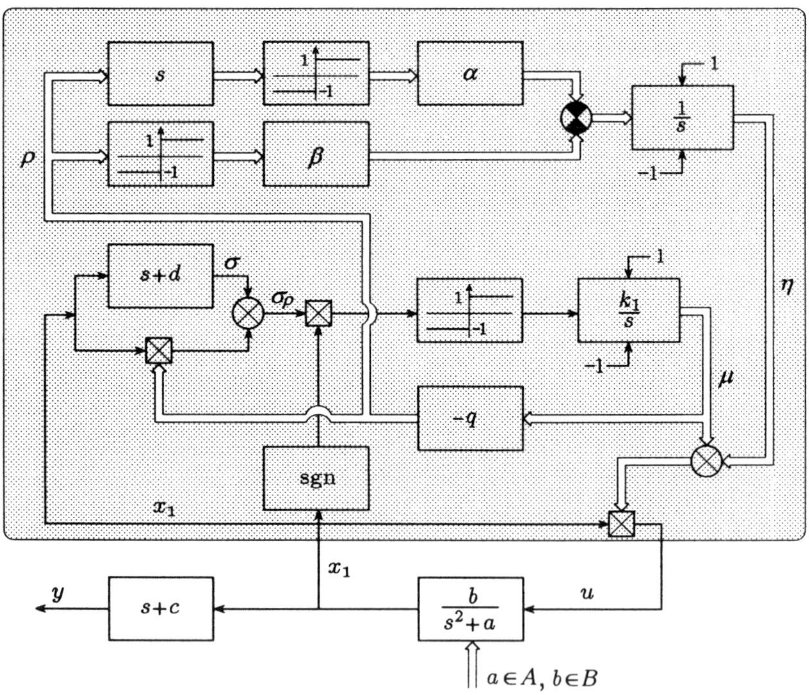

Figure 8.9.

To conclude this section, we give a resulting structure of a binary system (Figs. 8.9, 8.10) with all types of feedback and with constraints. Note that the resulting controller R_u is very complicated and it is not so easy to guess it.

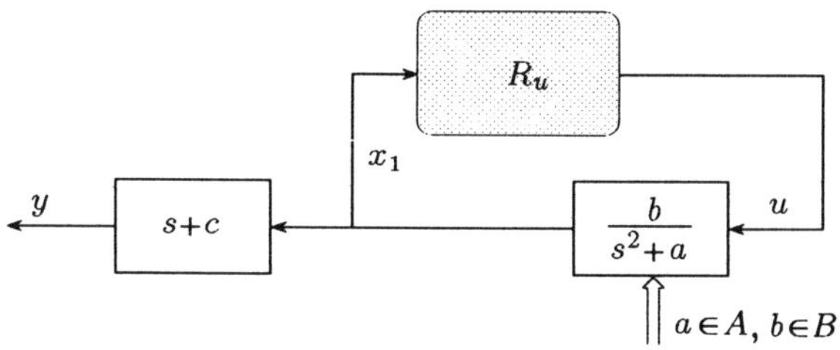

Figure 8.10.

8.2. On the Global Behavior of a Binary System

In this section we shall answer the first of the questions put at the beginning of this chapter, namely, the question concerning motion equations and the properties of a synthesized binary system under arbitrary initial conditions with respect to x, i.e., including those which do not belong to the set G_δ (Fig. 8.11). Let us consider, for definiteness, a binary system with C- and CO-feedbacks of the form

$$u = k\mu x_1, \quad \dot{\mu} = -k_1 \operatorname{sgn}\left[\xi + \mu|\xi|\right]. \tag{8.9}$$

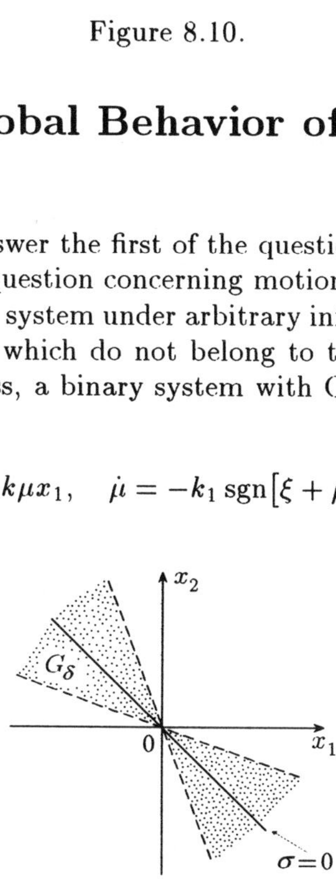

Figure 8.11.

If other types of feedback are also used, then the corresponding investigation is even simpler.

It follows from (8.9) that the variables μ and u are bounded:

$$|\mu| \leq 1, \quad |u| \leq k|x_1|.$$

Moreover, if the action of Eqs. (8.9) is extended beyond the limits of the set G_δ, and in order to do this it suffices to set

$$\xi = \operatorname{sat}\left(\frac{\sigma}{\delta x_1}\right)$$

instead of the standard

$$\xi = \sigma/x_1,$$

then, when the phase point is outside of the set G_δ, the variable μ assumes the value

$$\mu = -\operatorname{sgn}(\sigma x_1) \tag{8.10}$$

in a time internal smaller than $2/k_1$. Substituting (8.10) into the first relation of (8.9), we find that the control assumes the form $u = -k|x_1|\operatorname{sgn}\sigma$ which coincides with the feedback of VSS. And since this is the case, the phase trajectories are the same as in VSS.

Thus, if the parameter k in the VSS is chosen such that the phase point falls on the line of discontinuity $\sigma = 0$, then this also occurs in the binary system, i.e., the set G_δ is attracting. The necessary and sufficient condition for the phase point to fall on the line of discontinuity is the absence of real positive zeros of the characteristic polynomial of the system

$$\dot{x}_1 = x_2,$$
$$\dot{x}_2 = ax_1 - bkx_1$$

for any fixed $a \in A$, $b \in B$, and this is obviously valid when

$$kb^- > a^0. \tag{8.11}$$

Thus, when condition (8.11) is fulfilled, the set G_δ is attracting. If, in addition, we make it invariant (i.e., if all the motions that begin in it do not leave it), then G_δ will be an attractor, and the whole analysis from the preceding chapters will be suitable for the investigation of this situation. However, in the binary system (of the C-feedback + CO-feedback) in question the set G_δ is not an attractor but only conditionally invariant, namely, if, inside G_δ, there exists, for instance, a set of the same configuration $G_{\delta'}$ $(0 < \delta' < \delta)$ such that the motions that begin in $G_{\delta'}$ do not leave G_δ, then the latter is a conditionally invariant set. If, in addition, it is an attracting set, then we speak about a conditional attractor (Fig. 8.12). In Fig. 8.12a the set G_δ is an attractor and in Fig. 8.12b it is a conditional $G_{\delta'}$-attractor.

Since the straight line $\sigma = 0$ is also attracting when condition (8.11) is fulfilled, then it is most convenient to take it as $G_{\delta'}$. In order to obtain relationships between the coefficients of the system that would guarantee the σ-conditional invariance of the set G_δ, we shall consider the following motion equation with respect to the variable $\xi = \sigma/x_1$ which is known from the preceding chapters:

$$\dot{\xi} = 2d\xi + kb\mu + a_*, \quad a_* = a - d^2, \quad k = \text{const}, \tag{8.12}$$

and the equation of the CO-controller

$$\dot{\mu} = -k_1 \operatorname{sgn}(\xi + \mu|\xi|), \quad k_1 = \text{const}\,.$$

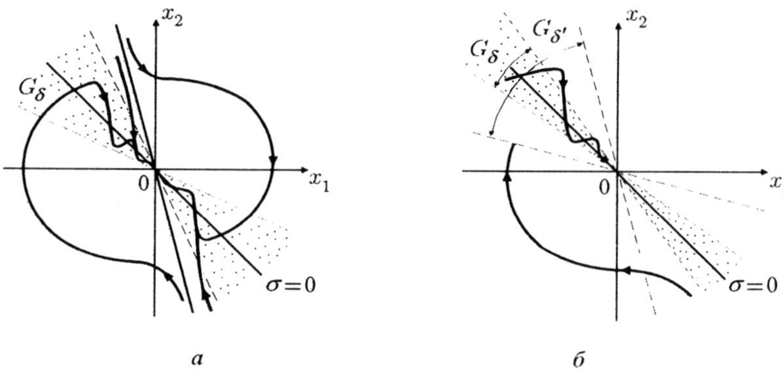

Figure 8.12.

It is clear that the set G_δ is σ-conditionally invariant if every motion that starts from the line $\sigma = 0$ (i.e., for $\xi = 0$) does not attain the boundary of G_δ (i.e., $|\xi| = \delta$).

From Eq. (8.12) we have in G_δ an estimate

$$|\dot{\xi}| \leq 2d\delta + a_*^0 + kb^+, \quad a_*^0 = a^0 - d^2,$$

and therefore, in time t, the variable ξ "grows" not more than by the value

$$|\xi| \leq (2d\delta + a_*^0 + kb^+)t.$$

The required condition $|\xi| \leq \delta$ yields the following estimate for the time of "growth":

$$t \leq \frac{\delta}{2d\delta + a_*^0 + kb^+}. \tag{8.13}$$

However, the growth is sure to terminate when the variable μ assumes its extreme value $\mu = -\operatorname{sgn}\xi$ before the variable $|\xi|$ assumes the value δ and, in addition, the inequality

$$2d\delta - kb^- + a_*^0 < 0, \quad a_*^0 = a^0 - d^2, \tag{8.14}$$

is satisfied.

The maximal time needed for the change of the variable μ from one extreme value to the other is less than $2/k_1$. Therefore the last inequality and (8.13) give the following final estimate for the parameter k_1:

$$k_1 \geq \frac{2}{\delta}(2d\delta + a_*^0 + kb^+), \tag{8.15}$$

where the parameter k is determined from (8.14) with the use of the inequality

$$k \geq \frac{2d\delta + a_*^0}{b^-}. \tag{8.16}$$

Relations (8.15) and (8.16) completely define the parameters of the C- and CO-feedbacks since (8.16) guarantees the attraction for the line $\sigma = 0$. It should also be pointed out that when we use the O-feedback, we can reduce the lower estimate of the factor k_1 by half.

8.3. Physical Foundations of the Compensation for Uncertainty

We shall now answer the third question concerning the physical foundations of the compensation for uncertainty.

Note, first of all, the obvious fact that every disturbance can be interpreted as a force whose effect can be compensated only by the effect of another force equivalent to it. For instance, for the direct compensation for the disturbance f in the control

$$\dot{\sigma} = d\sigma + bu + f$$

the control is formed as the sum

$$u = u_x + u_f,$$

where the component $u_f = -f/b$ is a compensating force and the component u_x is a stabilizing force.

When the disturbance f is unknown, use can be made of a high gain factor in the feedback $u = -k\sigma$, and then, in the feedback described by the equation

$$\frac{1}{k}\dot{\sigma} = -\left(b + \frac{d}{k}\right)\sigma + \frac{f}{k},$$

the compensation is achieved as $k \to \infty$ when the reduced disturbance $(f/k) \to 0$. In this case, the compensating force is formed as a result of the multiplication of a high k by a small signal σ.

The relay control

$$u = -u^0 \operatorname{sgn} \sigma, \quad u^0 = \text{const}$$

damps, also by the method of force, the action of the disturbance f:

$$\dot{\sigma} = d\sigma - u^0 b \operatorname{sgn} \sigma + f,$$

which, however, must now be uniformly bounded:

$$|f| < u^0.$$

In this case, there exists a sliding mode at the zero of the variable σ and the mean value of the discontinuous signal coincides, with an accuracy to within the sign, with the component u_f in the direct compensation method

$$u^0 \operatorname{sgn}_{\text{eq}} \sigma = f/b.$$

Thus, in all the indicated cases

- the idea of the force compensation for the disturbance is realized with the aid of a negative feedback.

In binary systems (and, of course, in VSS as well) the compensation can be achieved by other means. Indeed, in this case,

$$u = k\mu x_1,$$

and the product $k\mu$ plays the part of the gain factor of feedback. However, if we use a relay CO-controller

$$\mu = -\operatorname{sgn}\xi$$

or an integral-relay CO-controller

$$\dot{\mu} = -k_1 \operatorname{sgn}(\sigma + \mu|\sigma|),$$

then the variable μ varies from -1 to $+1$ and, hence, the feedback factor k_μ of the principal loop also changes sign. And this means that an alternating feedback is used in the principal loop and the behavior of the system can be described by the Σ^+-equations

$$\begin{aligned} \dot{x}_1 &= x_2, \\ \dot{x}_2 &= ax_1 + kbx_1, \end{aligned}$$

or by the Σ^--equations

$$\begin{aligned} \dot{x}_1 &= x_2, \\ \dot{x}_2 &= ax_1 - kbx_1. \end{aligned}$$

Since $kb^- > a^0$, the Σ^+-system is always unstable, and this is essential. Namely,

- the use of unstable structures makes it possible to increase, in finite time, in a system with bounded parameters, the norm of the phase vector to a value sufficient for damping the disturbance.

We already know that the use of discontinuous feedback for this purpose is not at all inevitable since any, not necessarily linear, control laws which change the sign of feedback are suitable.

8.4. On the Compensation for the Coordinate Disturbance

Let us now consider the technique of solving the fourth problem posed at the beginning of the chapter, namely, the problem concerning the algorithms of stabilization of forced motion.

Up till now we stipulated that $f(t) \equiv 0$. Suppose now that this is not the fact, namely, that there exists an additional term, i.e.,

$$\dot{\sigma} = d\sigma + bu + a^* x_1 + f, \quad a^* = a - d^2, \quad b \in B, \quad a \in A,$$

in the equation for the variation of the error in the CO-loop σ.

The presence of the term $f(t)$ on the right-hand side of the equation changes the situation qualitatively. Since the disturbance $f(t)$ is assumed to be known with an accuracy to within the inclusion

$$f \in F = \{f \mid |f| \leq f_M\},$$

where, in addition, the parameter $b \in B$ is unknown, it cannot be included into the parametric disturbance or eliminated by the well-known compensation methods, i.e., by a direct or an indirect measurement. Therefore the question arises as to the changes which must be introduced into the design of binary control systems in order to endow them with the stabilization property in the case of coordinate disturbance.

Note that for the unknown $b \in B$ the problem remains unsolved even for the known $f(t)$ since, in this situation, the standard stabilization methods are inapplicable. If the wave model of the disturbance f is known, i.e., if the annihilating operator K_f is known:

$$K_f f \equiv 0,$$

then, by the change of control

$$u = K_f^{-1} v$$

we can reduce the problem of stabilization of forced motion to the problem of stabilization of free oscillations since

$$\dot{\sigma} = d\sigma + bK_f^{-1} v + a^* x_1 + f,$$

and, by introducing the new variables

$$\tilde{\sigma} = K_f \sigma, \quad \tilde{x}_1 = K_f x_1$$

we obtain the standard stabilization problem

$$\dot{\tilde{\sigma}} = d\tilde{\sigma} + bv + a^* \tilde{x}_1.$$

In these calculations we assumed, for simplicity and without loss of generality, that $b = \text{const}$. Thus,

- the most general and difficult situation arises when only the fact of inclusion $f \in F$ is known as concerns the disturbance $f(t)$ and the parameters of the object are unknown, i.e., $a \in A$, $b \in B$.

In what follows, we shall deal precisely with this situation.

Suppose that the coordinate disturbance satisfies the matching condition. Then, using the standard change of variable

$$x_2 = \sigma - dx_1,$$

we reduce the initial problem of stabilization of the object P

$$\begin{aligned} \dot{x}_1 &= x_2, \\ \dot{x}_2 &= ax_1 + bu + f \end{aligned}$$

to the problem of stabilization of the scalar object P_σ:

$$\dot{\sigma} = d\sigma + bu + a_* x_1 + f, \quad a_* = a - d^2.$$

As usual, when controlling a forced motion, we choose the control as the sum

$$u = u_x + u_f, \tag{8.17}$$

where the component u_x stabilizes the free oscillations of the object P_1:

$$\dot{\sigma} = d\sigma + bu_x + a_* x_1,$$

and the component u_f stabilizes the forced oscillations of the object P_2:

$$\dot{\sigma} = bu_f + f.$$

As before, we assume that

$$a \in A, \quad b \in B, \quad f \in F.$$

When we use discontinuous controls, the stabilization of P_1- and P_2-objects implies the stabilization of the P_σ-object. Indeed, the discontinuous stabilizing controls u_x, u_f obviously satisfy the conditions

$$\sigma(d\sigma + bu_x + a_* x_1) \leq 0, \tag{8.18}$$
$$\sigma(bu_f + f) \leq 0. \tag{8.19}$$

Since this is the fact, their sum satisfies a similar condition

$$\sigma(d\sigma + b(u_x + u_f) + a_* x_1 + f) \leq 0,$$

which is equivalent to the inequality

$$\sigma\dot{\sigma} < 0$$

that implies the stabilization of the P_σ-object at zero.

The laws of control which resolve inequalities (8.18), (8.19) for any $a \in A$, $b \in B$, and $f \in F$ obviously have the form

$$u_x = -k|x_1| \operatorname{sgn}\sigma, \tag{8.20}$$
$$u_f = -lf_M \operatorname{sgn}\sigma, \tag{8.21}$$

where the constants k, l satisfy the relations

$$\begin{aligned} kb^- &> a_*^0 = a^0 + d^2, \\ lb^- &> 1. \end{aligned} \tag{8.22}$$

The control algorithms (8.20), (8.21) are standard for the VSS theory.

When passing from discontinuous stabilization algorithms to continuous ones, we can do the following. Using the standard notation

$$\mu = -\operatorname{sgn}(x_1\sigma) = -\operatorname{sgn}\xi, \quad \xi = \sigma/x_1, \tag{8.23}$$

we can write the stabilization algorithms for control (8.20), (8.21) in the form

$$u = u_x + u_f = (kx_1 + lf_M \operatorname{sgn} x_1)\mu. \tag{8.24}$$

Let us now take relation (8.24) as the basis when synthesizing a continuous law, but instead of a relay CO-controller we shall employ, for instance, an integral-relay CO-controller

$$\dot{\mu} = -k_1^0 \operatorname{sgn}(\xi + \mu|\xi|), \tag{8.25}$$

where k_1^0, as we shall see later, is no longer a coefficient but a function dependent on the phase vector and the disturbance majorant.

To find the function k_1^0 which guarantees the σ-conditional invariance of the set G_δ, i.e., the satisfaction of the inequality $|\xi| \le \delta$ if $\xi(0) = 0$, we shall use the equation for the variation of the O-variable ξ in the set G_δ. This equation can be derived in a standard way and has the form

$$\dot{\xi} = 2d\xi + b\frac{u}{x_1} + a_* + \frac{f}{x_1}.$$

Since, according to (8.24), we have

$$\frac{u}{x_1} = \left(k + \frac{lf_M}{|x_1|}\right)\mu,$$

we can write the required equation in the final form as

$$\dot{\xi} = 2d\xi + b\left(k + \frac{lf_M}{|x_1|}\right)\mu + a_* + \frac{f_1}{x_1}. \tag{8.26}$$

From (8.24) we have $|\mu| \le 1$ and, consequently, the estimate

$$|\dot{\xi}| \le 2d\delta + b^+\left(k + \frac{lf_M}{|x_1|}\right) + a_*^0 + \frac{f_M}{|x_1|}$$

is valid for the right-hand side of Eq. (8.26) in G_δ.

During the time t the variable ξ increases from zero ($\xi(0) = 0$) by not more than the value $|\xi_{\max}|$ (we can assume x_1 to be constant since the motion is in G_δ and δ and t are small), where

$$|\xi_{\max}| = \left[2d\delta + b^+k + a_*^0 + \frac{b^+l + 1}{|x_1|}f_M\right]t. \tag{8.27}$$

The obtained value will not exceed the admissible limit δ if, in time t, according to (8.22), the O-variable μ is sure to assume the value $\mu = -\operatorname{sgn}\xi$. However, this may happen only when

$$t \geq 2/k_1^0. \tag{8.28}$$

From relations (8.27), (8.28) and the inequality $|\xi_{\max}| \leq \delta$ we obtain a lower estimate for the function k_1^0:

$$k_1^0 \geq \left[2d\delta + b^+k + a^0 + \frac{b^+l+1}{|x_1|}f_M\right] = k_1 + \frac{2(b^+l+1)}{\delta|x_1|}f_M. \tag{8.29}$$

It follows from this estimate that the function k_1^0 differs from the constant k_1, which acts when a free motion is controlled, by the term

$$\frac{2(b^+l+1)}{\delta|x_1|}f_M$$

caused by the presence of the external force f. When this force is absent ($f_M \equiv 0$), this estimate coincides with the earlier established estimate.

The second conclusion from (8.29) is that in the neighborhood of zero of the variable x_1 the function k_1^0 increases indefinitely, and this is inadmissible. In order to restrict this function, we must take instead of (8.29) an estimate

$$k_1^0 \geq k_1 + \frac{2(b^+l+1)}{\delta\Delta}f_M = k_1 + k_1' f_M, \tag{8.30}$$

where Δ is a positive constant which defines the size of the ball of dissipativity

$$B_r(0) = \{x \mid x_1^2 + x_2^2 \leq r^2(\Delta)\}.$$

The last remark concerning the dissipativity of the system is due to the fact that for the bounded function k_1^0 from (8.30) for small x_1 ($|x_1| < \Delta$) the projections of the phase trajectories of the system onto the set G_δ may leave this

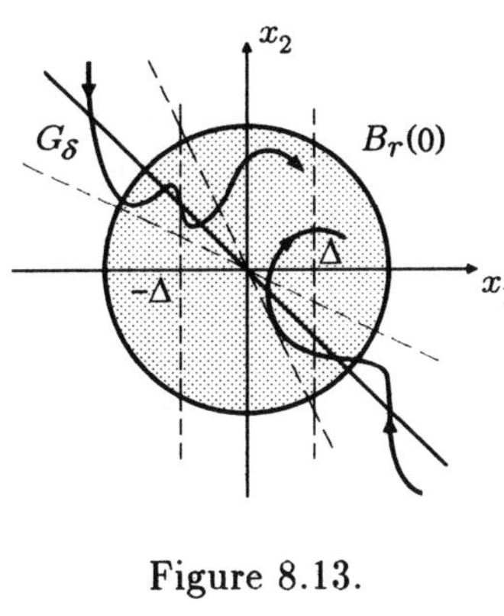

Figure 8.13.

set, and this cannot happen for the function k_1^0 from (8.29). Therefore condition (8.30) guarantees the stabilization of all solutions (of course, when the conditions of attraction for G_δ are fulfilled) in the neighborhood of zero immersed in the ball $B_r(0)$ (Fig. 8.13). Note that as $\Delta \to 0$, the ball degenerates into the point 0, and this is what must happen. We can diminish the dissipativity ball $B_r(0)$ and, hence, increase the precision of stabilization by introducing appropriate O- and OC-feedbacks. We do not give here a detailed justification of this statement.

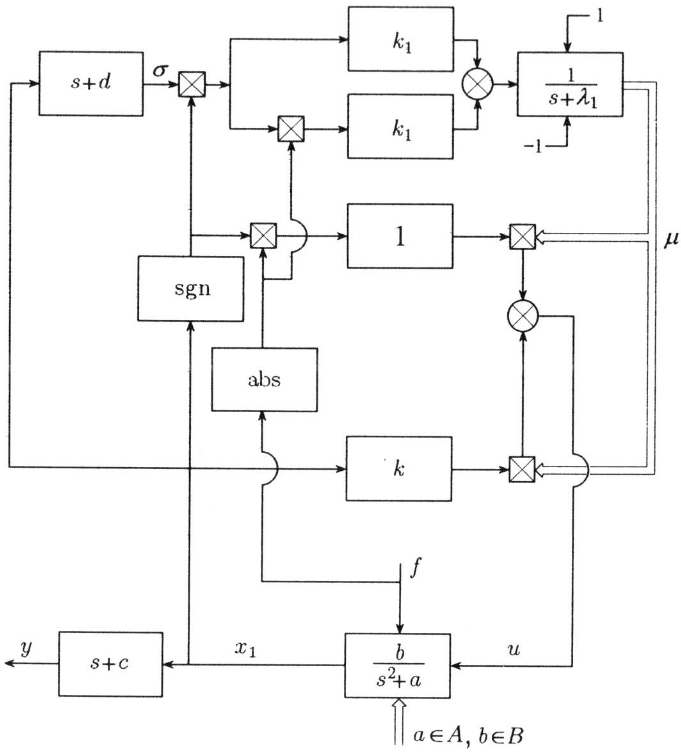

Figure 8.14.

The block diagram of the synthesized binary system of stabilization of a forced motion for the case where the disturbance is measurable is given in Fig. 8.14. Concluding the discussion of this theme, we want to point out that the problem of stabilization of a forced motion can also be solved without any available information about the majorant of the coordinate disturbance f_M and even when the matching condition is violated. In this case, the compensation can be achieved by the combined use of the indirect estimate of the disturbance in terms of the internal variables of the object and the alternating feedback.

Chapter 9

Signal Differentiation

Differentiation is interpreted in control theory as processing a signal for obtaining estimates of its derivatives. The main obstacles in the way to ideal differentiation is a physical unrealizability of an ideal differentiator and an unremovable contradiction between the precision of the differentiation operation and the operation of noise filtration. The differentiation theory is a division of the general theory of filtration, but it has an essential peculiarity which consists in the absence of a model of the signal being processed or an exact information concerning its parameters, which is typical of the filtration theory. Therefore differentiation remains one of the most difficult problems of the control theory. The a priori uncertainty in the statement of a differentiation problem presupposes the search for perfect differentiators in the class of nonlinear dynamical systems.

The aim of this chapter is to demonstrate the potentialities of the new types of nonlinear feedbacks in the differentiation problem.

9.1. Statement of the Differentiation Problem

When solving some natural science problems, we often have to know the estimates of the derivatives of functions. In what follows, we assume, for convenience, that these are time functions and will traditionally call them signals.

A differentiation problem may also arise in control theory in connection, for instance, with reconstruction of the complete phase vector of the system from its output. When this problem is solvable, we can make a change of variables in order to obtain the phase vector in arbitrary coordinates. The problem of signal differentiation is one of ill-posed problems, and this is due, in particular, to the fact that the relation

$$g_1(t) = g_2(t) \tag{9.1}$$

does not at all imply that

$$\frac{dg_1(t)}{dt} = \frac{dg_2(t)}{dt}. \tag{9.2}$$

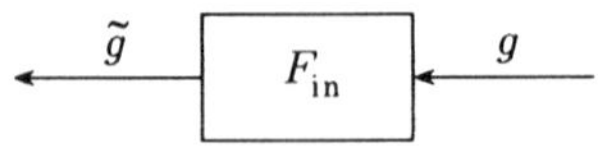

Figure 9.1.

Moreover, derivatives may not exist at all, and therefore we must first of all process the signal in such a way that the consequence (9.1)→(9.2) would take place. This can be achieved by the use of the filter F_{in} of the preliminary signal processing (Fig. 9.1) and its transformation into a smooth signal $\tilde{g}$. The differentiation problem is ill-posed also because the result of differentiation of two arbitrarily close signals may differ by an arbitrary value.

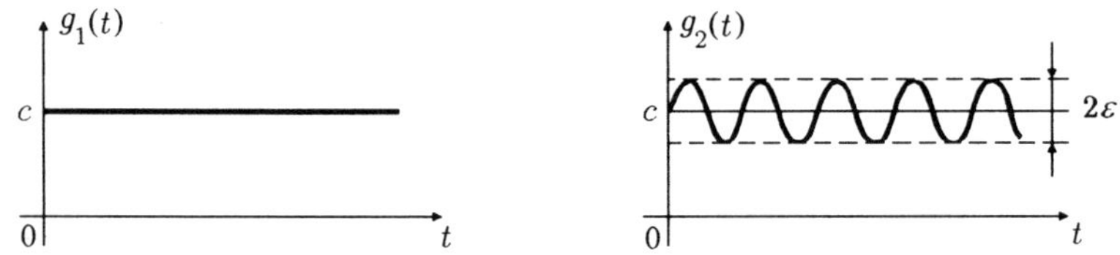

Figure 9.2.

Indeed, let us consider two signals

$$g_1(t) = c = \text{const}, \quad g_2(t) = g_1(t) + \varepsilon \sin \omega t, \tag{9.3}$$

where ε is an arbitrarily small constant and ω is the frequency of interference (Fig. 9.2). Differentiating the functions from (9.3), we have $\dot{g}_1 = 0$, $\dot{g}_2 = \varepsilon\omega \cos \omega t$. The graphs of the derivatives are given in Fig. 9.3. We can see that the differentiation error

$$\sup_t |\dot{g}_1 - \dot{g}_2| = \varepsilon\omega$$

may exceed any preassigned number. Therefore we cannot speak about "pure" differentiation but only about the estimates of the derivative $\hat{\dot{g}}$ measuring its closeness to the actual derivative in some metric. It is often convenient to distinguish the part in the structure of a differentiator which is responsible for

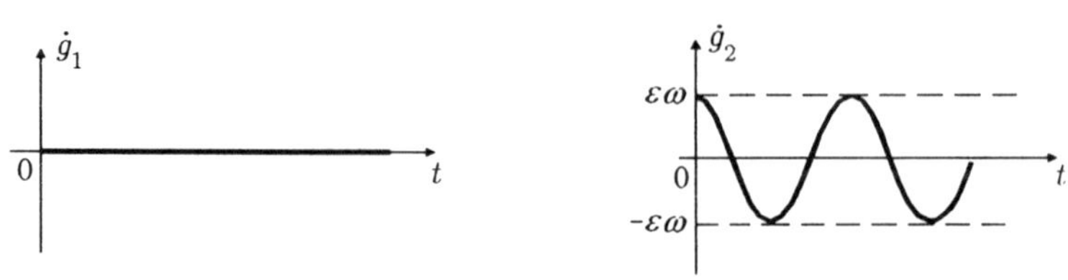

Figure 9.3.

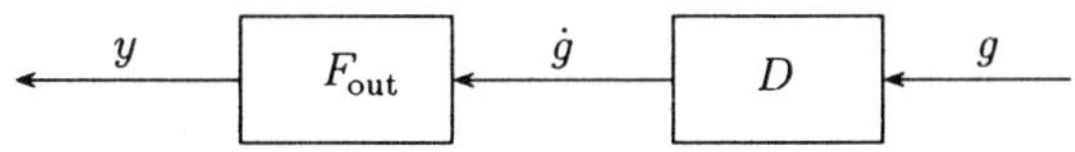

Figure 9.4.

the signal differentiation proper from the part meant for the formation of the estimate of the derivative. The latter operation can be realized with the aid of a certain filter F_{out}, and the indicated separation of functions is illustrated by Fig. 9.4. Thus, not only the signal being differentiated but also the obtained derivative must, generally speaking, be subjected to an additional processing in order to make the problem solvable and the estimate of the derivative admissible, i.e., the operators of the filters F_{in}, F_{out} and the differentiation operator D must be actually synthesized (Fig. 9.5). The choice of the filter depends on many circumstances which are not directly connected with the operation of "pure"

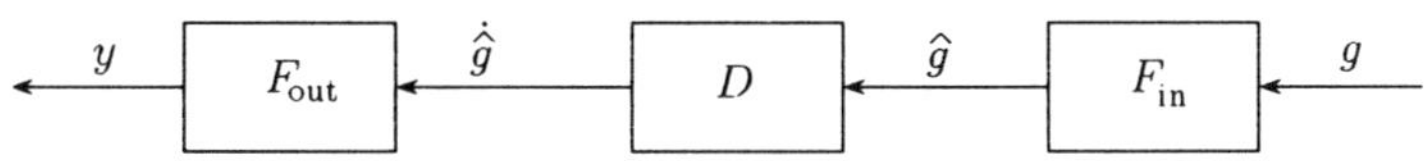

Figure 9.5.

differentiation (for instance, on the spectral composition of the noise, on the existence of amplitude constraints, etc.), and it is precisely these circumstances that make the decomposition of the approximate differentiation operator shown in Fig. 9.5 useful.

Now we assume that the signal $g(t)$ is "good", and therefore we can set $F_{\text{in}} \equiv 1$.

9.1.1. Filtration

Filtration is an operation of a separation of a useful signal from among the mixture of signals. Since such a separation is not unique, a quality indicator or a filtration criterion is introduced in order to compare the versions of a separation. The extremalization of this criterion gives an optimal filter.

In order to carry out a successful separation, we need a priori information about a useful signal in the interference. Most often dynamical systems that generate these signals serve as such an information, to put it otherwise, wave models of a signal and interference. In these cases, the problem of filtration can be reduced to a certain differentiation problem. At the same time, the differentiation problem can always be interpreted as a filtration problem, i.e., there is much in common between these problems but they are not identical.

The matter is not so much as the filtration problem is usually considered in a stochastic statement and the differentiation problem in a deterministic statement. The main difference is that a smaller volume of a priori information is usually required for solving a differentiation problem than that required for solving a filtration problem.

9.1.2. *RC*-circuit

Up till now the simplest differentiators realized on passive electric circuits have been used for approximate differentiation. The RC-circuit (Fig. 9.6) is the best known differentiator of this type. Here g is the input voltage, y is the output voltage, i is current, R and c are parameters of the circuit, namely, resistance and capacitance respectively. According to Ohm's law

$$g = R_i + \frac{1}{c}\int i\,dt, \quad y = R_i.$$

Differentiating this expression and excluding the current i, we obtain a differential equation

$$\dot{y} + (1/T)y = \dot{g}, \tag{9.4}$$

where $T = KC$ is the differentiation time constant. The general solution of Eq. (9.4) has the form

$$y = \underbrace{e^{-t/T}y(0)}_{\text{free motion}} + \underbrace{T\dot{g}\left(1 - e^{-t/T}\right)}_{\text{excited motion}}.$$

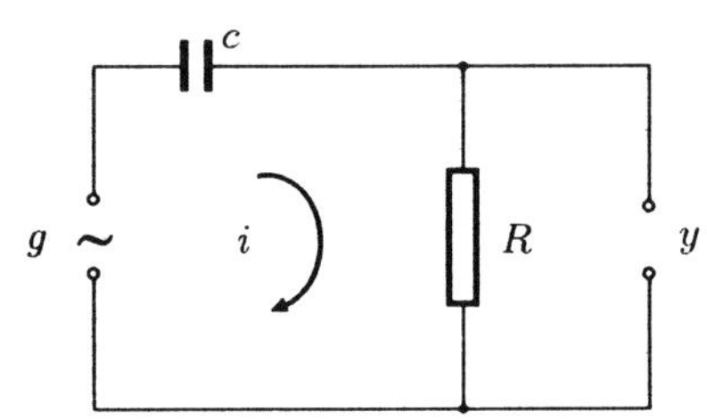

Figure 9.6.

Let us investigate the differentiating properties of the RC-circuit considering only the exited component ($y(0) = 0$) of the solution (output) since the free motion exponentially damps out. Suppose, for the beginning, that

$$\dot{g} = \text{const}.$$

In this case, we can see from the equation

$$y = T\dot{g}\left(1 - e^{-t/T}\right)$$

that $y \to T\dot{g}$ as $t \to \infty$, and, as T diminishes, the "rate" of convergence increases, but, at the same time, the coefficient in front of the derivative diminishes (Fig. 9.7 where $T_1 > T_2$). Thus, the RC-circuit realizes differentiation with an accuracy to within the transient process which, however, is unremovable.

Let now $g(t)$ be a harmonic signal, i.e.,

$$g(t) = e^{j\Omega t},$$

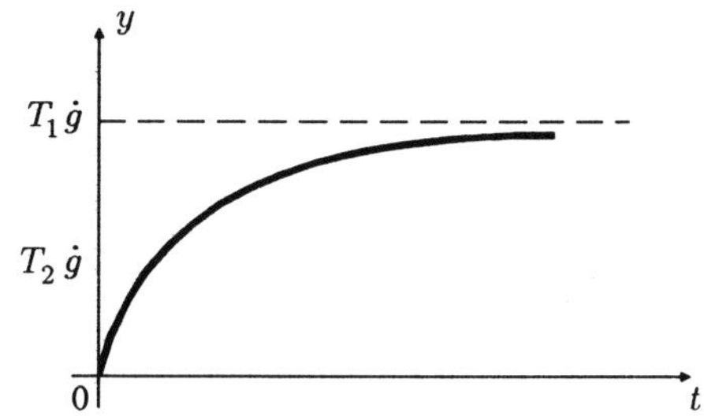

Figure 9.7.

where Ω is frequency and j is an imaginary unit. Then we seek the forced solution of the equation

$$\dot{y} + (1/T)y = j\Omega e^{j\Omega t} \tag{9.5}$$

in the form

$$y = Ae^{j(\Omega t+\varphi)}, \tag{9.6}$$

where A is the amplitude and φ is the phase of the output signal of the RC-circuit. Substituting (9.6) into (9.5), we get a relation

$$(j\Omega + 1/T)Ae^{j\varphi} = j\Omega,$$

from which we obtain an expressions for the amplitude and phase:

$$A = \frac{\Omega}{\sqrt{\Omega^2 + 1/T^2}}, \quad \varphi = \frac{\pi}{2} - \tan^{-1}(T\Omega).$$

Since the "pure" derivative is given by the expression

$$\dot{g} = j\Omega e^{j\Omega t} = \Omega e^{j(\Omega t+\pi/2)},$$

it is expedient to call the quantities $\Omega - A$ and $\pi/2 - \varphi$ distortions with respect to the amplitude and the phase respectively.

The cases of dependence of differentiation errors on the time constant T when $\Omega > 1$ are given in Fig. 9.8. We can see from the graphs that as the time

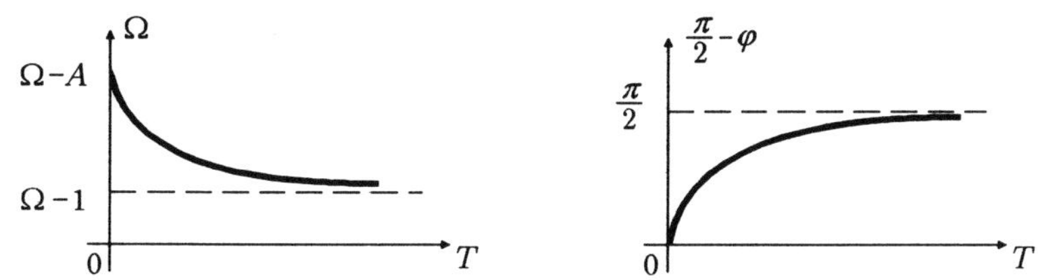

Figure 9.8.

constant increases, the amplitude distortions decrease but the phase distortions increase when a harmonic signal is differentiated. Consequently, the problem of choosing a time constant is not as simple as it may seem.

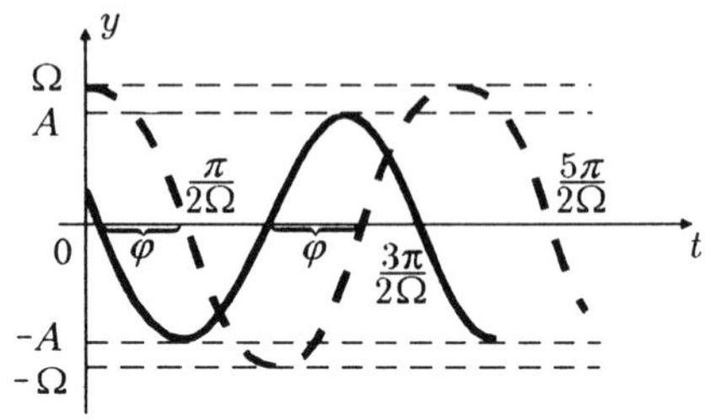

Figure 9.9.

In Fig. 9.9 the dash line shows the "pure" derivative $\Omega\cos(\Omega t)$ and the solid line shows the "real" derivative at the output of the RC-circuit.

It is also useful to know how the differentiation errors change upon the variation of the frequency of the signal being processed. The corresponding graphs for $T < 1$ are given in Fig. 9.10. We can see that at different frequencies signals are processed with different errors and that the RC-circuit has the best char-

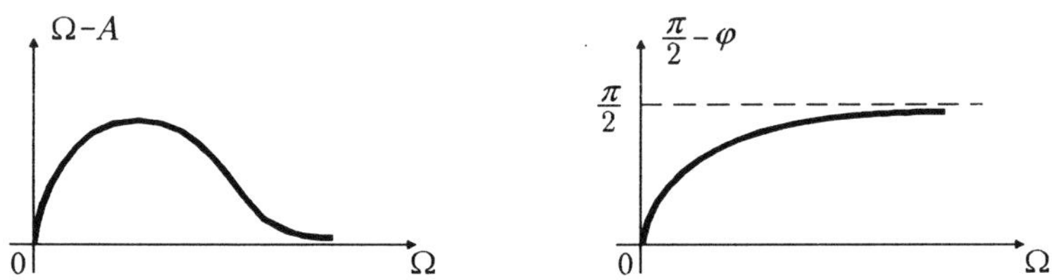

Figure 9.10.

acteristics as $\Omega \to 0$, i.e., when a constant signal is differentiated. When there exists an interference, i.e., when the signal $g(t) = e^{j\Omega t} + \varepsilon e^{j\omega t}$ is differentiated, where the interference frequency $\omega \gg \Omega$ and the amplitude $\varepsilon \ll 1$, the output of the RC-circuit for $y(0) = 0$ can be represented as a sum $y = y_\Omega(t) + y_\omega(t)$, where the first component is the estimate of the derivative of the useful signal and y_ω is the estimate of the derivative of the interference. Since formulas (9.5), (9.6) and other related formulas given above are valid for the component $y_\omega(t)$, it follows that the additional contribution made by this component to the differentiation error is defined by its amplitude A_ω and phase φ_ω:

$$A_\omega = \frac{T\omega}{\sqrt{1 + (T\omega)^2}}, \qquad \varphi_\omega = \tan^{-1}(-T\omega).$$

It follows from the expressions for the amplitude and phase that in order to eliminate the harmful effect of the interference, we must increase the time constant of the RC-circuit, but then phase distortions in the useful component increase. Therefore a compromise is possible here. The best way, however, is to find differentiation schemes such that there would exist several free parameters that would make it possible to solve independently the problem of interference "filtration" and the problem of diminishing the amplitude and phase distortions.

Let us now consider differentiation schemes of this kind, but note that the transfer function of the RC-circuit has the form

$$W_{RC}(s) = \frac{s}{s + 1/T}$$

which can be easily determined from its differentiation equation

$$\dot{y} + \frac{1}{T}y = \dot{g}.$$

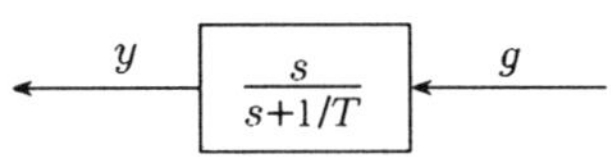

Figure 9.11.

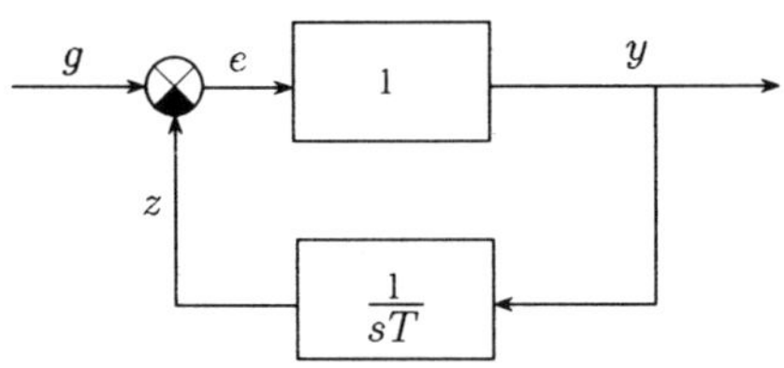

Figure 9.12.

This transfer function is associated with the simplified block diagram shown in Fig. 9.11 which can be developed as a system given in Fig. 9.12, where z is the output of the tracking system and e is its error. In these block diagrams there is only one adjustable parameter T, and this predetermines their potentialities. If, instead of the identity transformation in the direct channel, we use some other laws of feedback, then we can hope to obtain better differentiators.

9.1.3. Discrete-difference approximations

The improved differentiation methods are methods based on the use of discrete differences of various orders of the function being differentiated. For instance, the difference

$$\tilde{g} = \frac{g(t) - g(t-h)}{h} \cong \frac{dg(t)}{dt}, \tag{9.7}$$

where $h = \text{const} > 0$, can serve as the estimate of the derivative of the function $g(t)$. Note that for a linear function estimate (9.7) of the derivative coincides with its exact value. In the general case, we have a relation

$$g(t) = g(t-h) + \dot{g}(t)h + \ddot{g}(t+dh)\frac{h^2}{2},$$

where d is a number from the interval $[0, 1]$, and, consequently, estimate (9.7) has an error of order $o(h^2)$.

The use of discrete values at the $(r+1)$th point for forming the estimate

$$\hat{\dot{g}} = \sum_{k=0}^{r} \alpha_k g(t+kh)$$

under a proper choice of the parameters α_k makes it possible to raise the accuracy of approximation to $o(h^r)$. It seems that the problem of differentiation is exhausted by tending h to zero, $h \to 0$. However, this is not the fact, and when there is an interference, some problems arise similar to those described in the preceding section. Namely, suppose that instead of $g(t)$ the signal $g_\varepsilon = g(t) + \varepsilon \sin \omega t$ is differentiated according to the scheme (9.7). Then the estimate of its derivative $\hat{\dot{g}}_\varepsilon$ is given by the expression

$$\begin{aligned} \hat{\dot{g}}_\varepsilon &= \frac{g_\varepsilon(t) - g_\varepsilon(t-h)}{h} \\ &= \frac{g(t) - g(t+h)}{h} + \frac{\varepsilon}{h}\big[\sin \omega t - \sin \omega(t-h)\big]. \end{aligned} \tag{9.8}$$

However, $\sin \omega(t-h) = \sin \omega t \cos \omega h - \cos \omega t \sin \omega h$, and, for $\omega h \ll 1$, we have

$$\cos \omega h \cong 1 - \omega h, \quad \sin \omega h \cong \omega h,$$

and therefore the approximate relation

$$\sin \omega(t-h) \cong \sin \omega t - \omega h(\sin \omega t + \cos \omega t)$$

is valid. Consequently, the interference introduces into estimate (9.8) an additional and, generally speaking, arbitrarily large error

$$\hat{\dot{g}}_\varepsilon \cong \hat{\dot{g}} - \varepsilon\omega(\sin \omega t + \cos \omega t),$$

which is not removed as $h \to 0$.

In some situations, the quality of this differentiation can be raised by the use of the idea of filtration which in difference schemes is equivalent to the operation of averaging. Namely, suppose that the signal g_ξ being differentiated is an additive mixture of a useful signal g and a random noise ξ with zero mean, i.e., $M\xi = 0$. To calculate the derivative, we shall use N differentiators D^i connected in parallel (Fig. 9.13) of the form

$$\hat{\dot{g}}^i_\xi(t) = \frac{g(t) - g(t-h)}{h} + \frac{\xi^i(t) - \xi^i(t-h)}{h}. \tag{9.9}$$

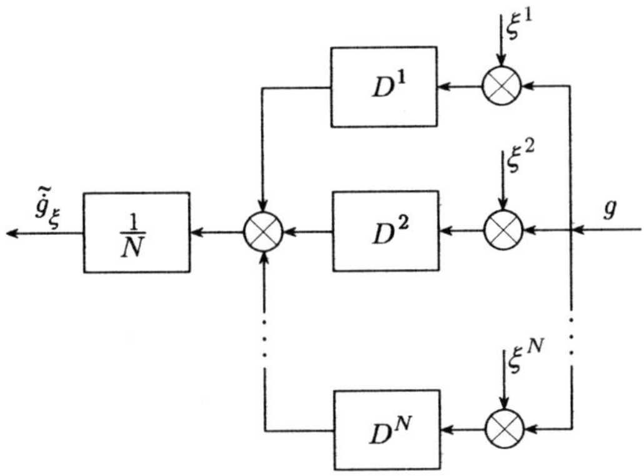

Figure 9.13.

The subscript i $(1, 2, \ldots, N)$ in the figure shows the output of the ith differentiator D^i; the existence of a subscript in the interference means that its "own" realization of the stochastic process ξ acts at the input of every D^i.

In order to obtain the desired estimate, we shall use the following simple averaging operation:

$$\hat{g}_\xi(t) = \frac{1}{N}\sum_{i=1}^{N} \hat{g}^i_\xi(t). \tag{9.10}$$

Then, from (9.9) and (9.10) we get

$$\hat{g}_\xi = \hat{g} + \frac{1}{hN}\sum_{i=1}^{N} \Delta\xi^i(t).$$

As $N \to \infty$, the last term tends to zero under our assumptions. It stands to reason that in order to decrease the regular error in this case (recall that the error is of order $o(h^2)$), we have to diminish h, and this leads to an increase in the number N and, hence, makes the differentiator more complicated.

Moreover, it is easy to see that the block diagram of the differentiator of the first difference

$$\tilde{g} = \frac{g(t) - g(t-h)}{h}$$

has the form shown in Fig. 9.14 and includes a delay element whose realization is not a simple matter either. The transfer function of differentiator (9.10) is shown in the block diagram in Fig. 9.14 and is given by the expression $W_D(s) = h/(1 - e^{-hs})$. For small values of $|hs|$ the decomposition $1 - e^{-hs} = hs - (h^2/2)s^2 + o(|hs|^3)$ is valid, and therefore $W_D(s)$ can be approximated by the expression

$$W_D(s) \cong s - \frac{h}{2}s^2 = s\left(1 - \frac{h}{2}s\right) \cong \frac{s}{1 + (h/2)s}.$$

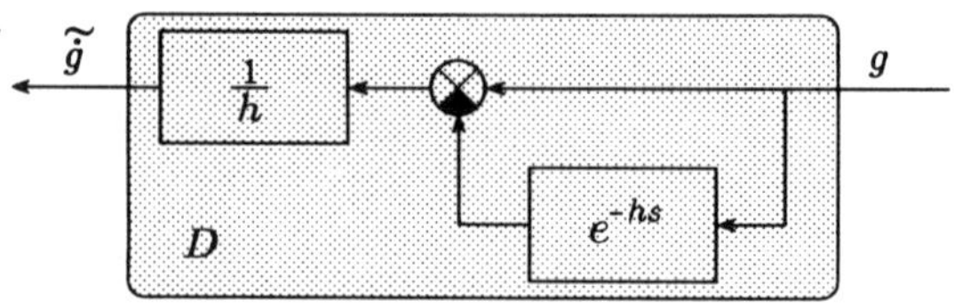

Figure 9.14.

In other words, if we denote $T = h/2$, then the transfer function of this differentiator will be close to the transfer function of the RC-circuit $W_D(s) \cong s/(Ts+1)$, and consequently, their properties will also be close. Therefore, without repeating ourselves, we shall consider "tracking" differentiators.

9.2. Tracking Differentiating Systems

The block diagram of differentiators of the type being considered is given in Fig. 9.15 in which F_{out} is the output filter and R_y is a control which stabilizes at zero the tracking error $e = g - z$.

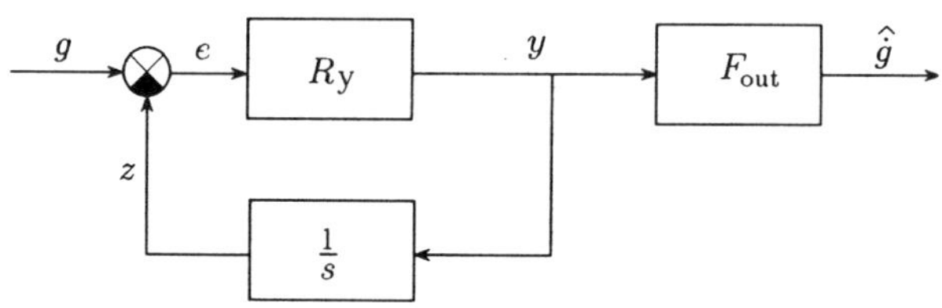

Figure 9.15.

In the case where the signals $g(t)$ and $z(t)$ are smooth, the relations $e = 0$, $\dot{z} = y$ immediately yield the required $y = \dot{g}$. Now if $z(t)$ does not have an ordinary derivative, then we can use the filter F_{out} to form the required estimate of the derivative

$$\hat{\dot{g}} = F_{\text{out}} y.$$

Thus, the problem of differentiation can be formulated as a problem of tracking, and the versions arising in this case are defined by the choice of the pair $\{R_y, F_{\text{out}}\}$. Let us consider some of them.

9.2.1. A linear differentiator

We can consider the linear differentiator for which

$$\{R_y, F_{\text{out}}\} = \{k, 1\}, \quad k = \text{const} > 0,$$

to be the simplest tracking differentiator. Its block diagram is shown in Fig. 9.16.

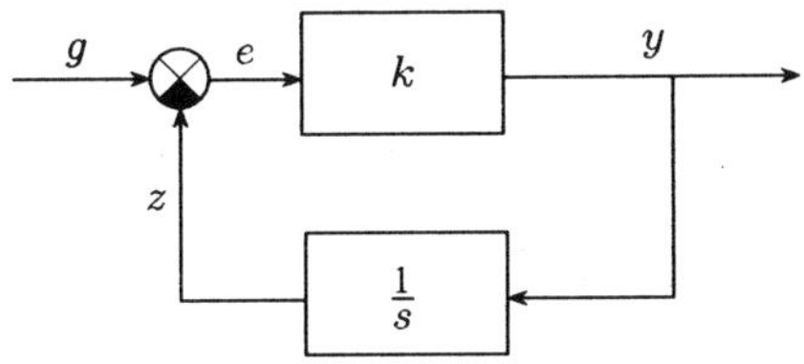

Figure 9.16.

The equations of the differentiator have the form

$$e = g - z, \quad \dot{z} = y, \quad \hat{\dot{g}} = y = ke. \tag{9.11}$$

Let us analyze its operation. The motion equation relative to the tracking error from (9.11) has the form

$$\dot{e} = -ke + \dot{g}, \quad t \geq 0,$$

and, consequently, the total error consists of two components, a free motion $e_c(t) = e(0)e^{-kt}$ and an excited one,

$$e_f(t) = e^{-kt} \int_0^t e^{k\tau} \dot{g}(\tau)\, d\tau. \tag{9.12}$$

We shall investigate only $e_f(t)$ since only this component contains the information about the derivative whereas $e_c(t)$ exponentially tends to zero as $t \to \infty$ and does not depend on $\dot{g}$.

Let $\dot{g} = \text{const}$, and then (9.12) can be easily integrated and

$$e_f(t) = \frac{\dot{g}}{k}\left(1 - e^{-kt}\right).$$

Since $y = ke$, the error of the derivative is defined by the relation

$$\hat{\dot{g}} = \dot{g}\left(1 - e^{-kt}\right).$$

The time graphs of the estimates for different k ($k_1 > k_2$) are shown in Fig. 9.17. We can see from these graphs that an increase in the gain factor and feedback improves the differentiating properties of this scheme. However, we must always remember that we deal with high gain feedback, namely, as $k \to \infty$, we must take into account the effect produced by constraints and singular perturbations on the robustness of the system. If $\dot{g} \neq \text{const}$, then, to diminish the tracking error, we must tend k to infinity, $k \to \infty$.

To analyze the consequences of singular perturbations, we must assume that in the block diagram shown in Fig. 9.16 pure integration is replaced by an integration with a small time delay $\tau = \text{const} > 0$. The situation being analyzed is illustrated by Fig. 9.18 where the equation with errors has the form

$$\dot{e}(t) = -ke(t - \tau) + \dot{g}. \tag{9.13}$$

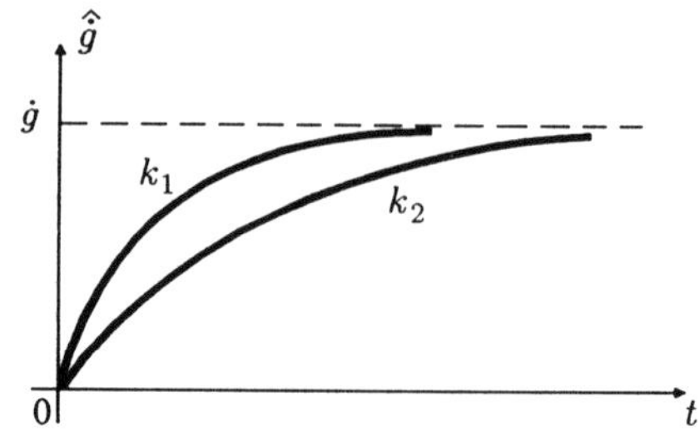

Figure 9.17.

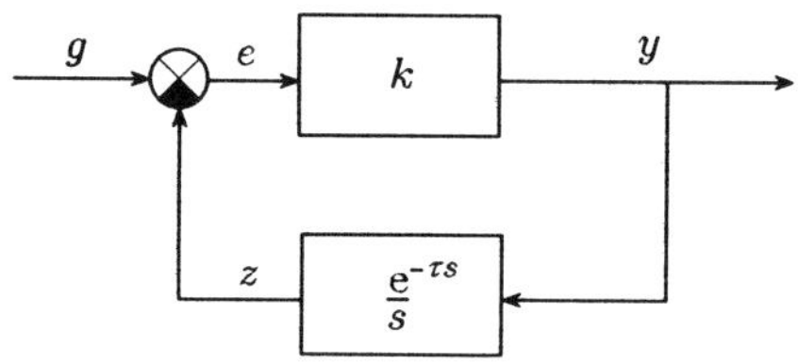

Figure 9.18.

Now the question of whether we must take into account the free component of the solution of Eq. (9.13) or not depends on the stability of the quasipolynomial

$$\varphi(s) = s + ke^{-\tau s}.$$

However, as $k \to \infty$, its zeros tend to the zeros of the equation $e^{-\tau s} = 0$ which has zeros at infinity of the right complex half-plane of the variable s ($\mathrm{Re}\, s > 0$). Thus, the polynomial being investigated is unstable, and therefore the system being differentiated is also unstable. Consequently, the stability of a system under a singular perturbation is restricted by the feedback factor of the value $k \leq k_{\mathrm{cr}}$.

The presence of a critical gain factor also determines the maximally attainable phase distortions when a sinusoidal signal is processed. Indeed, the gain factor of the linear differentiator

$$W_D(s) = \frac{s}{s/k + 1}, \tag{9.14}$$

and, consequently, the minimally attainable time constant of the differentiator $T = 1/k_{\mathrm{cr}}$, and this, as was already indicated in the analysis of an RC-circuit, defines phase distortions. Note the validity of the remarks made above concerning the properties of differentiators with transfer functions of the form (9.14) in this case as well.

Let us investigate now the effect produced by the amplitude constraint at the output of the differentiator shown in Fig. 9.19. The equations of this differentiator differ from those of the differentiator that we have considered only by the output equation

$$\hat{g} = \mathrm{sat}(y),$$

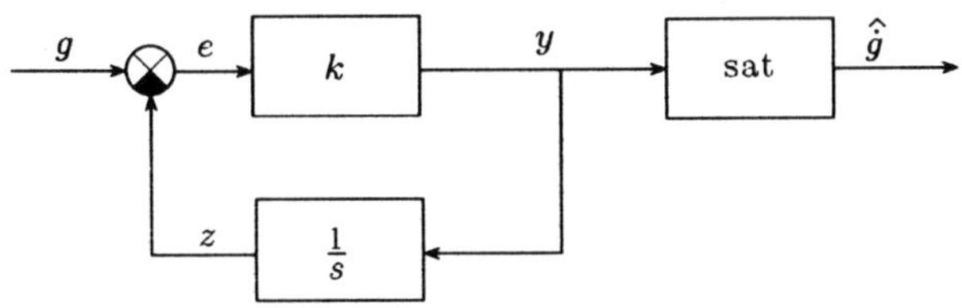

Figure 9.19.

where sat(·) is a saturation function. We can see from the block diagram in Fig. 9.19 that $\hat{g} = \text{sat}(ke)$, but, as $k \to \infty$,

$$\text{sat}(ke) = \text{sat}_{1/k}(e) \to \text{sgn}\, e.$$

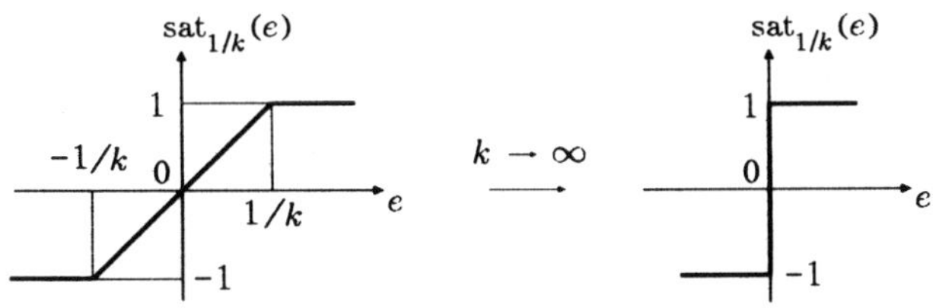

Figure 9.20.

Thus, in the limit, as $k \to \infty$, the presence of constraints leads to a discontinuous signal (Fig. 9.20). Therefore, in order to obtain a plausible estimate, it is necessary to use an output filter F_{out} which properly "averages" the discontinuous signal sgn y and inevitably introduces additional phase distortions (Fig. 9.21).

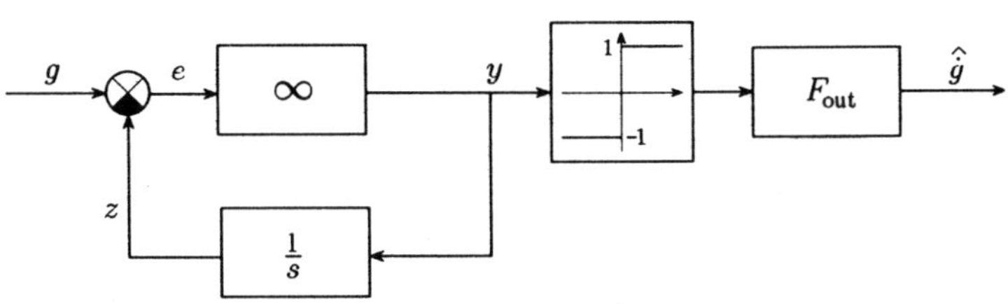

Figure 9.21.

In what follows, we investigate the averaging methods in greater detail, and here we must only point out that the initial problems arise when there is an amplitude constraint in a feedback circuit (Fig. 9.22) where the limiting process $k \to \infty$ leads to a relay differentiator. Now we pass to the analysis of the properties of a relay differentiator.

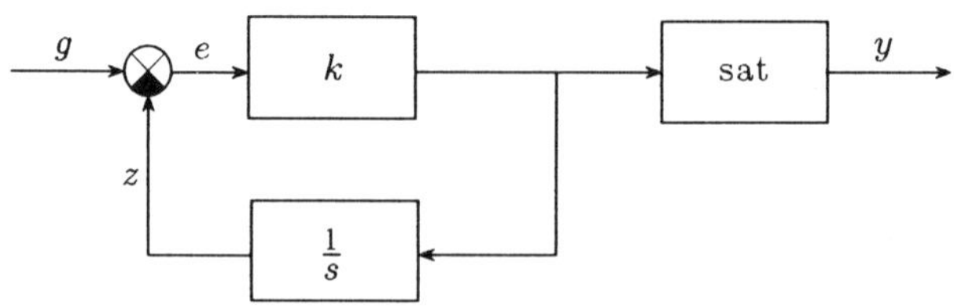

Figure 9.22.

9.2.2. Relay differentiator

The block diagram of a relay differentiator characterized by the pair

$$\{R_y, F_{\text{out}}\} = \left\{ k \operatorname{sgn} e, \frac{1}{\tau s + 1} \right\}$$

is shown in Fig. 9.23. We assume the output of the system to be the estimate of the signal, i.e., set $\hat{\dot{g}} = x$. To analyze the processes in a feedback circuit, we write the motion equation with respect to the error e:

$$\dot{e} = -k \operatorname{sgn} e + \dot{g}, \tag{9.15}$$

which easily results from the equations of the differentiator

$$\begin{aligned} &e = g - z, \quad y = k \operatorname{sgn} e, \quad k = \text{const} > 0, \\ &\dot{z} = y, \qquad T\dot{x} + x = y, \quad T = \text{const} > 0. \end{aligned} \tag{9.16}$$

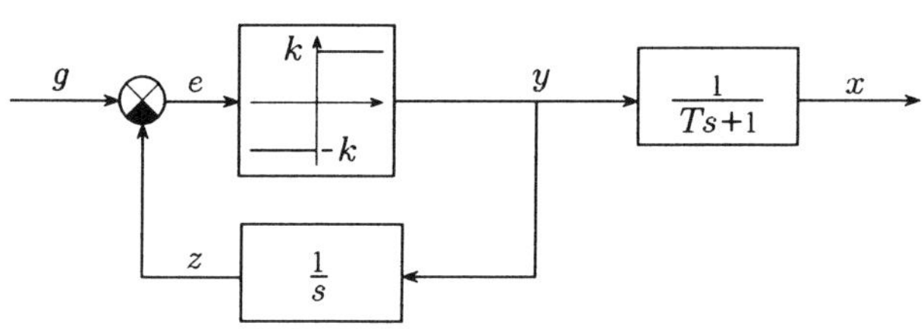

Figure 9.23.

We see that if $k > |\dot{g}|$, then, in a finite time, the process stabilizes at zero $e = 0$ where a sliding mode appears (Fig. 9.24). In the sliding mode, the mean value of the discontinuous signal can be found from the relations $e = 0$, $\dot{e} = 0$ and is given by the expression $k \operatorname{sgn}_{\text{eq}} e = \dot{g}$. The output filter is needed precisely in order to distinguish this mean value.

Indeed, (9.16) and (9.15) give the equation of the filter

$$T\dot{x} + x = \dot{g} - \dot{e},$$

the forced component of whose solution is given by the well-known formula

$$x_{\text{f}}(t) = \frac{1}{T} e^{-t/T} \int_0^t e^{\tau/T} \left(\dot{g}(\tau) - \dot{e}(\tau) \right) d\tau. \tag{9.17}$$

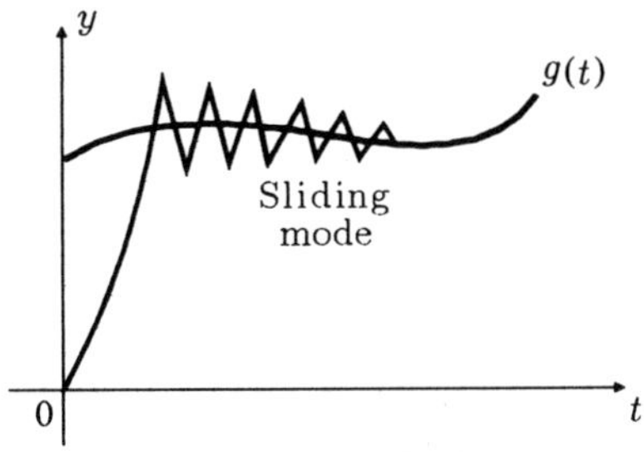

Figure 9.24.

Suppose that there exists a second derivative of the signal g and that this derivative is uniformly bounded, i.e.,

$$|\ddot{g}| \leq M, \quad M = \text{const} > 0, \tag{9.18}$$

and that the sliding mode appeared at $t = 0$. Then the following chain of equalities is valid:

$$\int_0^t e^{\tau/T}\dot{e}(\tau)\,d\tau = \int_0^t e^{\tau/T}de(\tau) = e(\tau)e^{\tau/T}\Big|_0^t - \int_0^t e(\tau)e^{\tau/T}d\left(\tau/T\right) = 0. \tag{9.19}$$

Consequently, we can simplify (9.17) up to the expression

$$x_{\mathrm{f}}(t) = \frac{1}{T}e^{-t/T}\int_0^t e^{\tau/T}\dot{g}\,d\tau, \tag{9.20}$$

for whose further transformation we shall again use integration by parts. As a result we obtain

$$\begin{aligned}\int_0^t e^{\tau/T}\dot{g}d(\tau/T) &= \int_0^t \dot{g}de^{\tau/T} = \dot{g}e^{\tau/T}\Big|_0^t - \int_0^t e^{\tau/T}\ddot{g}d\tau \\ &= \dot{g}e^{\tau/T} - \dot{g}(0) - \int_0^t e^{\tau/T}\ddot{g}d\tau.\end{aligned} \tag{9.21}$$

The last term in this relation can be estimated with due account of (9.18) as follows:

$$\left|\int_0^t e^{\tau/T}\ddot{g}\,d\tau\right| \leq \int_0^t e^{\tau/T}|\ddot{g}|\,d\tau \leq M\int_0^t e^{\tau/T}d\tau = TM(e^{t/T}-1). \tag{9.22}$$

Substituting (9.21) into (9.20), we find that

$$x_{\mathrm{f}}(t) = \dot{g}(t) - \dot{g}(0)e^{-t/T} - e^{t/T}\int_0^t e^{\tau/T}\ddot{g}\,d\tau = \dot{g}(t) + \varepsilon(t). \tag{9.23}$$

For the error of differentiation of $\varepsilon(t)$ we have from (9.22) and from the inequality $|\dot{g}(0)| < k$ an estimate

$$|\varepsilon(t)| \leq TM + (k + TM)e^{-t/T}. \tag{9.24}$$

The final result is

$$|x_{\mathrm{f}}(t) - \dot{g}(t)| \leq TM + (k + TM)e^{-t/T}. \tag{9.25}$$

To diminish the regular $\{TM\}$ and asymptotically vanishing $\{(k+TM)e^{-t/T}\}$ components of the differentiation error, we must (as we see from (9.25)) diminish the time constant T of the filter and then, in the limit, for $T = 0$, we shall obtain an absolutely exact differentiation. However, it is physically unattainable because switchings are always nonideal and the choice of the parameter T must be matched with these nonideal properties of switching elements.

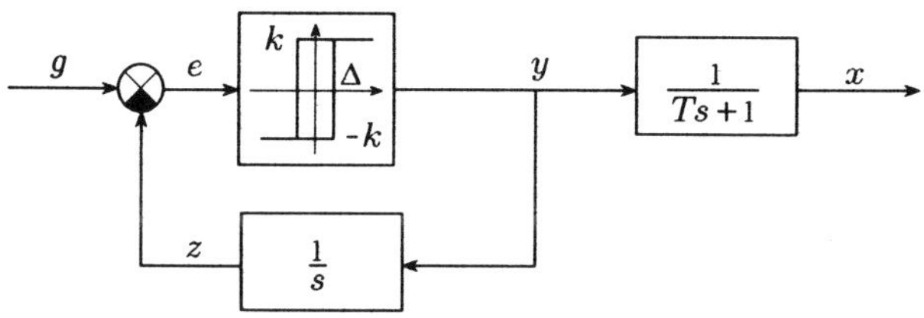

Figure 9.25.

Suppose, for instance, that there exists a space delay $\Delta = \text{const}$ in switchings (Fig. 9.25). Then, all equations of the relay differentiator from the block diagram shown in Fig. 9.23 are valid for describing the differentiator with nonideal switchings when the ideal relay in them is replaced by a relay with a hysteresis $\operatorname{sgn}_\Delta e$, in particular, we have the following equation in errors:

$$\dot{e} = -k \operatorname{sgn}_\Delta e + \dot{g}.$$

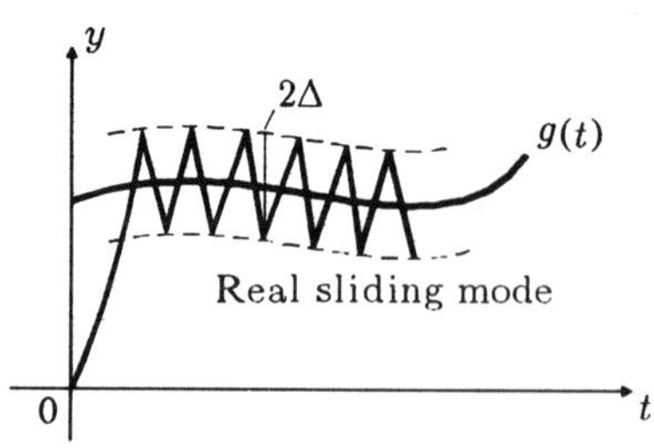

Figure 9.26.

A real sliding mode arises in the neighborhood of $|e| \leq \Delta$ (Fig. 9.26). The estimate for the error (9.24) introduced by the dynamics of the filter is also valid, but now $|e(t)| \le \Delta$ and estimate (9.22) does not hold. Using the expression from (9.22), we find an estimate which is suitable for this case. We have

$$\left|\int_0^t e^{\tau/T}\dot{e}(t)\,d\tau\right| \le \left|e(t)e^{t/T} - e(0)\right| + \Delta(e^{t/T}-1) \le 2\Delta\left(e^{t/T}+1\right).$$

Substituting the obtained estimate (9.23) and (9.24) into (9.17), we establish the validity of the estimate of the differentiation error:

$$|x_{\mathrm{f}}(t) - \dot{g}(t)| \le TM + \frac{2\Delta}{T} + \left(k + TM + \frac{2\Delta}{T}\right)e^{-t/T}. \tag{9.26}$$

We can see now that as $T \to 0$, the error does not decrease, as could be expected, but even increases.

The accuracy of differentiation will, in fact, increase if the condition

$$\lim_{T\to 0}\frac{\Delta}{T} = 0$$

is fulfilled. From the point of view of physics, the last relation implies that the high-frequency component of the signal $y(t)$ has been really filtered out and only the useful component $\dot{g}$ remains. The investigation that we have carried out shows that

- in a relay differentiator we cannot use only the choice of the unique parameter T in order to diminish both the errors introduced by the dynamics of the filter (TM) and those introduced by the nonideal properties of switchings (Δ/T).

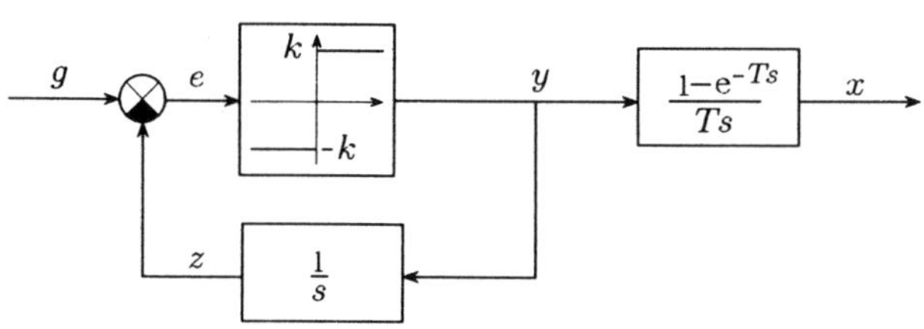

Figure 9.27.

We have a similar result when instead of an inertial filter shown in Fig. 9.23 we use a filter of a "sliding" mean shown in Fig. 9.27. In addition, exponential, polynomial, and high-frequency harmonic signals are not subjected to the constraint $|\dot{g}| < k$. Therefore it is interesting to investigate the possibilities that appear when we use a variable-structure control in the feedback of the differentiator.

9.2.3. Variable-structure differentiator

The block diagram of a variable-structure differentiator associated with a pair

$$\{R_y, F_{\text{out}}\} = \left\{\psi(z,e), \frac{1}{Ts+1}\right\}$$

is shown, in standard form, for the VSS theory in Fig. 9.28. The equations of a variable-structure differentiator (called a VSS-differentiator in the sequel) have the form

$$e = g - z, \quad y = \psi(z,e) = \begin{cases} kz, & ez < 0, \\ -kz, & ez > 0, \end{cases}$$
$$\dot{z} = y, \quad T\dot{x} + x = y, \quad k, T = \text{const} > 0;$$

the output x is assumed to be the estimate of the derivative $\hat{\dot{g}}$. Since this ψ-cell can be described by the relation

$$y = \psi(z,e) = k|z| \operatorname{sgn} e,$$

the motion equation of the feedback of the differentiator has the form

$$\dot{e} = -k|z| \operatorname{sgn} e + \dot{g}.$$

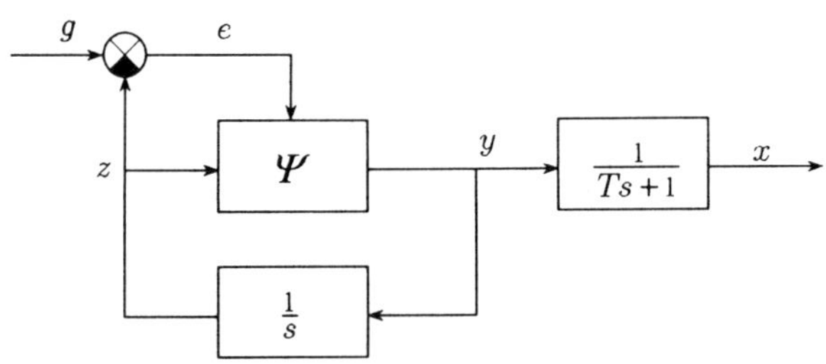

Figure 9.28.

If the signs of $g(0)$ and $z(0)$ coincide and the signal $g(t)$ belongs to the class of signals Γ, i.e.,

$$g(t) \in \Gamma \stackrel{\text{def}}{=} \{g \mid |\dot{g}| \leq \gamma|g|, \ |\ddot{g}| \leq M, \ M, \gamma = \text{const} > 0\},$$

then, if the condition $k > \gamma$ is fulfilled, the signal

$$z(t) = z(0)e^{\pm kt}$$

increases (decreases) exponentially and "overcomes", in a finite time, the signal $g(t)$, and a sliding mode appears at zero $z = 0$ (Fig. 9.29).

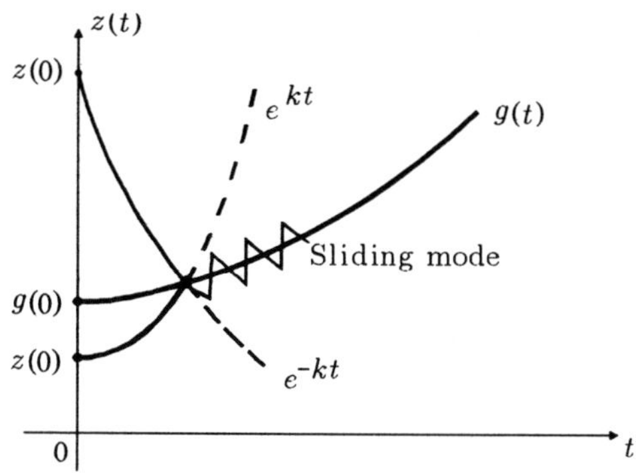

Figure 9.29.

In the sliding mode $e = 0$, i.e., $z = g$, the mean value of the signal y is defined by the expression

$$y_{\text{eq}} = \dot{g}$$

and, consequently, at the output of the inertial filter we have an estimate $x_{\text{f}}(t)$ of the derivative which is similar to that established in the preceding section, namely, under a similar assumption $|\ddot{g}| \leq M$ and for ideal switchings the estimate

$$|x_{\text{f}}(t) - \dot{g}(t)| \leq TM + (\gamma|g(0)| + TM)e^{-t/T}$$

is valid.

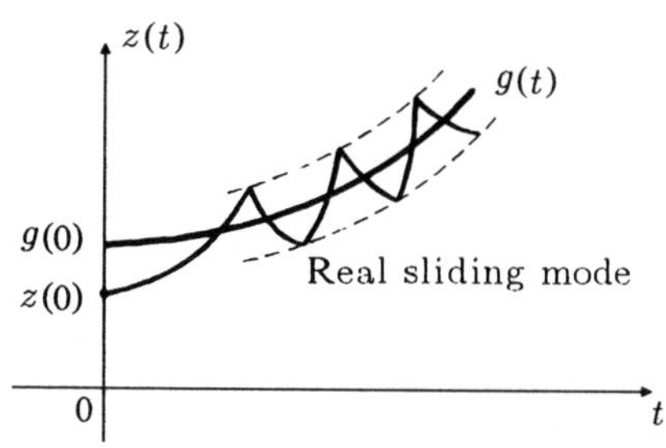

Figure 9.30.

When there is a space delay $\Delta > 0$ in switchings, we have a real sliding mode (Fig. 9.30). The accuracy of differentiation in a real sliding mode is characterized by the estimate

$$|x_{\text{f}}(t) - \dot{g}(t)| \leq TM + \frac{2\Delta}{T} + \left(\gamma|g(0)| + TM + \frac{2\Delta}{T}\right) e^{-t/T}.$$

In spite of the fact that the class of admissible signals includes now exponents, no significant changes have occurred.

- As before, it is difficult to make a compromise between the inertial property of the filter and its filtrating properties in the real sliding mode. In addition, when the signs of $g(0)$ and $z(0)$ do not coincide, the differentiator does not operate.

As a result, it is inefficient to use it for differentiating sinusoidal signals.

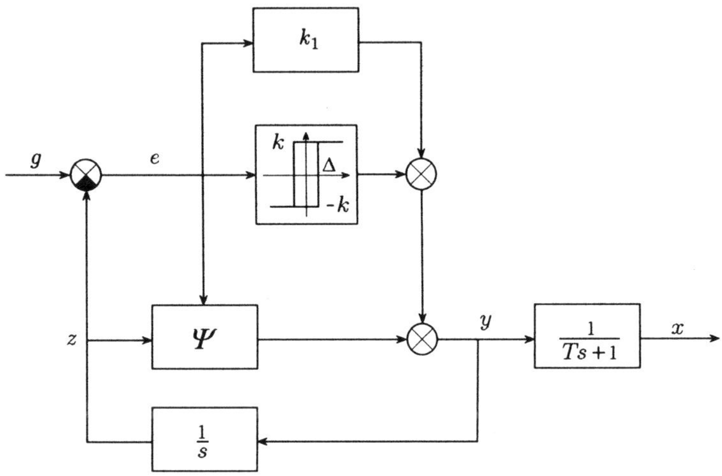

Figure 9.31.

In order to overcome these difficulties, we can combine in a feedback control a relay (linear) controller and a VSS-controller. The block diagram of such a system is given in Fig. 9.31. In this case, the equation of feedback has the form

$$\dot{e} = -(k_3|z| + k_2)\,\mathrm{sgn}\,e - k_1 e + \dot{g}$$

and the class of signals being differentiated satisfies the condition

$$|\dot{g}| \leq k_2 + k_3|g|, \quad k_1, k_2, k_3 = \mathrm{const} > 0.$$

In all other respects the situation is the same.

The indicated difficulties can be overcome, but the problem of accuracy of differentiation remains, and some other schemes are needed for its solution. Different approaches are also needed for the realization of repeated differentiation since the derivative x is discontinuous and a necessity arises to have an additional filter or to use some other principle of the differentiator design.

9.3. Tracking Asymptotic Binary Differentiator

The block diagram of a binary differentiator is shown in Fig. 9.32 where the form of the CO-controller R_μ is not yet defined. We must choose the controller R_μ such that when the output signal y is smooth, which fact makes the use

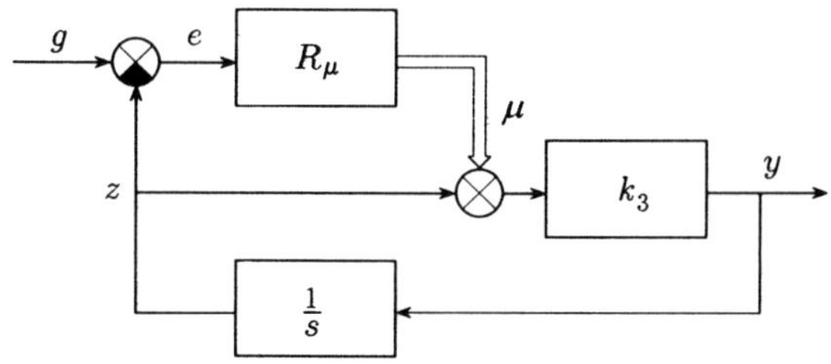

Figure 9.32.

of the output filter F_{out} unnecessary and makes it possible to use a repeated differentiation, we would be able, in addition, to ensure the robustness of the system of differentiation.

The equations of a binary differentiator are

$$e = g - z, \quad \mu = R_\mu e, \quad y = k_3 \mu z, \quad \dot{z} = y. \tag{9.27}$$

Let the signal g satisfy the condition $g \in \Gamma$, where

$$\Gamma = \{g \mid |\dot{g}| \le \gamma |g|, \ \gamma = \text{const} > 0\}.$$

In the further investigation it is convenient to use the O-variables

$$\xi = e/g, \quad a = \dot{g}/g. \tag{9.28}$$

Now we can replace the stabilization of the error e at zero by the stabilization of the function ξ at zero. In this case, a will play the part of a uniformly bounded disturbance ($|a| \le \gamma$). When seeking the motion equation of a feedback relative to the O-error ξ, we first have from (9.27) and (9.28)

$$\begin{aligned}\dot{\xi} &= \frac{\dot{e}}{g} - \frac{e\dot{g}}{g^2} = \frac{\dot{g} - y}{g} - \xi\frac{\dot{g}}{g} \\ &= (1-\xi)\frac{\dot{g}}{g} - \frac{k_3\mu z}{g}\bigg|_{z=g-e} = (1-\xi)\frac{\dot{g}}{g} - k_3\mu(1-\xi),\end{aligned}$$

or, after the change $\dot{g}/g$, we obtain from (9.28) the required equation

$$\dot{\xi} = (1-\xi)(a - k_3\mu), \quad |a| \le \gamma. \tag{9.29}$$

If a is an unknown function, then the stabilization in the small of the error ξ at zero is guaranteed for $k_3 > \gamma$ by the discontinuous law

$$\mu = \operatorname{sgn} \xi. \tag{9.30}$$

When switchings are ideal, this law leads to a variable-structure differentiator since in this case we have

$$\mu = \operatorname{sgn} \frac{e}{g} = \operatorname{sgn} e \operatorname{sgn} g = \operatorname{sgn} e \operatorname{sgn} z$$

and then

$$y = k_3 z \mu = k_3 |z| \operatorname{sgn} e,$$

and this in precisely the output of a VSS-differentiator.

When switchings are not ideal, say, when there is a space delay Δ, we have $|\xi| \leq \Delta$ and, hence,

$$|e| \leq \Delta |g|,$$

and not $|e| \leq \Delta$ as in a VSS-differentiator, and this means that

- when small signals are differentiated, the binary differentiator being considered is more preferable than a VSS-differentiator.

Nevertheless, a relay CO-controller leads to a discontinuous output signal y and, as a consequence, to the necessity of using an output smoothing filter, and this does not answer the goal that we have announced. Therefore we shall now consider an integral-relay CO-law

$$\dot{\mu} = k_2 \operatorname{sgn}[\xi + \mu |\xi|], \tag{9.31}$$

which, in conjunction with Eq. (9.29), defines the feedback system that we have to analyze.

Since we consider stabilization in the small ($|\xi| \leq \delta$, $\delta \ll 1$), we can use a change of variable

$$\varphi = -\ln(1 - \xi) = \xi + o(\xi^2)$$

and get an equivalent problem of stabilization of the Σ_1^{φ}-system

$$\begin{aligned} \dot{\varphi} &= a - k_3 \mu, \\ \dot{\mu} &= -k_2 \operatorname{sgn}[\varphi + \mu |\varphi|]. \end{aligned} \tag{9.32}$$

This is a standard scheme whose detailed investigation was carried out in the CO-theory (Chap. 4). On the basis of the results established in the CO-theory, we infer that the Σ^{φ}-system is dissipative for $\dot{a} = \text{const}$ (Fig. 9.33) and, consequently, this differentiator does not give an asymptotically exact differentiation. This difficulty can be overcome in two ways.

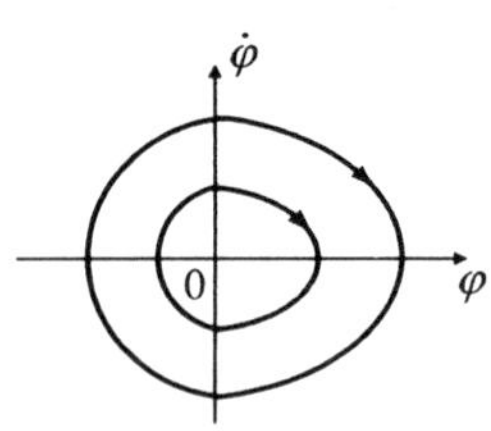

Figure 9.33.

The first way is to use an integral-relay CO-controller

$$\begin{aligned} \mu &= \mu_1 + \mu_2, \\ \mu_1 = k_1 \xi, \quad & \dot{\mu}_2 = k_2 \operatorname{sgn}[\xi + \mu_2 |\xi|]. \end{aligned} \tag{9.33}$$

Using the scheme described above, we can obtain now equations for the feedback Σ_2^{φ}-system

$$\dot{\varphi} = a - k_3 k_1 \varphi - k_3 \mu_2, \tag{9.34}$$

$$\dot{\mu}_2 = k_2 \operatorname{sgn}[\varphi + \mu_2 |\varphi|]. \tag{9.35}$$

Differentiating Eq. (9.34) with respect to t and substituting $\dot{\mu}_2$ from (9.35) into the result, we get for $|\mu| \leq 1$ an equation

$$\ddot{\varphi} + k_3 k_1 \dot{\varphi} + k_3 k_2 \operatorname{sgn} \varphi = \dot{a}. \tag{9.36}$$

For $k_3 k_2 > \sup\limits_{t \leq 0} |\dot{a}|$ this equation has zero as an asymptotically stable equilibrium position. Consequently, the O-variable $\varphi \to 0$ as $t \to \infty$, and, hence, the initial tracking error e diminishes.

In accordance with what was proved above, $e \to 0$ as $t \to \infty$, and, hence, $z \to g$ as $t \to \infty$, and, therefore, by virtue of smoothness, $\dot{z} \to \dot{g}$. However,

$$\dot{z} = u = y + k_1 e,$$

and this implies the desirable

$$\dot{g}(t) - y(t) \to 0, \quad t \to \infty.$$

Figure 9.34 shows the block diagram of binary differentiator with a linear integral-relay stabilization law.

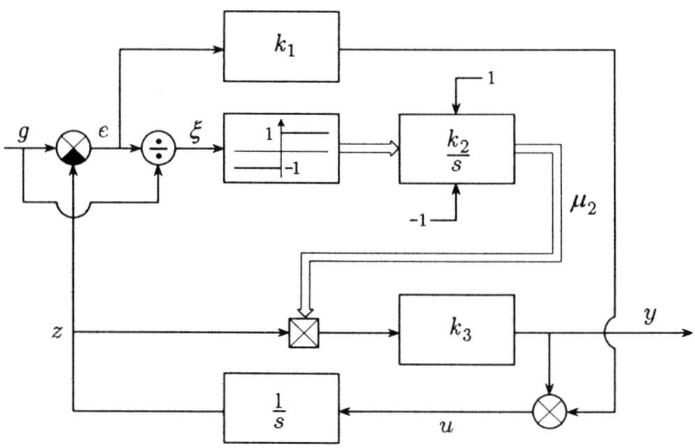

Figure 9.34.

The second technique of achieving the asymptotic accuracy of differentiator is based on the use of an inertial-relay CO-controller (here we have already made a change of $\varphi(\xi)$)

$$\dot{\mu} + k_2 \mu = k_2 \operatorname{sgn} \varphi. \tag{9.37}$$

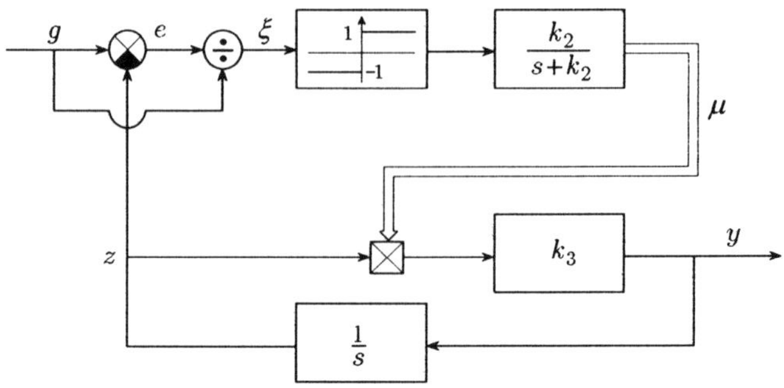

Figure 9.35.

This equation and the equation of the O-object

$$\dot{\varphi} = a - k_3\mu$$

define the feedback Σ_3^{φ}-system.

The block diagram of a binary differentiator with an inertial-relay controller is shown in Fig. 9.35. To analyze the stability of the feedback of the Σ_3^{φ}-system, we shall differentiate the equation $\dot{\varphi} = a - k_3\mu$ with respect to t, substitute into it $\dot{\mu}$ from (9.37), and in the resulting equation replace μ by the expression $(a - \dot{\xi})/k_3$. As a result, we obtain an equation

$$\ddot{\varphi} + k_2\dot{\varphi} + k_3k_2 \operatorname{sgn} \varphi = \dot{a} + k_2 a. \tag{9.38}$$

It is easy to make this equation asymptotically stable at zero, for which purpose it suffices to ensure the fulfilment of the condition

$$k_3k_2 > \sup_{t \geq 0} |\dot{a} + k_2 a|.$$

Thus, $\varphi \to 0$ as $t \to \infty$, and, hence, this system also guarantees asymptotically exact differentiation

$$y(t) - \dot{g}(t) \to 0, \quad t \to \infty.$$

We shall now describe a finite binary differentiator.

9.4. Finite Binary Differentiator

The generalized block diagram of a finite controller does not differ from the standard block diagram of a binary differentiator. Its peculiarity is that use is made of a C-controller stabilizing the binary CO-controller R_μ which guarantees the appearance of a second-order sliding mode (Fig. 9.36a), i.e., a mode where not only e but also $\dot{e}$ is continuous, and only $\ddot{e}$ suffers discontinuities. When this sliding begins, the relations

$$e = 0, \quad \dot{e} = 0$$

are certainly identically satisfied, and from this instant we have $y(t) = \dot{g}(t)$ (Fig. 9.36*b*). We can use any algorithm described in Chap. 5 as such a stabilization algorithm, namely, a finite twisting algorithm, an algorithm optimal with respect to time or drift. For details we refer the reader to the indicated chapter.

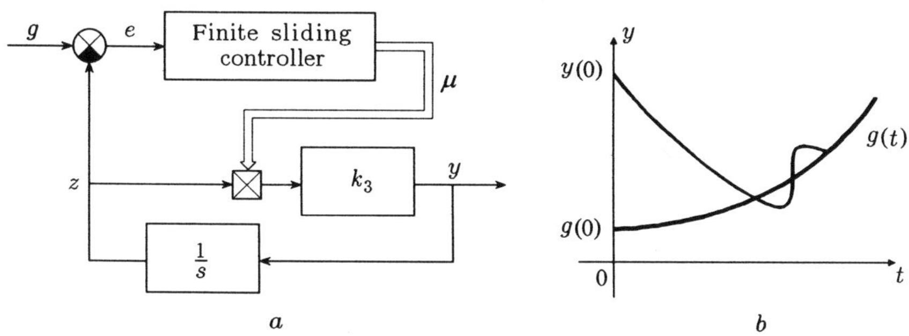

Figure 9.36.

By way of example, we shall describe here the principle of construction of the finite twisting algorithm. As was established above, the problem of stabilization of feedback can be reduced to the problem of stabilization at zero of a scalar object

$$\dot{\varphi} = a - k_3\mu, \quad |a| \leq \gamma, \quad \gamma = \text{const}\,. \tag{9.39}$$

Let us set, for simplicity, $a = \text{const}$ and assume that the signs of the variable φ and its derivative $\dot{\varphi}$ are known. Then the algorithm

$$\dot{\mu} = -k_1 \operatorname{sgn}\varphi - k_2 \operatorname{sgn}\dot{\varphi}, \quad k_1, k_2 = \text{const} > 0 \tag{9.40}$$

finitely stabilizes the object (9.39) at zero. Indeed, from (9.39) and (9.40) we have a second-order equation

$$\ddot{\varphi} + k_3k_2 \operatorname{sgn}\dot{\varphi} + k_3k_1 \operatorname{sgn}\varphi = 0$$

whose phase trajectories are parabolas defined by the equations

$$\ddot{\varphi} = \pm\alpha = \pm k_3(k_2 + k_1),$$
$$\ddot{\varphi} = \pm\beta = \pm k_3(k_2 - k_1),$$

and if $\alpha > \beta > 0$, then, in the twisting mode, the phase point gets to zero in a finite time.

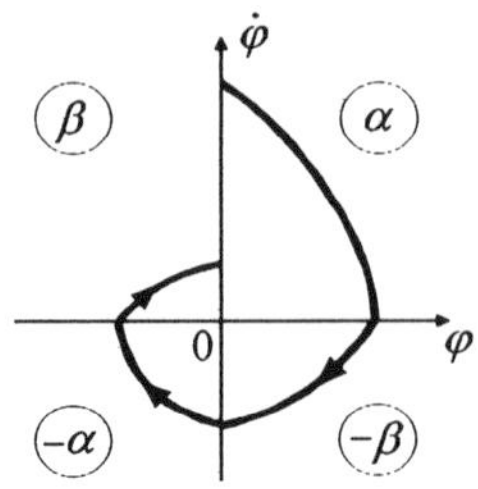

Figure 9.37.

Note that a second-order sliding mode arises at zero, and this means that when there is a small delay τ in the switching, then the deviation of φ from zero has the second order of smallness with respect to τ, i.e., $|\varphi| \sim O(\tau^2)$, and not the first order ($|\varphi| \sim O(\tau)$), as in the case of regular sliding.

9.5. Nonstandard Differentiating Systems

In tracking differentiating systems (Fig. 9.38) the choice of the operator of C-feedback R_y was aimed at ensuring the validity of the relation

$$z = g, \tag{9.41}$$

from which, under a proper smoothness, we get an estimate for the derivative $y = \dot{z} \cong \dot{g}$.

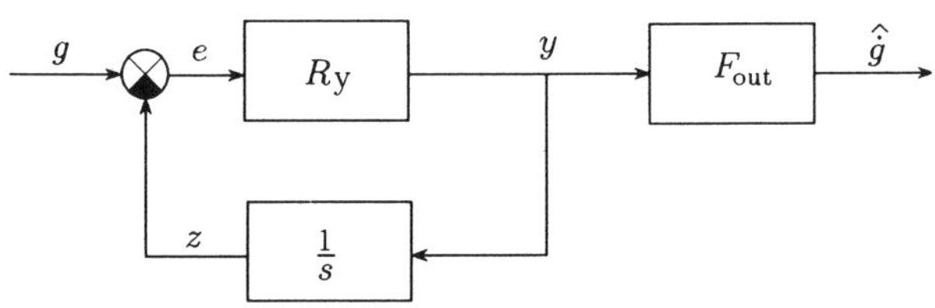

Figure 9.38.

The exact equality (9.41) can only be ensured with the aid of high gain ($k \to \infty$) or discontinuous feedback. In all cases, the estimate of the derivative was formed by the output signal y of the C-controller R_y. When there exist an additive interference in the signal being processed and an amplitude constraint at the input of the C-controller, the use of high gain or discontinuous feedback lowers the noise protection of the differentiator and, as a consequence, deteriorates the estimate of the derivative.

Moreover, it follows from the estimates given above that the guaranteed quality of differentiation in systems with discontinuous C-controller R_y becomes noticeably worse as the amplitude of the discontinuous (commuted) signal grows. Consequently, it is necessary to find new means and methods of constructing differentiating systems in which the effect produced by these factors would be weaker. Let us consider two new types of differentiating systems.

9.5.1. Differentiator with a "small" amplitude of discontinuities

The principle of construction of a differentiator with a small amplitude of discontinuities is based on the following, sufficiently obvious, idea which we shall explain by way of an example of a relay differentiator. If we distinguish the mean component of a discontinuous signal at the input of an integrator, say, with the aid of an inertial element and add the result to the output signal of

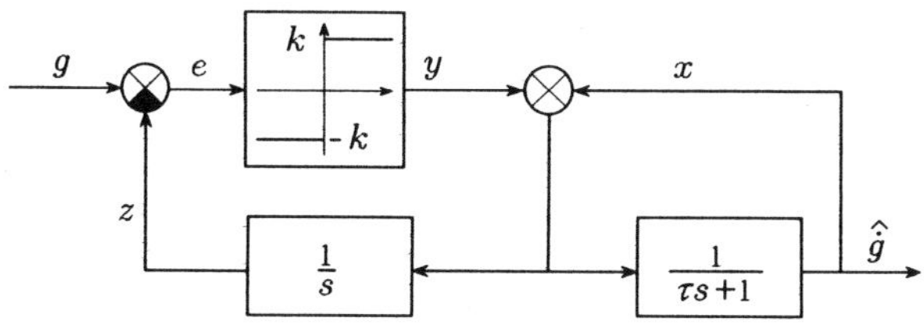

Figure 9.39.

the relay controller (Fig. 9.39), then the mean value of the discontinuous signal y in the stabilization mode, when $e = 0$, will be zero, and this means that, in this case, the sliding mode can be sustained by an arbitrarily small magnitude of the commuted signal. Formally this follows from the equations

$$\begin{aligned} \dot{e} = \dot{g} - u, \quad & u = y + x, \\ \tau\dot{x} + x = u, \quad & y = k \operatorname{sgn} e. \end{aligned} \tag{9.42}$$

If a sliding mode arises at the point $e = 0$, then $\dot{e} = 0$, and the equivalent value of the control is given by the expression

$$u_{\text{eq}} = \dot{g}.$$

However, $\tau\dot{x} + x = u$, and, hence, the signal $x(t)$ provides an asymptotic estimate for the derivative $\dot{g}$. Consequently, the mean value of the signal

$$(y_{\text{eq}} = u_{\text{eq}} - x) \to 0, \quad t \to \infty.$$

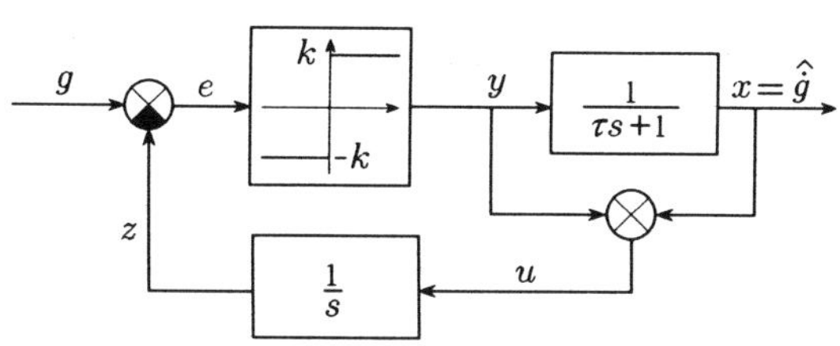

Figure 9.40.

In order to carry out a more exact investigation of this differentiator, it is convenient to reduce its block diagram to the form shown in Fig. 9.40. In addition to the tracking error e we introduce an error of differentiation

$$\varepsilon = \dot{g} - x \tag{9.43}$$

and write the motion equations of the differentiator in the variables (e, ε). From (9.42), (9.43), and Fig. 9.40 we have

$$\dot{\varepsilon} = \ddot{g} - \dot{x} = \ddot{g} - \frac{y}{\tau} = \ddot{g} - \frac{k}{\tau}\,\mathrm{sgn}\, e, \tag{9.44}$$

$$\dot{e} = \varepsilon - y = \varepsilon - k\,\mathrm{sgn}\, e. \tag{9.45}$$

To analyze the stability of zero of system (9.44), (9.45), we shall use a Lyapunov function of the form

$$v = (k/\tau)|e| + \varepsilon^2/2.$$

By virtue of system (9.44), (9.45), its derivative for $e \neq 0$ has the form

$$\begin{aligned}\dot{v} &= \frac{k}{\tau}(\mathrm{sgn}\ e)\dot{e} + \varepsilon\dot{\varepsilon}\\ &= \frac{k}{\tau}\,\mathrm{sgn}\ e(\varepsilon - k\,\mathrm{sgn}\, e) + \varepsilon\left(\ddot{g} - \frac{k}{\tau}\,\mathrm{sgn}\, e\right) = \varepsilon\ddot{g} - \frac{k^2}{\tau}.\end{aligned}$$

It follows from the last expression that if

$$|\ddot{g}| \leq M, \quad M = \mathrm{const}, \tag{9.46}$$

then, at least for the initial conditions

$$\varepsilon(0) \leq \frac{k^2}{\tau M}, \tag{9.47}$$

a sliding mode arises in finite time at the point $e = 0$. In this case, the equivalent value of the discontinuous signal is defined by the relation

$$k\,\mathrm{sgn}_{\mathrm{eq}}\, e = \varepsilon,$$

substituting which into Eq. (9.44), we obtain an equation for the variation of the differentiation error

$$\tau\dot{\varepsilon} + \varepsilon = \tau\ddot{g}.$$

If (9.46) is satisfied, then the following estimate of the differentiation error is valid:

$$|\varepsilon(t)| \leq \tau M. \tag{9.48}$$

Estimates (9.47) and (9.48) imply an inequality which bounds from below the magnitude of discontinuities:

$$k > \tau M.$$

Inferences.

- In an ideal case, the amplitude of discontinuities can be made arbitrarily small.
- In the real scheme the constant of the filter $\tau \geq \tau_{\rm cr}$, where $\tau_{\rm cr}$ is defined by nonideal switchings, and therefore the amplitude of discontinuities cannot be smaller than $k_{\rm cr} \geq \tau_{\rm cr} M$.
- In the differentiator that we have considered the class of admissible signals is extended to signals with bounded $\ddot{g}$.

9.5.2. Nonstandard binary differentiator

Whereas the preceding nonstandard differentiator was based on a tracking system, the differentiator that we consider below is based on an idea which is new in principle. Its essential features can be seen from Fig. 9.41 which shows that we propose to use the "tracking" error

$$e = g - z$$

as the estimate $g(t)$ of the derivative of the signal.

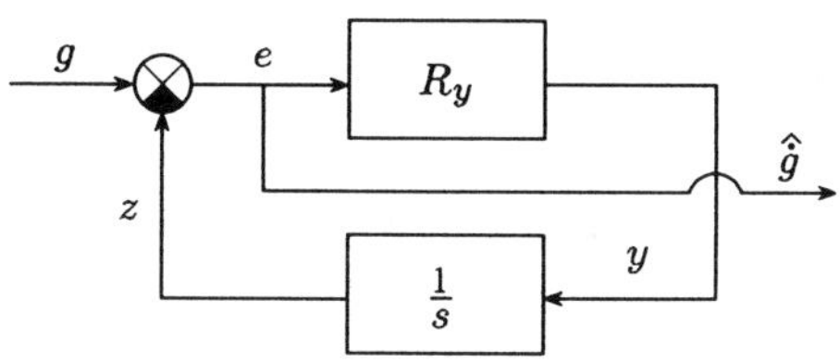

Figure 9.41.

We can explain the gist of this idea as follows. In a standard tracking system $e = 0$ and the output of the controller R_y coincides with the derivative $\dot{g}$. If an interference $q \sin \omega t$ is added to the signal $g(t)$, then we have a signal $\dot{g} + q\omega \cos \omega t$ at the output of the controller, and it may be that $q\omega \gg 1$. Since there are always amplitude constraints at the output of the controller, they will be constantly violated and the estimate of the derivative will be distorted. Therefore an idea arises of using the signals at the input of the controller in order to evaluate the derivative of the signal since then we can expect a decrease of the negative influence of the amplitude constraints.

Let us carry out a preliminary analysis of the block diagram given in Fig. 9.41. It is described by the operator relation

$$e = g - z = g - \frac{R_y}{s} e,$$

where s is the symbol of the differentiation operation. Consequently,

$$e = \frac{s}{s + R_y} g. \tag{9.49}$$

For the differentiator to give arbitrarily small phase and amplitude distortions, the relation

$$e = \frac{s}{\tau s + 1} g \tag{9.50}$$

must be satisfied for the arbitrarily assigned time constant $\tau = \text{const} > 0$. Comparing (9.49) and (9.50), we find the expression for the operator

$$R_y = (\tau - 1)s + 1 \tag{9.51}$$

required for this purpose. We can see from relation (9.51) that in order to solve this problem (i.e., the problem of differentiation) by linear means, we need an operator of a "pure" derivative. We get a vicious circle which can be broken only by the use of the effects which are observed in nonlinear systems.

It is expedient to recall here that in Chap. 6 we had a similar situation (the O-theory) where, as a result of using three types of feedback (C-, CO-, and O-feedbacks), we managed to obtain an operator of the form (9.51), i.e., an operator which transforms a control error x_1 into a weighted sum of the error and its derivative:

$$u = (-k/\delta)\sigma = (-k/\delta)(\dot{x}_1 + dx_1).$$

However, $u = \mu x_1$, and, comparing these expressions, we find that the action of μ is equivalent to the action of the operator

$$R_u = (-k/\delta)(s + d).$$

Let us use the recommendations given by the CO- and O-theories. As a result we obtain a block diagram shown in Fig. 9.42. In this diagram we must define

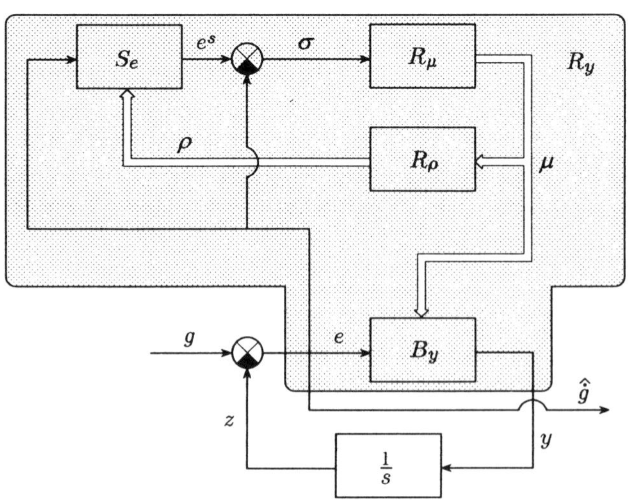

Figure 9.42.

the operators R_μ and R_ρ of the CO- and O-controllers respectively, the binary operation B_y, and the reference device S_e. The toned part is the controller (regulator) R_y.

If we are interested only in the principal potentialities of the scheme under investigation, then it is natural to use the results of the O-theory and choose the operators R_μ, R_ρ, S_e and the binary operation B_y in the form

$$\begin{aligned} &R_\mu : \mu = k \operatorname{sgn} \sigma e, \quad \sigma = e^s - e, \quad k = \text{const} > 0, \\ &R_\rho : \rho = q\mu, \quad q = \text{const} > 0, \\ &S_e : T\dot{e}^s + c_\rho e^s = e, \quad c_\rho = c + \rho, \quad c = \text{const} > 0, \\ &B_y : y = B_y e = \mu e. \end{aligned}$$

This choice is dictated by the following reasons. If the error $\sigma = 0$, then $e = e^s$, and from the equation of the reference device S_e we get

$$T\dot{e} + c_\rho e = e.$$

However, $c_\rho = c + q\mu$, and therefore the last equation can be rewritten in the form

$$T\dot{e} + (c-1)e = -q\mu e.$$

This means that the "action" of the operator variable μ is equivalent, in its effect, to the "action" of the operator

$$R_y = -\frac{Ts + (c-1)}{q}.$$

Comparing this expression to the required expression

$$R_y = (\tau - 1)s + 1,$$

we obtain relations for calculating the parameters of the scheme

$$1 - \tau = \frac{T}{q}, \quad 1 = \frac{1-c}{q}, \tag{9.52}$$

which can be easily satisfied. As a result, we get the required equation

$$\tau\dot{e} + e = \dot{g}.$$

In accordance with this choice, the block diagram of the differentiator becomes more concrete and assumes the form shown in Fig. 9.43.

Let us analyze the differentiating system. The indicated consequences are observed only if $\sigma \equiv 0$ from some time moment. To obtain the conditions that would guarantee this fact, we write an equation for the variation of the error σ.

Since $\sigma = e^s - e$, we first have

$$\dot{\sigma} = \dot{e}^s - \dot{e}\big|_{S_e} = \frac{e - (c + q\mu)e^s}{T} - \dot{g} + \mu e.$$

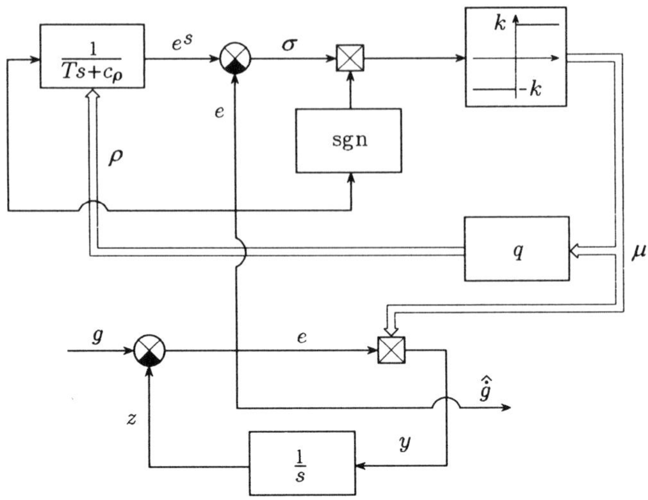

Figure 9.43.

However, $e_s = \sigma + e$, and therefore

$$\dot{\sigma} = \frac{e - c(\sigma + e)}{T} - \dot{g} + \mu e\left(1 - \frac{q}{T}\right) - \frac{q\mu\sigma}{T}$$
$$= \frac{1-c}{T}e - \frac{c + q\mu}{T}\sigma - \dot{g} + \mu e\left(1 - \frac{q}{T}\right). \tag{9.53}$$

Relations (9.52) imply

$$\frac{1-c}{T} = \frac{1}{1-\tau}, \quad \frac{c}{T} = \frac{1-q}{q(1-\tau)}, \quad 1 - \frac{q}{T} = -\frac{\tau}{1-\tau}, \quad \frac{q}{T} = \frac{1}{1-\tau},$$

and therefore we can rewrite (9.53) as

$$\dot{\sigma} = \frac{e}{1-\tau} - \frac{(1 - q + q\mu)\sigma}{q(1-\tau)} - \dot{g} - \mu e\frac{\tau}{1-\tau}.$$

Since $\mu = k\,\mathrm{sgn}(\sigma e)$, we finally have an equation

$$\dot{\sigma} = \frac{e}{1-\tau} - \frac{(1 - q + q\mu)\sigma}{q(1-\tau)} - \dot{g} - k|e|\frac{\tau}{1-\tau}\,\mathrm{sgn}\,\sigma. \tag{9.54}$$

Note now that nothing hinders the choice of q as a sufficiently small parameter, say, such that

$$q(k+1) < 1. \tag{9.55}$$

Then the coefficient (recall that $|\mu| \leq k$)

$$\frac{1 - q + q\mu}{q(1-\tau)} > 0.$$

For the sliding mode to exist at the point $\sigma = 0$, the inequality

$$\sigma(e - \dot{g} + \tau\dot{g} - k\tau|e|\operatorname{sgn}\sigma) \leq 0$$

must be satisfied as it follows from (9.54). At least in a small neighborhood of the point $\sigma = 0$ this condition is fulfilled if

$$|\dot{g}| < k, \tag{9.56}$$

since, in this case, $e \cong \dot{g}$.

Thus, if relations (9.55)–(9.56) are satisfied, then, under the initial conditions, from some neighborhood of $\sigma = 0$ in a finite time a sliding mode appears for which the required equality $\sigma = 0$ is satisfied. As a result, the favorable consequences described above are observed, namely,

- the use of the discontinuous stabilizing feedback does not entail the necessity of using a "smoothing" output filter in this case whereas this is obligatory in other discontinuous differentiating systems; however, to obtain higher derivatives, these "smoothing" filters may be necessary;
- constraints imposed on the class of signals being differentiated,

  $$|\dot{g}| \leq \text{const}, \quad |\ddot{g}| \leq \text{const},$$

  may be weakened up to

  $$|\ddot{g}| \leq \text{const}$$

  if use is made of a combination of the ideas which serve as the basis for the construction of the nonstandard differentiating system that we have considered (Fig. 9.44).

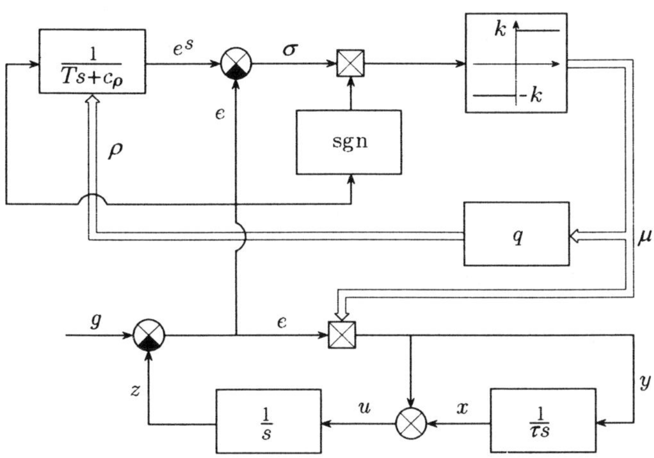

Figure 9.44.

This technique makes it possible to diminish the amplitude of the commuted signal. We omit the details.

To complete this theme, we give the results of discrete simulation of a nonstandard binary differentiator.

9.5.3. The results of discrete simulation of a nonstandard binary differentiator

The signal $g(t) = \sin t + \varepsilon \sin(1000t)$ being differentiated contains a useful signal $\sin t$ and an interference $\varepsilon \sin(1000t)$. The parameters in the block diagram of the differentiator shown in Fig. 9.43 are the following:

$$T = 1/20, \quad q = 1/19.8, \quad c = 0.95, \quad k = 1.$$

At these values $\tau = 0.01$. The difference scheme is constructed with the use of Euler's method with a pitch $h = 0.001$. The results of this simulation are given in the figures.

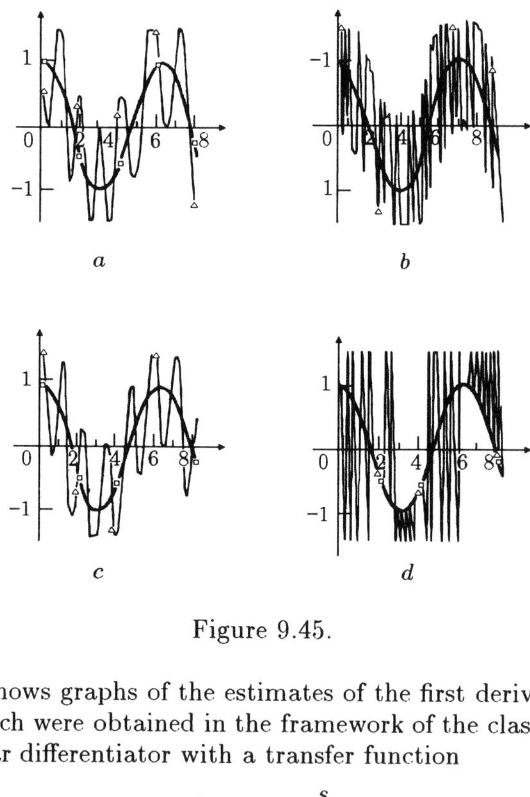

Figure 9.45.

Figure 9.45 shows graphs of the estimates of the first derivative in the presence of noise which were obtained in the framework of the classical scheme with the aid of a linear differentiator with a transfer function

$$W(s) = -\frac{s}{\tau s + 1}.$$

Figure 9.45*a* shows the results of differentiation of a signal with an interference level $\varepsilon \sim 10^{-2}$ and Fig. 9.45*b* shows graphs obtained as a result of differentiation of the same signal with an interference level $\varepsilon \sim 10^{-1}$.

Figure 9.46 gives the results of differentiation in the framework of the classical differentiation scheme with the aid of a nonstandard binary differentiator. It shows the graphs of the estimates of the second, third, fourth, and fifth derivatives, respectively, when the differentiation is without interference ($\varepsilon = 0$).

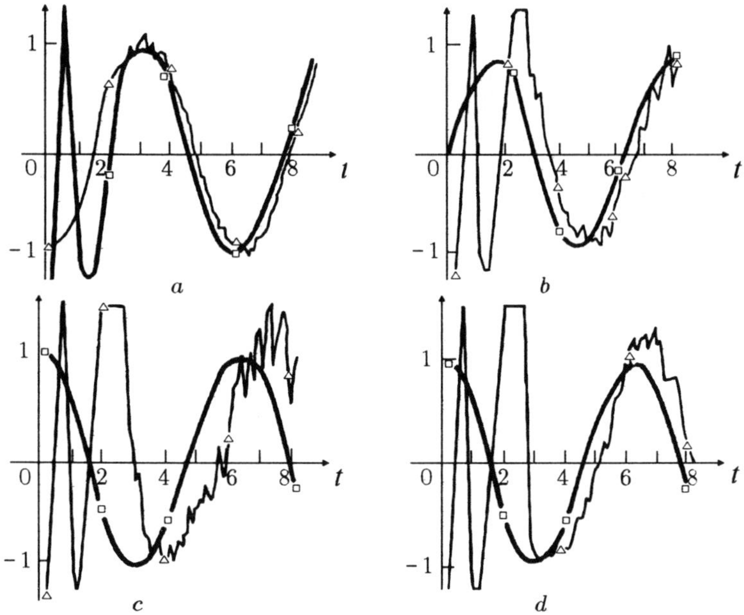

Figure 9.46.

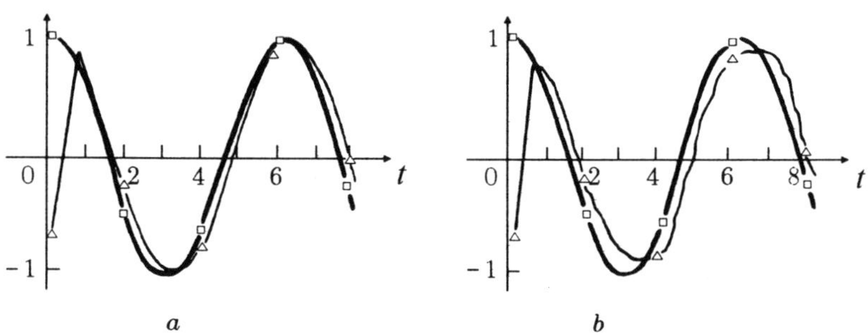

Figure 9.47.

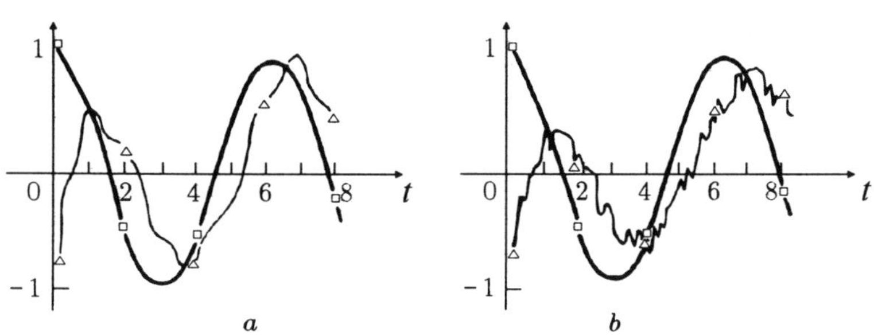

Figure 9.48.

Figure 9.47 shows the graphs of estimates of the first derivative in the presence of noise which were obtained in the process of a successive differentiation by a nonstandard binary differentiator with a noise level $\varepsilon \sim 10^{-2}$. The graphs of estimates of the first derivative in the presence of noise which were obtained in the process of a successive differentiation by nonstandard binary differentiator with a noise level $\varepsilon \sim 10^{-1}$ are given in Fig. 9.48.

The squares in Figs. 9.45–9.48 denote the derivatives and triangles denote the estimates of the derivatives.

Chapter 10

Suboptimal Stabilization of an Uncertain Object

Let us consider cases where the problem of optimal stabilization of an uncertain object can be solved exactly or approximately. The approach to the solution is based on the combination of ideas of optimal stabilization, asymptotic invariance, and the binary control theory. The invariance of the field of extremals relative to the uncertainty factors serves as the support element of the construction being developed. In this case, uncertainty factors affect only the optimal feedback. New types of feedback and binary control methods are used for its approximate realization. The proposed approach is compared to such well-known methods as the method of averaging, the method of guaranteed result, the method of feedback with high gain, and others.

10.1. Statement of the Optimal Stabilization Problem

The formation of optimal stabilization as a self-sustained division of the optimal control theory in a deterministic statement dates back to the works of A.A. Letov, H. Bellman, A.A. Krasovskii, and, in a stochastic statement, to the works of R. Kalman. As a result of long-term efforts, an ASOC theory (analytic synthesis of optimal controllers) was created.

The central place in the ASOC theory is occupied by the problem of synthesis of optimal feedback which stabilizes a determined or stochastic object under the conditions where there exists full information concerning the behavior of the object and the characteristics of external forces. Under real conditions, the information about the object and the disturbances acting on it is always incomplete, and since the optimal solution is, as a rule, very sensitive to the variations of the conditions of the problem, the recommendations of this theory can only serve as the basis and are seldom used in practice.

The potentialities of the ASOC theory become somewhat extended when its methods are combined with the methods of adaptive control. The internal constraints of the theory of adaptive control could not lead to universal methods of optimal stabilization under uncertainty, and therefore the problem being considered remains urgent. A new viewpoint is required concerning the optimal control problem when there is uncertainty. In particular, it is of interest to distinguish situations where it is possible to obtain an optimal or arbitrarily close to optimal (suboptimal) solution of a stabilization problem under a considerable uncertainty in the description of control objects or external forces. Here and in what follows, we understand a considerable uncertainty as the availability of information only about the majorants of functions which describe the disturbances and, perhaps, about the channels of their action on the object.

In this section we shall restrict the consideration to optimal stabilization problems in the following formulation: for a finite-dimensional object

$$\dot{x} = f(t, x, u, a), \quad t \geq t_0,$$

which contains a compact uncertainty

$$a \in A,$$

we have to synthesize a realizable optimal feedback with respect to the state which transfers the object from any initial position x_0 to the preassigned state x_1 by a time moment $t_1 > t_0$. This feedback keeps the object in the position x_1 for all $t > t_1$ and minimizes the functional

$$J(a) = \int_{t_0}^{t_1} F(t, x, u, a)dt.$$

Here $x \in \mathbb{R}^n$ is a phase vector and $u \in \mathbb{R}$ is control. The functions f and F are such that the existence of optimal feedback is guaranteed for every $a \in A$. The essence of this problem in that the perturbation a is not known exactly and cannot be identified. We have to describe the class of objects and functionals for which the formulated statement of the problem of optimal stabilization admits a physically sensible solution. The positive answer to this range of questions constitutes the principal content of this section.

We can briefly explain the gist of the proposed approach as follows. Suppose that the perturbation a in known. Then the function

$$u_{\text{opt}}(t, x, a) = \operatorname{argmin}_u J(a)$$

defines the optimal feedback and the equation of the field of extremals assumes the form

$$\dot{x}_{\text{opt}} = f_{\text{opt}}(t, x_{\text{opt}}) = f(t, x_{\text{opt}}, u(t, x_{\text{opt}}, a), a).$$

The first obvious possibility of realizing the optimal system is when u_{opt} does not depend on the perturbation a. The second possibility is less trivial and arises when the field of extremals does not depend on the perturbation a. In this case, of course, u_{opt} depends on the perturbation and cannot be exactly realized. The use of new types of feedback and the methods of binary control theory opens up a possibility of an arbitrarily close approximation to the optimal control law. We preface the general consideration of the problem by an illustrative example which demonstrates the peculiarities of the approach that we develop.

10.2. Example of an Optimal Stabilization Problem under Uncertainty

For simplicity, we shall consider a scalar object

$$\dot{x} = ax + u, \quad t \geq 0,$$

with an unknown parameter a satisfying the constraint

$$|a| \leq A,$$

where A is a known constant. The optimal feedback must transfer the object from the arbitrary state $x_0 \in \mathbb{R}$ to zero in an infinite time providing a minimum to the quadratic "full energy" functional

$$J(a) = \int_0^\infty \left(x^2 + \frac{1}{\gamma^2}\dot{x}^2\right) dt.$$

Here γ is a known positive parameter. Following the tradition, we use the Bellman function

$$v(t) = \int_t^\infty \left(x^2 + \frac{1}{\gamma^2}\dot{x}^2\right) d\tau, \quad t \geq 0,$$

when calculating u_{opt}.

In accordance with the optimality principle, the function $v(t)$ satisfies the differential Bellman equation

$$\min_u \left\{\frac{dv}{dt} + x^2 + \frac{1}{\gamma^2}\dot{x}^2\right\} = 0.$$

Setting $v = kx^2/2$, we find an expression for the optimal control in the form

$$u_{\text{opt}} = -ax - \frac{\gamma^2 k}{2}x.$$

Substituting this expression into Bellman's equation, we can determine the Bellman function parameter

$$k = 2/\gamma$$

and, together with it, the finite form of the optimal feedback

$$u_{\text{opt}} = -(\gamma + a)x.$$

The field of extremals in this problem is generated by the differential equation independent of the parameter a:

$$\dot{x}_{\text{opt}} + \gamma x_{\text{opt}} = 0,$$

and the minimal value of the functional $J(a)$ does not depend on a either and is defined by the relation

$$J_{\text{opt}} = \frac{x_0^2}{\gamma}.$$

It stands to reason that it is impossible to realize the obtained feedback, and therefore we must find the possibilities of its approximate realization. Let us consider the possibilities that are based on the removal of the uncertainty in the problem being considered as early as the initial stage. The first possibility which we consider is connected with the idea of averaging and presupposes the availability of information about the density of distribution of the parameter a.

10.3. Optimal Stabilization "in the Mean"

Suppose, for definiteness, that the density of distribution of the parameter a is uniform over the interval $[-A, A]$, i.e.,

$$p(a) = \frac{1}{2A}.$$

Then the mean value of the function,

$$\bar{J}(A) = \int_{-A}^{A} p(a)\, J(a)\, da,$$

must be minimized, and therefore we must consider the averaged Bellman function

$$\bar{v} = \int_{-A}^{A} v(t)\, p(a)\, da$$

if, of course, Bellman's principle of optimality is used to synthesize the feedback. In this case, the optimality equation has the form

$$\min_u \left\{ \frac{d\bar{v}}{dt} + \overline{x^2} + \frac{1}{\gamma^2}\overline{\dot{x}^2} \right\} = 0.$$

If we set $\bar{v} = \bar{k}x^2/2$ and carry out the averaging, then the optimality equation assumes the form

$$\min_u \left\{ \frac{A^2 x^2}{3\gamma^2} + \bar{k}xu + \frac{u^2}{\gamma^2} \right\} = 0.$$

All the quantities from the preceding equation are known, i.e., the situation is completely definite and, consequently, the equation can be solved. We begin with finding

$$u_{\text{opt}} = -\frac{\gamma^2 \bar{k}}{2} x.$$

Then we substitute the obtained expression into the optimality equation and determine the parameter of Bellman's averaged function

$$\bar{k} = \frac{2}{\gamma}\sqrt{1 + \frac{A^2}{3\gamma^2}}.$$

Then the final form of the optimal feedback is defined by the expression

$$u_{\text{opt}} = -\sqrt{\gamma^2 + \frac{A^2}{3}}\, x,$$

the field of extremals satisfies the differential equation

$$\dot{x}_{\text{opt}} + \left(\sqrt{\gamma^2 + \frac{A^2}{3}} - a\right) x_{\text{opt}} = 0,$$

and the minimal value of the averaged functional is defined by the relation

$$\bar{J}_{\text{opt}} = \sqrt{1 + \frac{A^2}{3\gamma^2}}\, \frac{x_0^2}{\gamma}.$$

Thus, in the averaging method the optimal feedback is determined, the field of extremals depends on the uncertain parameter, and the optimal value $\bar{J}_{\text{opt}}$ of the functional exceeds the minimally possible value J_{opt} since

$$\Delta \bar{J} = \bar{J}_{\text{opt}} - J_{\text{opt}} = \left(\sqrt{1 + \frac{A^2}{3\gamma^2}} - 1\right) \frac{x_0^2}{\gamma} > 0.$$

This is precisely the inequality that defines the losses in optimality when the stabilization is carried out "in the mean".

10.4. Minimax Optimal Stabilization

If there is no information about the density of distribution, then it is possible to reduce the problem with uncertainty being considered to a problem without uncertainty with the use of the idea of a guaranteed result where the "worse" value of the functional must be minimized. Formally, this procedure reduces to the well-known problem of minimax and a search for a saddle point. We shall follow Bellman's optimality principle, and, to determine Bellman's function $v = kx^2/2$, shall use the differential minimax equation

$$\min_u \max_{a \in A} \left\{ kx(ax + u) + \frac{1}{\gamma^2}(ax + u)^2 + x^2 \right\} = 0.$$

A direct analysis of the expression in braces allows us to state that its maximum with respect to $a \in A$ is achieved for

$$a^* = A \operatorname{sgn}\left(\frac{\gamma^2 k}{2} x^2 + ux\right).$$

Therefore the choice of the optimal feedback must now be made on the basis of a minimization of a completely determined function, i.e.,

$$\min_u \left\{ \left(1 + \frac{A^2}{\gamma^2}\right) x^2 + A \left| kx^2 + \frac{2xu}{\gamma^2} \right| + \frac{u^2}{\gamma^2} + kxu \right\} = 0.$$

Hence we obtain the optimal feedback in the form

$$u_{\text{opt}} = -\frac{\gamma^2 k}{2} x,$$

where the Bellman function coefficient

$$k = \frac{2}{\gamma}\sqrt{1 + \frac{A^2}{\gamma^2}},$$

i.e., the feedback which is optimal according to the minimax principle is linear and is given by the expression

$$u_{\text{opt}} = -\sqrt{\gamma^2 + A^2}\, x.$$

The field of extremals is described by the differential equation

$$\dot{x}_{\text{opt}} + \left(\sqrt{\gamma^2 + A^2} - a\right) x_{\text{opt}} = 0,$$

and the minimax value of the functional is given by the expression

$$J^M_{\text{opt}} = \frac{1}{\gamma}\sqrt{1 + \frac{A^2}{\gamma^2}}\, x_0^2.$$

Under this approach to the synthesis of feedback the absolute losses in optimality are defined by the relation

$$\Delta J^M = J^M_{\text{opt}} - J_{\text{opt}} = \sqrt{1 + \frac{A^2}{\gamma^2}}\,\frac{x_0^2}{\gamma}.$$

10.5. Stabilization with the Use of the Standard Model and an Error Feedback with High Gain

It is well known that feedback with high gain is an efficient means for suppressing disturbances. However, its direct application in the problem being considered is inefficient. Therefore we shall, first of all, pay attention to the fact that the field of extremals does not depend on the uncertainty factor and, consequently, the corresponding differential equation can be taken as the equation of a standard model. Then we can introduce feedback with respect to the error signal

$$e = x - x_{\text{opt}}$$

and use it to eliminate the error in accordance with the block diagram shown in Fig. 10.1. Suppose, for instance, that a feedback with high gain is linear and is realized by increasing the gain factor $k > 0$ to infinity, i.e.,

$$u = -ke, \quad k \to \infty.$$

Then the equation of a system with deviations has the form

$$\dot{e} = -(\gamma + k)e + (a + \gamma)x.$$

As $k \to \infty$, the error diminishes to zero, and the behavior of the feedback system does not differ in anything from the optimal behavior.

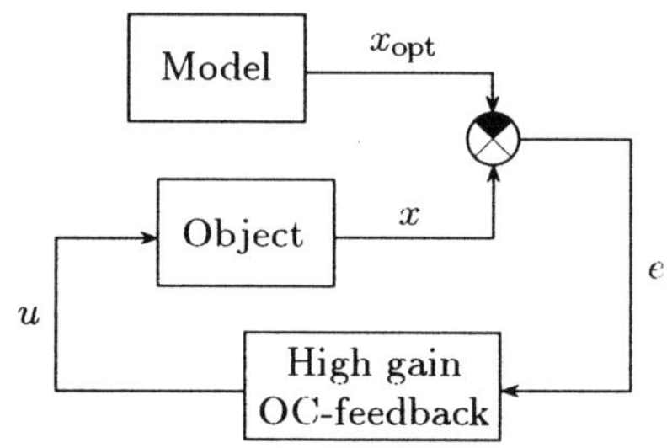

Figure 10.1.

Formally, the obtained solution completely exhausts the problem under consideration since the losses in optimality are zero. However, if we are interested in a solution which would have a pragmatic sense, then we have to take into account the presence of physical constraints and investigate the robustness of such an optimal system with respect to regular and singular perturbations.

It is well known that systems with a very high gain factor in feedback are not robust with respect to singular perturbations, and therefore it is forbidden to use high gain factors that exceed a certain critical value k_{cr}. However, if $k < k_{\text{cr}}$, then the error $e \not\equiv 0$ and losses in optimality occur for whose evaluation we write an equation with deviations

$$\dot{e} = -k_\gamma e + a_\gamma x,$$

where $k_\gamma = k + \gamma$, $a_\gamma = a + \gamma$. Since $e(0) = 0$, the estimate

$$|e| \le \frac{A_\gamma}{k_\gamma}|x|, \quad A_\gamma = A + \gamma,$$

is valid for all $t > 0$. This relation and the obvious inequality

$$|x| \le |x_{\text{opt}}| + |e|$$

imply, for a sufficiently large value of k (i.e., $k > A$), the estimate

$$|x| \le m|x_{\text{opt}}|, \quad \bar{m} = \frac{k_\gamma}{k - A}$$

which is necessary for the sequel.

The integrant of the functional of the problem

$$F = x^2 + \frac{1}{\gamma^2}(ax - ke)^2$$

is estimated by the inequality

$$F \leq Mx^2, \quad M = 1 + \frac{1}{\gamma^2}\left(A + \frac{kA_\gamma}{k_\gamma}\right)^2.$$

Since on the extremal the integrant is defined by the relation

$$F_{\text{opt}} = 2x_{\text{opt}}^2,$$

the inequality

$$F \leq \frac{m^2 M}{2} F_{\text{opt}}$$

holds for $t > 0$, and therefore the absolute loss in optimality satisfies the inequality

$$\Delta F^k = J_{\text{opt}}^k - J_{\text{opt}} \leq \left(\frac{m^2 M}{2} - 1\right)\frac{x_0^2}{\gamma}.$$

This estimate is, of course, overstated. To obtain a more "delicate" result, we must introduce into consideration a function $\xi(t)$ as a solution of the differential equation

$$\dot{\xi} = -(k_\gamma + a)\xi + k\xi^2 + a_\gamma,$$

which corresponds to the trivial initial condition $\xi(0) = 0$.

The error e is now defined by the relation $e = \xi x$ and the integrant of the functional satisfies the relation

$$F = x^2\left[1 + \frac{1}{\gamma^2}(a - k_\gamma)^2\right] \leq M_\xi x^2,$$

where

$$M_\xi = \max_\xi \left[1 + \frac{1}{\gamma^2}(a - k_\gamma)^2\right].$$

If $m_\xi^2 = \max\limits_\xi [1/(1-\xi)^2]$, then the final estimate of the absolute losses has the form established above but with more "exact" constants:

$$\Delta J^{k,\xi} = J_{\text{opt}}^{k,\xi} - J_{\text{opt}} \leq \left(\frac{m_\xi^2 M_\xi}{2} - 1\right)\frac{x_0^2}{\gamma}.$$

This estimate is remarkable because it decreases to zero as $k \to \infty$. Note that this property does not hold for the preceding estimate.

Thus, an increase in the gain factor in feedback raises the quality of a control system, but only to a certain limit which depends on the critical gain factor k_{cr}. However, the question concerning the existence of other methods which minimize the losses in optimality or eliminate them remains urgent. If $a = \text{const}$, then we can pin our hopes on the ideas of identification and adaptation. The process of identification of a parameter causes losses in optimality which are comparable with losses caused by the preceding methods and, consequently, does not lead to a qualitatively new result. The possibilities connected with this approach are also limited by the fact that the parameter a must be fixed.

10.6. Stabilization by the Methods of the Binary Control Theory

The methods of the binary control theory are characterized by a fundamentally different, for instance, from the method of a high gain factor, application of the independence of the field of extremals in the problem of the uncertainty factor.

Whereas in the case of a standard feedback we first determine a control law and then reveal the effect produced by its parameters on the quality of the problem solution, in the case of binary control the form of feedback is not fixed but is determined automatically, with the aid of the new type of error feedback.

The task of the new loop of feedback is to ensure the coincidence of the qualitative behavior of the control object with a certain dynamical element (the reference device of dynamical properties) which defines this behavior. It stands to reason that the feedback in this case is nonlinear.

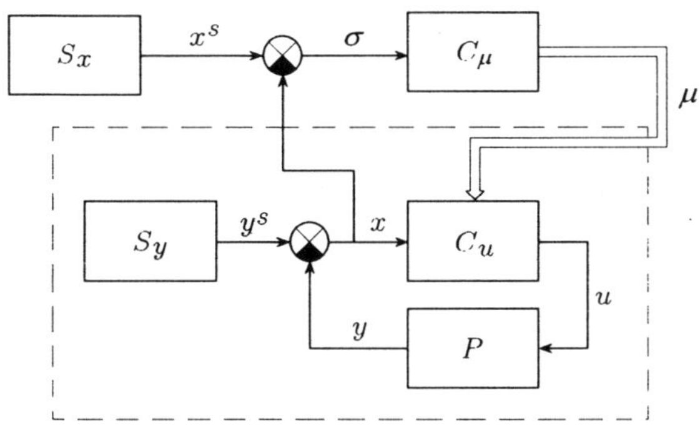

Figure 10.2.

The block diagram of the control system in which this idea is realized is shown, in a generalized form, in Fig. 10.2 in which P is an object, S_y and S_x are the main and the auxiliary reference device with output reference actions y^S and x^S, respectively, C_u and C_μ are the operators of the main and the binary feedback, x and σ are the errors of control of the main and the coordinate-operator feedback, u and μ are coordinate and operator control signals. In this diagram, we must choose the operator C_μ and the binary operation in C_u for which the error σ of the coordinate-operator feedback loop becomes zero.

Specifically, for the problem under consideration, this diagram assumes the form shown in Fig. 10.3. For convenience and simplicity, the block diagram of the control system shown in this figure reflects the choice of the operator C_u and the binary operation for which

$$u = \mu x.$$

In a binary control system of this kind only the operator C_μ must be chosen from the optimality conditions of the whole system. This should be done only when the error $\sigma \equiv 0$. The motion equation for this error has the form

$$\dot{\sigma} = -\gamma\sigma - a_\gamma x - \mu x, \quad \sigma(0) = 0,$$

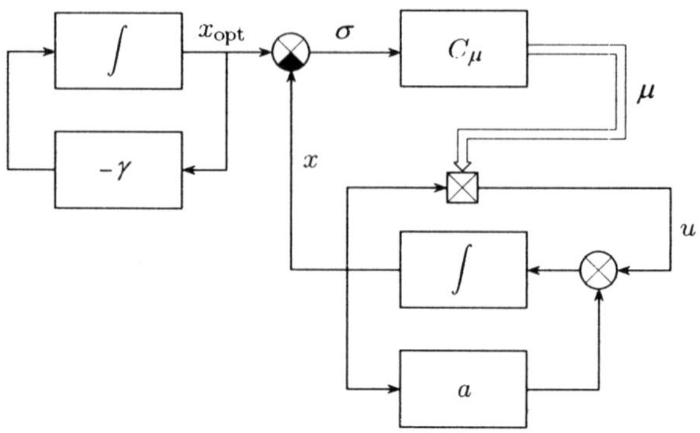

Figure 10.3.

and the problem consists in a "correct" choice of the algorithm of the variation of the operator variable μ. Let us consider the possibilities that exist for the realization of this idea.

10.6.1. A variable structure system

In the classical theory of variable structure systems use is made of a discontinuous feedback which ensures a sliding mode on the surface of discontinuity. In our case, this idea leads to the use of a relay coordinate-operator feedback

$$\mu = k \operatorname{sgn}(\sigma x), \quad k = \text{const} > 0.$$

As a result, when the inequality

$$k > A_\gamma$$

is satisfied in a feedback control system described by the equation

$$\dot{\sigma} = -\gamma\sigma - a_\gamma x - k|x| \operatorname{sgn} \sigma,$$

a sliding mode occurs on the "surface" of discontinuity $\sigma = 0$, and, from the initial moment (since $\sigma(0) = 0$), the motion equations of the control system are the same as the motion equation of the optimal system for any a. However, the question is whether the properties of such a control system are invariant, in this case, relative to the parameter a.

To evaluate the quality of a control system, it is natural to use the functional of the problem

$$J = \int_0^\infty \left(x^2 + \frac{1}{\gamma^2} \dot{x}^2 \right) dt.$$

If we act formally and do not pay attention to the discontinuity of $\dot{x}$, then, to calculate the functional J, we must replace

$$\dot{x} = ax + k|x| \operatorname{sgn} \sigma$$

in the integrant by

$$\dot{x} = ax + k|x| \operatorname{sgn}_{\text{eq}} \sigma.$$

However, $k|x| \operatorname{sgn}_{\text{eq}} \sigma = -a_\gamma x$, and therefore $\dot{x} = -\gamma x$ and, hence, the optimal value of the functional obtained in this way is invariant relative to the parameter a.

Now if we act more accurately, we may notice that

$$\dot{x} = a^2x^2 + 2akx|x| \operatorname{sgn}_{\text{eq}} \sigma + k^2x^2 = \left(k^2 - a^2 - 2a\gamma\right)x^2,$$

and if $a = \text{const}$, then the optimal value of the functional depends on the parameter a and is given by the expression

$$J_{\text{VSS}}^{\text{opt}}(a) = \frac{1}{2}\left[1 + \frac{1}{\gamma^2}(k^2 - a^2 - 2a\gamma)\right]\frac{x_0^2}{\gamma}.$$

Consequently, when use is made of the methods of classical theory of variable structure systems, the minimum of the functional in the problem under consideration is not fixed and is defined by a specific value of the parameter a. In this sense, the optimal control problem being considered cannot be solved by the VSS methods because the integrant of the functional is nonlinear with respect to discontinuous control. In this connection, of a certain interest are methods of optimal stabilization which ensure the motion along the extremal by means of a continuous control.

10.6.2. Binary stabilization with an integral CO-feedback

A motion arbitrarily close to optimal can be ensured by a continuous binary control with an integral CO-feedback

$$\dot{\mu} = \alpha \operatorname{sgn}\left\{\left[\sigma + \frac{M}{k}|\sigma|\right]x\right\},$$

where $\alpha = \text{const} > 0$ is a sufficiently large number and $k > A_\gamma$ as before.

To verify this fact, we make a change of variable $\sigma = \xi x$ and obtain motion equations

$$\dot{\xi} = -(1+\xi)(a_\gamma + \mu), \quad \dot{\mu} = \alpha \operatorname{sgn}\left[\xi + \frac{M}{k}|\xi|\right].$$

Since $\mu \to \mu_{\text{eq}} = k \operatorname{sgn}_{\text{eq}} \xi$ as $\alpha \to \infty$ and the process $\xi(t)$ is stabilized at zero, it is easy to see that for a finite but sufficiently large α the process $\xi(t)$ will be stabilized in a certain neighborhood of zero

$$|\xi| \le \delta_1$$

with a constant $\delta_1 = \delta_1(\alpha)$ such that $\delta_1(\alpha) \to 0$ as $\alpha \to \infty$.

Let $\eta = \mu - \mu_{\text{eq}}$. Then it follows from what we had before that

$$|\eta| \le \delta_2,$$

where the constant $\delta_2 = \delta_2(\alpha)$ and $\delta_2(\alpha) \to 0$ as $\alpha \to \infty$. In this notation the estimate

$$F = x^2 + \frac{1}{\gamma^2}\dot{x}^2 = x^2\left[1 + \left(\frac{\gamma+\eta}{\gamma}\right)^2\right] \le \frac{\gamma^2 + (\gamma^2+\delta_2)^2}{2\gamma^2(1-\delta_1)^2}2x^2_{\text{opt}}$$

is valid for the integrant of the functional F and, consequently, the inequality

$$J^B_{\text{opt}} \le \frac{\gamma^2 + (\gamma^2+\delta_2)^2}{2\gamma^2(1-\delta_1)^2}\frac{x_0^2}{\gamma}$$

holds for the optimal value of the functional J^B_{opt}. It follows from this inequality that $J^B_{\text{opt}} \to x_0^2/\gamma$ as $\alpha \to \infty$, and we can now speak about a suboptimal stabilization in the class of binary feedbacks. Since binary systems are robust relative to regular and singular perturbations, the constructed suboptimal control system is also robust, and this distinguishes it advantageously from a stabilization system with a high gain factor.

10.6.3. Stabilization with the use of a second-degree sliding mode

For a suboptimal binary stabilization we have $x(t) \not\equiv x_{\text{opt}}(t)$. However, the exact equality $x(t) = x_{\text{opt}}(t)$ can be ensured by a continuous control from a finite moment t_* if we use a second-degree sliding mode. Let us again consider a binary control system in coordinates (ξ, μ) :

$$\dot{\xi} = -(1+\xi)(a_\gamma + \mu).$$

Then, for $0 < \rho < 1/2$ and sufficiently large constants α and λ, the control algorithm of the form $\mu = \mu_1 + \mu_2$, where

$$\dot{\mu}_1 = \alpha \operatorname{sgn}\left(\xi + \frac{M}{k}|\xi|\right), \quad \mu_2 = \lambda|\xi|^\rho \operatorname{sgn}\xi,$$

ensures the stabilization of ξ at zero in a finite time t_* which can be made arbitrarily small if $\xi(0) = 0$. In this case, the loss for search if defined by the relation

$$\Delta J^S = J^S_{\text{opt}} - J_{\text{opt}} = \int_0^{t_*}(F - F_{\text{opt}})\,dt,$$

and, consequently, second-degree sliding modes also open up a possibility of a suboptimal control of an uncertain object. If we introduce an index of suboptimality (with respect to the loss in optimality) according to the relation

$$I = \frac{\Delta J}{J_{\text{opt}}},$$

then we can demonstrate the advantages of each of the stabilization methods that we have considered (Fig. 10.4). We have $I^S < I^B$ in this figure since for

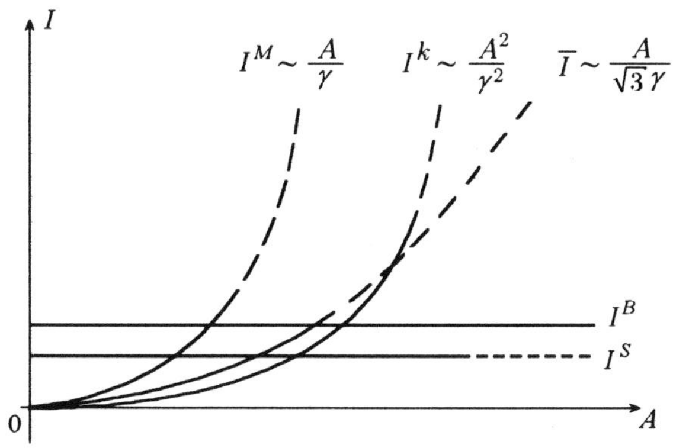

Figure 10.4.

the same values of the factors α and k the losses in optimality are smaller for the second-degree sliding mode. The sign "$\sim$" denotes the rate of growth of the suboptimality index as the "degree" of uncertainty rises.

Let us consider now the general theory following the works [29, 30, 43].

10.7. Reduction of the Suboptimal Stabilization Problem to the Problem of Asymptotic Invariance

The generalization of the results presented above is the approach to suboptimal stabilization based on the reduction of the initial problem to the problem of asymptotic invariance. The theory of asymptotic invariance is a universal tool for solving various control problems under the conditions of uncertainty, and this reduction makes it possible to extend the field of application of the ideas and approaches that we have considered.

10.7.1. Main concepts of the theory of asymptotic invariance

Let A be a compact set and suppose that the pair of functions $\{\hat{x}, u_{\text{opt}}\}$ for each element $a \in A$ minimizes the functional in the Lagrange form

$$J(u, a) = \int_{t_0}^{t_1} L(t, x, u, a)\, dt + l(x_0, x_1)$$

under the constraints

$$\dot{x} = f(t, x, u, a), \quad x(t_0) = x_0, \ x(t_1) = x_1, \ u \in \mathcal{U}.$$

Suppose, in addition, that the optimal value of the functional does not depend on the element $a \in A$, i.e.,

$$J(u_{\text{opt}}, a) = J(u_{\text{opt}}) \quad \text{for any } a \in A.$$

Here and in the sequel $x \in \mathbb{R}^n$ and u is a scalar control. It is assumed that the functions f, L, l and the classes A and $\mathcal{U}$ are such that there exists a solution of the optimization problem. It follows from the assumption that we have made that the optimal control depends on a, i.e., $u_{\text{opt}} = u_{\text{opt}}(t, x, a)$, and its realization is impossible if the perturbation a is unknown. Therefore it is relevant to pose the question concerning the approximation of the optimal control u_{opt} in the class of admissible feedbacks. To evaluate the quality of such an approximation, we can use the suboptimality index (relative losses in optimality)

$$I(u, a) = \frac{J(u, a) - J(u_{\text{opt}})}{J(u_{\text{opt}})}.$$

We introduce the concept of the quality of approximation by means of the following definition: if, for any $\varepsilon > 0$, there exists an admissible feedback $u \in \mathcal{U}$ such that $\sup_{a \in A} I(u, a) \leq \varepsilon$, then the feedback system is suboptimal in the class of controls $\mathcal{U}$.

The concept of suboptimal stabilization system can be connected with the concept of asymptotic invariance.

Let ε and γ be arbitrary positive numbers. For a certain function $h(t, x)$ we introduce a notation

$$h_{u,a} = h\big(t, x(t; t_0, x_0, u, a)\big), \quad \sigma_u = h_{u,a_1} - h_{u,a_2}.$$

We say that the dynamical system $\dot{x} = f(t, x, u, a)$ is (ε, γ)-exponentially h-invariant relative to the perturbations $a \in A$ in the class of controls $\mathcal{U}$ if there exist a feedback $u \in \mathcal{U}$ and a function $p(x)$ ($p(0) = 0$), which is positive for all $x \neq 0$, such that

$$|\sigma_u| \leq \varepsilon p(x_0) e^{-\gamma(t-t_0)}, \quad t \geq t_0,$$

for any elements a_1, $a_2 \in A$ and every $x_0 \in \mathbb{R}^n$.

The following statement reveals the relationship between the suboptimality and exponential invariance.

Statement 10.1. *Let the following proposition be fulfilled for certain positive numbers γ, q_1, q_2, q_3 and an arbitrary $\varepsilon > 0$:*

(1) *the dynamical system $\dot{x} = f(t, x, u, a)$ is $(\varepsilon q_1, \gamma)$-exponentially L-invariant relative to the perturbations $a \in A$ in the class of controls $\mathcal{U}$,*

(2) *for any control $u \in \mathcal{U}$ there exists an element $a_* \in A$ such that*

$$I(u, a^*) \leq \varepsilon q_2,$$

(3) *for any $u \in \mathcal{U}$, $a \in A$*

$$\big|l\big(x_0, x(t_1; t_0, x_0, u, a)\big) - l\big(x_0, x(t_1; t_0, x_0, u, a_*)\big)\big| \leq \varepsilon q_3.$$

Then the dynamical system $\dot{x} = f(t, x, u, a)$ is suboptimal in the class $\mathcal{U}$ relative to the Lagrangian L and there exists a realizable feedback u_* such that

$$I(u^*, a) \le \varepsilon \left(q_2 + \frac{q_1 p(x_0) + q_3 \gamma}{\gamma J(u_{\text{opt}})} \right).$$

In this statement conditions are indicated under which the problem of suboptimization reduces to the problem of exponential invariance. For the latter problem, constructive methods for synthesizing a feedback have been worked out, and therefore its solution is essentially simplified. In the theory of asymptotic invariance use is made of universal sufficient conditions of exponential invariance which constitute a unique basis for obtaining laws of control which can be employed for solving a specific problem. These sufficient conditions can be formulated as the following statement.

Statement 10.2. *Let $\eta(t) = \varepsilon p(x_0) e^{-\gamma(t-t_0)}$ $(\gamma > 0)$ and suppose that there exist a constant $q > 0$ and a continuous positive function $v(t) = v(t, x_0) > 0$ such that*

(1) *$v(t)$ is differentiable with respect to t when $v(t) \ge \eta(t)$,*

(2) *$v(t_0) < \eta(t_0)$,*

(3) *the condition*

$$\dot{v} + \gamma v \big|_{v(t) \ge \eta(t)} \le 0$$

is fulfilled for a certain $u \in \mathcal{U}$,

(4) *$|\sigma_u(t)| \le q v(t, x(t))$ for all $t \ge t_0$, $a \in A$.*

Then

$$|\sigma_u(t)| \le \varepsilon q p(x_0) e^{-\gamma(t-t_0)}$$

uniformly with respect to t and $a \in A$, i.e., for $h = L$ there exists a (ε, γ)-exponential L-invariance.

We shall show how the sufficient conditions for exponential invariance formulated above can be used in specific situations by way of an example of suboptimal linearly quadratic stabilization of an uncertain system.

10.7.2. Suboptimal linearly quadratic stabilization

For a linear object and a quadratic functional the scheme of synthesis of suboptimal control proposed above can be completed. Indeed, let us consider in $\mathbb{R}^n$ for $t \ge 0$ a linear stationary control object

$$\dot{x} = Ax + b(u + a), \quad x(0) = x_0,$$

where $a \in A$ is an unknown perturbation such that

$$|a| \le a_1 |x|, \quad |\dot{a}| \le a_2 |x|.$$

Here the constants a_1, a_2 are known. We have to construct a realizable feedback which minimizes the "full" energy functional

$$J = \int_0^\infty L(t)\,dt, \quad L = x^\top \tilde{P} x + \dot{x}^\top \tilde{Q} \dot{x}, \quad \tilde{P} > 0, \quad \tilde{Q} > 0.$$

Using the notation

$$P = \tilde{P} + (A - bd^\top)^\top \tilde{Q}(A - bd^\top), \quad Q = b^\top \tilde{Q} b, \quad d = \frac{A^\top \tilde{Q} b}{b^\top \tilde{Q} b},$$

$$v = u + d^\top x + a, \quad A_d = A - bd^\top, \quad d \in \mathbb{R}^n,$$

we reduce the functional and the feedback equation to the form

$$J = \int_0^\infty (x^\top P x + Qv^2)\,dt,$$

$$\dot{x} = A_d x + bv.$$

The optimal solution of the transformed problem is known and is given by the expression

$$v_{\text{opt}} = -\hat{k}x,$$

where $\hat{k} = Q^{-1} b^\top R$ and R is the solution of Riccati equation

$$P - RbQ^{-1}bR + RA_d + a_d R = 0.$$

In this case, the field of extremals of the problem is described by the equation

$$\dot{\hat{x}} = (A_d - b\hat{k})\hat{x},$$

and the optimal value of the functional can be found from the relation

$$J_{\text{opt}} = x_0^\top R x_0.$$

In order to obtain an optimal stabilization system, we must realize the feedback

$$u_{\text{opt}} = -kx - a, \quad k = \hat{k} + d^\top,$$

and this is impossible. Instead of this nonrealizable feedback, we use a feedback of the form

$$u = -kx + u_\varepsilon,$$

where, to generate the component u_ε, we use binary control methods and the principles of suboptimal control and exponential invariance presented above.

Suppose that in the block diagram shown in Fig. 10.2 the dynamical reference device S_x is defined by the equation of the field of extremals of the problem

$$\dot{x}^s = A_k x^s, \quad x^s(0) = x_0, \quad A_k = A - bk.$$

It is known that the spectrum $\sigma(A_k)$ of the matrix A_k lies in the left open complex half-plane, i.e., there exists a number $\gamma_0 > 0$ such that the condition $\operatorname{Re}\lambda \le -\gamma_0$ is fulfilled for any $\lambda \in \sigma(A_k)$. Consequently, for a certain $N > 1$ we have

$$\|e^{A_k t}\| \le N e^{-\gamma_0 t}.$$

We take a function

$$L = x^\top P x + Q(u_\varepsilon + a)^2$$

as $h_{u,a}$.

Using the sufficient conditions of exponential invariance formulated in the preceding section, we find that the controlled system is exponentially L-invariant in the class of feedbacks being considered if there exists a feedback u_a such that

$$\lim_{t\to\infty} \left| L(t) - \hat{x}^\top(t)\big(P + \hat{k}^\top Q\hat{k}\big)\hat{x}(t) \right| = 0.$$

We set

$$u_\varepsilon = -k_0\|x\|\mu, \qquad \dot{\mu} = -\alpha \operatorname{sgn}\big[\sigma + \mu|\sigma|\big],$$

where k_0 and α are positive parameters and the function σ is a solution of the system of differential equations

$$\dot{\sigma} + l\sigma = a - k_0\|x\|\mu, \qquad \dot{\mu} = -\alpha \operatorname{sgn}\big[\sigma + \mu|\sigma|\big], \qquad |\mu(t_0)| \le 1.$$

Here l is an arbitrarily assigned positive number. It is obvious that the function $\sigma(t)$ is also unknown, but if we abstract ourselves from this circumstance for the time being, then it becomes clear that by an appropriate choice of k_0, α we can easily ensure the validity of the inequalities

$$|\sigma| \le \varepsilon\|x_0\|e^{-\gamma t},\ t \ge 0, \qquad |\dot{\sigma}| \le \varepsilon q_1\|x_0\|e^{-\gamma t},\ t \ge \hat{t},$$

$$\operatorname{Var}\Big|_0^{\hat{k}} \sigma \le \varepsilon\|x_0\|, \qquad \sup_{[0,\hat{t}]} |\dot{\sigma}| \le q_2\|x_0\|,$$

for certain positive constants ε, q_1, q_2, and γ.

These inequalities and the equation

$$\dot{x} = A_k x + b(\dot{\sigma} + l\sigma)$$

imply the asymptotic invariance of the system and, hence, its suboptimality. Thus everything has reduced to the evaluation of the function $\sigma(t)$, and to carry out this evaluation use can be made of an asymptotic observer of the form

$$\hat{\sigma} = c^\top x - \xi,$$
$$\dot{\xi} = c^\top A_k x + l(c^\top x - \xi),$$

where $c^\top b = 1$. It is easy to find out that the evaluation error $e = \sigma - \hat{\sigma}$ satisfies the equation $\dot{e} + le = 0$ and, consequently, the function $\hat{\sigma}$ exponentially converges (with an arbitrarily assigned exponent l) to the function σ. It follows that the suboptimality property will be preserved if instead of σ in feedback we use its estimate $\hat{\sigma}$, i.e.,

$$u = -kx - k_0\|x\|\mu,$$
$$\dot{\mu} = -\alpha\,\mathrm{sgn}\big[\hat{\sigma} + \mu|\hat{\sigma}|\big], \quad |\mu(0)| \leq 1.$$

Thus, the use of methods of asymptotic invariance for the optimal stabilization under uncertainty makes it possible to synthesize feedbacks which robustly stabilize uncertain objects with a quality arbitrarily close to the optimal quality. The solution of problems of optimal control in the framework of the theory of asymptotic invariance (the latter being universal) allows us to extend the classes of uncertain objects subject to optimization.

In the general case, the problem of optimal stabilization under the conditions of uncertainty cannot be solved by classical methods, but their combination with the theory of new types of feedback and the idea of asymptotic invariance makes it possible to solve the problem for a sufficiently large class of uncertain objects. It is precisely this fact that we substantiate in this chapter. The development and generalization of these propositions can be found in the works [29, 30, 31, 43, 44, 83, 93].

Conclusion

Feedback “permeates” our environment, it serves as a key element of the biological evolution and natural selection, it ensures a regulating mechanism in equilibrium systems, in particular, in natural ecological systems, and is a necessary element of efficient economical constructions, and, finally, it serves as the basis for self-controlled and self-sustained biological systems. This enumeration can be easily continued. However, we still know very little about the mechanism of feedback since, actually, it was never an object of investigation in itself, for which there are certain reasons.

Indeed, the idea of feedback is almost obvious, it is easily perceived, and in simple situations its application does not cause any problems. However, it is not easy, as a rule, to synthesize feedback in nonstandard situations, its synthesis requires nontrivial solutions. This is due to the absence of a theory which would explain the mechanism of feedback formation. As a rule, these mechanisms escape the researcher since they are very complicated. The situation here resembles that in other fields of natural science. The physicist Richard Feynman used to say about the law of gravitation that the action of the law was complicated but its key idea was simple and that this circumstance made all our laws related [89].

To say the truth, regular attempts to study feedback are made in the theory of automatic control, in bionics, and in theories of economics. However, in these disciplines almost always the emphasis is laid on the use of feedback rather than on the mechanism of its formation. This is natural since, according to Henon, science works out rules for solving a problem but not the techniques of choosing these rules.

The absence of rules of the second level makes it necessary to guess the laws governing feedback every time when we deal with a nonstandard problem, but it is very difficult to think of new ideas and invent new principles since it requires rich imagination. It is not by chance that there were very little revelations in the history of the control theory. As a result, the control theory contains, in fact, records of solutions of standard problems whereas it is desirable to have rules of synthesis which can be extended to new situations.

In this monograph we made an attempt to advance a hypothesis concerning the structure of the mechanism of feedback formation. Our conjecture is based on the “hierarchy” of complexity of feedback. This idea of the structure of feedback seems to us quite natural since it allows us to reduce the problem of synthesis of a complicated linear control to a solution of a sequence of single-type and well-studied problems. To put it otherwise, the complexity of a problem that we can see is recurrently generated by the internal simplicity that we do not see.

In order to realize the principle of hierarchy, we had to introduce a concept of signal-operator which is new for control theory. This term reflects the dual nature of signals in nonlinear dynamical systems. In conjunction with the principle of feedback, the signal-operator gives the necessary possibilities for the transition from a direct solution of a problem to finding the algorithm for solving the problem and, if the necessity arises, to the algorithm which defines the algorithm of solving the problem, and so on. The remarkable feature of "hierarchical structure" of feedback obtained in this way is that at each level of hierarchy the control mechanisms are simple, of the same type, and can be obtained by the methods which are standard for the classical control theory.

Pay attention to the fact that the idea of the dual nature of variable quantities is ordinary for natural sciences and very fruitful. Thus, for instance, Max Born introduced the concept of an operator of a physical quantity which turned out to be very productive in quantum mechanics which deals with objects which are dual in their fundamental principle, and the concept of operator-time introduced by Il'ya Prigogine proved to be very useful in the physics of irreversible processes.

We are sure that the appearance of binary principle and new types of feedback is quite natural for the contemporary stage of development of the general theory of feedback. To justify this thesis, we considered in detail the evolution of the most important principles and methods of the control theory arising with the growth of the uncertainty factors in control problems, which is of interest by itself and can serve as a brief introduction to the classical feedback theory.

It should also be pointed out that from the point of view of mathematics the proposed approach can be regarded as the technique of synthesizing nonlinear dynamical systems with preassigned properties of their solutions such, for instance, as stability, low sensitivity to the variations of the problem parameters, etc. It stands to reason that the application of the proposed principle is not restricted by the problems of stabilization, filtration, and optimization. Even today we already know of its other applications, in particular, for

- separation of close signals of unknown frequency in finite time, whereas linear schemes of separation require, in general, an infinite period of time,
- robust solution of inverse problems of dynamics which refer to problems of synthesis of standard trajectories and dynamic processing of measurement results,
- solution of stabilization problems under uncertainty and search for saddle positions of equilibrium.

We take the liberty to state that in this monograph we propose a hierarchical principle of formation of feedback which is based on the dual nature of the variables of a nonlinear system and which endows an automatic control system with elements of perfect behavior under complicated conditions of uncertainty. Possibly, some other mechanisms of synthesis of a complicated feedback also exist, but the advantage of the proposed mechanism is that it is simple in the methods of its realization, makes it possible to do without the use of large gain factors or discontinuous control elements. On the contrary, the needed behavior of an automatic control system can be achieved with the use of bounded gain factors and smooth control signals, which is especially important since in natural control systems the perfect behavior of control mechanisms is achieved under precisely these conditions.

Bibliography

[1] Aizerman, M.A., *Automatic Control Theory*, Nauka, Moscow (1966).

[2] Andreev, Yu.N., *Control of Finite-dimensional Linear Objects*, Nauka, Moscow (1976).

[3] Athans, M. and Falb, P.L., *Optimal Control* (Russian translation), Mashinostroenie, Moscow (1968).

[4] Bellman, R., *Dynamic Programming*, Princeton (1957).

[5] Bellman, R. and Kalaba, R., *Dynamic Programming in Modern Control Theory* (Russian translation), Nauka, Moscow (1969).

[6] Besekerskii, V.A. and Popov, E.P., *Theory of Automatic Control Systems*, Nauka, Moscow (1972).

[7] Bozhukov, V.M., Kukhtenko, V.I., Levitin, V.F., and Shumilov, B.F., "Relay automatic control systems for objects with large ranges of variation of dynamic characteristics," *Izv. Akad. Nauk SSSR, Tekhn. Kibern.*, **6**, 154–163 (1966).

[8] Boltyanskii, V.G., *Mathematical Methods of Optimal Control*, Nauka, Moscow (1969).

[9] Bromberg, P.V., *Matrix Methods in the Theory of Relay and Momentum Control*, Nauka, Moscow (1967).

[10] Butkovskii, A.G., *Theory of Optimal Control of Systems with Distributed Parameters*, Nauka, Moscow (1965).

[11] Voronov, A.A., *Fundamentals of Automatic Control Theory*, Part I. *Linear Systems of Control of One Quantity*, Energiya, Moscow-Leningrad (1965).

[12] Voronov, A.A., *Fundamentals of Automatic Control Theory*, Part II. *Special Linear and Nonlinear Systems of Automatic Control of One Quantity*, Energiya, Moscow-Leningrad (1966).

[13] Vyshnegradskii, I.A., *On Direct Action Controllers*, Moscow (1876).

[14] Gelig, A.Kh., Leonov, G.A., and Yakubovich, V.A., *Stability of Nonlinear Systems with a Nonunique Equilibrium Position*, Nauka, Moscow (1978).

[15] Emelyanov, S.V., *Variable-structure Automatic Control Systems*, Nauka, Moscow (1967).

[16] Emelyanov, S.V. (ed.), *Theory of Variable-structure Systems*, Nauka, Moscow (1970).

[17] Emelyanov, S.V., *Binary Automatic Control Systems*, International Research Institute for Management Sciences, Moscow (1984).

[18] Emelyanov, S.V. and Korovin, S.K., "Application of the principle of deviation control for extending the set of types of feedbacks," *Dokl. Akad. Nauk SSSR*, **258**, 5, 1070–1074 (1981).

[19] Emelyanov, S.V. and Korovin, S.K., "Extension of the set of types of feedbacks and their application in the process of constructing feedback dynamical systems," *Izv. Akad. Nauk SSSR, Tekh. Kibern.*, **5**, 173–183 (1981).

[20] Emelyanov, S.V. and Korovin, S.K., "Theory of nonlinear feedback under uncertainty," in: *Univers. Rosii, MGU*, **I**. *Mathematical Modeling*, 214–278 (1993).

[21] Emelyanov, S.V. and Korovin, S.K., "New types of feedback. Synthesis of nonlinear control under the conditions of uncertainty," Jubilee Collection of Papers of the Division of Informatics, Computer Science and Automation of the Russian Academy of Sci., **1**, 115-137 (1993).

[22] Emelyanov, S.V., Korovin, S.K., and Levantovskii, L.V., "Higher-order sliding modes in binary control systems," *Dokl. Akad. Nauk SSSR*, **287**, 6, 1338-1342 (1986).

[23] Emelyanov, S.V., Korovin, S.K., and Levantovskii, L.V., "Second-order sliding modes in the control of uncertain systems," *Izv. Akad. Nauk SSSR, Tekhn. Kibern.*, **1**, 112-118 (1986).

[24] Emelyanov, S.V., Korovin, S.K., and Levantovskii, L.V., "A new class of algorithms of the second-order sliding," *Matemat. Model.*, **2**, 3, 89-100 (1990).

[25] Emelyanov, S.V., Korovin, S.K., and Levant, A., "Higher-order sliding modes in controlled systems," *Diff. Uravn.*, **29**, 11, 1877-1899 (1993).

[26] Emelyanov, S.V., Korovin, S.K., and Mamedov, I.G., "Quasisplitting method and its application for synthesizing automatic control systems," *Dokl. Akad. Nauk SSSR*, **286**, 2, 311-315 (1986).

[27] Emelyanov, S.V., Korovin, S.K., and Mamedov, I.G., "Structural transformations and spatial decomposition of discrete controlled systems, the quasisplitting method," *Tekhn. Kibern.*, 6, 118-128 (1986).

[28] Emelyanov, S.V., Korovin, S.K., Mamedov, I.G., and Nersisyan, A.L., "Binary algorithms for control of one class of delayed uncertain dynamical systems," *Diff. Uravn.*, **25**, 10, 1670-1679 (1989).

[29] Emelyanov, S.V., Korovin, S.K., Mamedov, I.G., and Nosov, A.P., "Asymptotic invariance in problems of control of uncertain objects," *Dokl. Akad. Nauk SSSR*, **311**, 1, 44-49 (1990).

[30] Emelyanov, S.V., Korovin, S.K., Mamedov, I.G., and Nosov, A.P., "Asymptotic invariance and robust stabilization of hereditary systems by continuous control," *Dokl. Akad. Nauk SSSR.*, **311**, 2, 296-300 (1990).

[31] Emelyanov, S.V., Korovin, S.K., Memedov, I.G., and Nosov, A.P., "Asymptotic invariance of delayed control systems," *Diff. Uravn.*, **27**, 3, 415-427 (1991).

[32] Emelyanov, S.V., Korovin, S.K., and Nersisyan, A.L., "On asymptotic properties of state observers for uncertain systems with a separated stationary linear part," *Dokl. Akad. Nauk SSSR.*, **311**, 4, 807-811 (1990).

[33] Emelyanov, S.V., Korovin, S.K., and Nersisyan, A.L., "Stabilization of uncertain neutral objects by a variable structure control," *Dokl. Akad. Nauk SSSR*, **312**, 4, 801-806 (1990).

[34] Emelyanov, S.V., Korovin, S.K., Nersisyan, A.L., and Nisenzon, Yu.E., "Stabilization of uncertain systems with respect to the output by discontinuous control," *Dokl. Akad. Nauk SSSR*, **311**, 3, 544-549 (1990).

[35] Emelyanov, S.V., Korovin, S.K., Nersisyan, A.L., and Nisenzon, Yu.E., "Stabilization of multidimensional uncertain objects with respect to the output," *Dokl. Akad. Nauk SSSR*, **311**, 5, 1062-1067 (1990).

[36] Emelyanov, S.V., Korovin, S.K., Nersisyan, A.L., and Nisenzon, Yu.E., "Asymptotic observers for a class of nonlinear dynamical objects," *Dokl. Akad. Nauk SSSR*, **313**, 5, 1052-1056 (1990).

[37] Emelyanov, S.V., Korovin, S.K., and Sizikov, V.I., "On the synthesis of nonlinear control of free motion of nonstationary systems," *Dokl. Akad. Nauk SSSR*, **265**, 2, 297-301 (1982).

[38] Emelyanov, S.V., Korovin, S.K., and Sizikov, V.I., "The principles of design and the general methods of synthesis of binary systems of control of uncertain nonlinear objects," *Dokl. Akad. Nauk SSSR*, **281**, 4, 810-814 (1985).

[39] Emelyanov, S.V., Korovin, S.K., and Ulanov, B.V., "On the synthesis of control systems with the application of coordinate-parametric and parametric feedbacks," *Dokl. Akad. Nauk SSSR*, **266**, 5, 1077-1081 (1982).

[40] Emelyanov, S.V., Korovin, S.K., and Ulanov, B.V., "Control of nonstationary dynamical systems with the application of coordinate-parametric feedback," *Tekh. Kibern.*, 6, 201-212 (1982).

[41] Emelyanov, S.V., Korovin, S.K., and Ulanov, B.V., "Control of linear stationary objects under external actions with the application of various types of feedback," *Tekh. Kibern.*, 1, 174-182 (1984).

[42] Emelyanov, S.V., Korovin, S.K., and Ulanov, B.V., "On the control of nonstationary dynamical systems," *Diff. Uravn.*, **20**, 10, 1683-1691 (1984).

[43] Korovin, S.K., Mamedov, I.G., and Nosov, A.P., "Suboptimality of asymptotically invariant control systems," *Diff. Uravn.*, **28**, 11, 1932-1945 (1992).

[44] Korovin, S.K., Mamedov, I.G., and Nosov, A.P., "Stabilization of uncertain systems on rings," in: *Univer. Rossii, MGU*, **1**. *Mathematical Modeling*, 279-295 (1993).

[45] Korovin, S.K., Nersisyan, A.L., and Nisenzon, Yu.E., "Control of linear uncertain objects with respect to the output," *Tekh. Kibern.*, 1, 67-73 (1990).

[46] Krasovskii, A.A., *Dynamics of Continuous Self-adjusting Systems*, Fizmatgiz, Moscow (1963).

[47] Krasovskii, A.A., *Systems of Automatic Control of Flight and their Analytic Design*, Nauka, Moscow (1973).

[48] Krasovskii, A.A., Bukov, V.P., and Shendrik, V.S., *Universal Algorithms for Optimal Control of Continuous Processes*, Nauka, Moscow (1977).

[49] Krasovskii, A.A. and Pospelov, G.S., *Fundamentals of Automatics and Technical Cybernetics*, Gosenergoizdat, Moscow-Leningrad (1962).

[50] Krasovskii, N.N., "Theory of optimal controlled systems," in: *Mechanics in the USSR for 50 years*, **1**, 179-244, Nauka, Moscow (1968).

[51] Krasovskii, N.N., *Control of a Dynamical System: Problem of the Minimum of a Guaranteed Result*, Nauka, Moscow (1985).

[52] Kulebakin, V.S., "Theory of invariance of automatically controlled systems," Proc. of the 1st Congress of the FAC in Automatic Control, 1, 247-255, USSR Acad. Sci. Publ. (1960).

[53] Letov, A.M., "Analytic design of controllers," *Avtom. Telemekh.*, 4, 436-441; 5, 561-568; 6, 661-665; 4, 425-435; 11, 1405-1413 (1962).

[54] Lee, R., *Optimal Estimates, Determination of the Characteristics and Control*, Nauka, Moscow (1966).

[55] Lee, Y. and Van der Velde, Y., "Theory of nonlinear self-adjusting systems," in: *Self-adjusting Automatic Systems*, Nauka, Moscow (1964).

[56] Lyapunov, A.M., *General Problem on the Stability of Motion*, Gostekhizdat, Moscow-Leningrad (1950).

[57] Meerov, M.V., *Synthesis of Structures of High-Precision Automatic Control Systems*, Nauka, Moscow (1967).

[58] Mikhailov, A.V., "Method of harmonic analysis in control theory," *Avtomat., Telemekh.*, 3 (1938).

[59] Moiseev, N.N., *Numerical Methods in the Theory of Optimal Systems*, Nauka, Moscow (1971).

[60] Pavlov, A.A., *Synthesis of Time Optimal Relay Systems*, Nauka, Moscow (1966).

[61] Petrov, B.N., "On the application of optimality conditions," in: Proc. IInd All-Union Conference in Automatic Control Theory, **2**, 241-246, USSR Acad. Sci Publ., Moscow-Leningrad (1955).

[62] Petrov, B.N., Rutkovskii, V.R., Krutova, I.N., and Zemlyakov, S.D., "Principles of design and construction of self-adjusting control systems," Mashinostroenie, Moscow (1972).

[63] Pontryagin, L.S., Boltyanskii, V.G., Gamkrelidze, R.V., and Mishchenko, E.F., *Mathematical Theory of Optimal Processes*, Nauka, Moscow (1969).

[64] Pugachev, V.S. (ed), *Fundamentals of Automatic Control*, Nauka, Moscow (1968).

[65] Samarskii, A.A., *Theory of Difference Schemes*, Nauka, Moscow (1977).

[66] Tikhonov, A.N. and Arsenin, V.Ya., *Methods of Solving Ill-Posed Problems*, Nauka, Moscow (1979).

[67] Ulanov, G.M., "Statistical and information problems of disturbance control," Energiya, Moscow (1970).

[68] Utkin, V.I., *Sliding Modes in Optimization and Control Problems*, Nauka, Moscow (1981).

[69] Fel'dbaum, A.A., "Problems concerning self-adjusting systems," in: *Self-adjusting Systems*, 5-22, Nauka, Moscow (1963).

[70] Fel'dbaum, A.A., *Fundamentals of the Theory of Optimal Automatic Systems*, Nauka, Moscow (1966).

[71] Filippov, A.P., "Differential equations with discontinuous right-hand side," *Mat. Sb.*, **5**, 99-126 (1960).

[72] Filippov, A.P., *Differential Equations with Discontinuous Right-hand Side*, Nauka, Moscow (1985).

[73] Fomin, V.N., Fradkov, A.L., and Yakubovich, V.A., *Adaptive Control of Dynamical Objects*, Nauka, Moscow (1981).

[74] Tsypkin, Ya.Z., *Adaptation and Training in Automatic Systems*, Nauka, Moscow (1968).

[75] Tsypkin, Ya.Z., *Relay Automatic Systems*, Nauka, Moscow (1974).

[76] Csaki, F., *Modern Control Theories. Nonlinear Optimal and Adaptive Systems*, Akademlai Kiado, Budapest (1972).

[77] Black, H., "Stabilized feedback amplifiers," *Bell Syst. Tech. J.*, **13**, 1-18 (1934).

[78] Bode, H., *Network Analysis and Feedback Amplifier Design*, Van Nostrand, New York (1945).

[79] Emelyanov, S.V. and Korovin, S.K., "Development of feedback types and their application to design of closed-loop dynamic systems," in: *Problems of Control and Inform. Theory*, **10**, 3, 161-174, Hungarian Acad. Sci. (1981).

[80] Emelyanov, S.V., Korovin, S.K., and Levantovskii, L.V., "A drift algorithm in control of uncertain processes," in: *Problems of Control and Inform. Theory*, **15**, 6, 425-438, Hungarian Acad. Sci. (1986).

[81] Emelyanov, S.V., Korovin, S.K., and Mamedov, I.G., "Structural transformations and spatial decomposition of control systems: the quasi-decoupling method," in: *Problems of Control and Inform. Theory*, **16**, 3, 155-168, Hungarian Acad. Sci. (1987).

[82] Emelyanov, S.V., Korovin, S.K., and Mamedov, I.G., "Stabilization of uncertain dynamic delayed processes by binary control systems," in: *Problems of Control and Inform. Theory*, **18**, 3, 135-149 (1989).

[83] Emelyanov, S.V., Korovin, S.K., Mamedov, I.G., and Nosov, A.P., "Asymptotic invariance and stabilization on uncertain delay systems," in: *Dynamics and Control*, **4**, 39-58, Kluwer Academic Publ., Boston (1994).

[84] Emelyanov, S.V., Korovin, S.K., Nersisyan, A.L., and Nisenzon, Y.Y., "Discontinuous output stabilizing an uncertain MIMO plant," *Int. J. Control*, **55**, 1, 63-107 (1992).

[85] Emelyanov, S.V., Korovin, S.K., and Sizikov, V.I., "Use of coordinate-parametric feedback in design of control systems," in: *Problems of Control and Infrom. Theory*, **10**, 4, 237-251, Hungarian Acad. Sci. (1981).

[86] Emelyanov, S.V., Korovin, S.K., and Sizikov, V.I., "Control of nonstationary plants with coordinate-parametric and parametric feedbacks," in: *Problems of Control and Inform. Theory*, **11**, 4, 259-269 (1982).

[87] Emelyanov, S.V., Korovin, S.K., and Ulanov, B.V., "Control of nonstationary dynamic systems with quasicontinuous generation of the control signal," in: *Problems of Control and Inform. Tneory*, **12**, 1, 11-32, Hungarian Acad. Sci. (1983).

[88] Emelyanov, S.V., Korovin, S.K., and Ulanov, B.V., "Control of dynamic systems in face of exogenous signals with feedbacks of various types," in: *Problems of Control and Inform. Theory*, **14**, 1, 3-16 (1985).

[89] Feynman, R.P. and Hibbs, A., *Quantum Mechanics and Path Integral*, McGraw-Hill Book Company, New York (1965).

[90] Flugge-Lotz, I., *Discontinuous Automatic Control*, Princeton Univ. Press, Princeton, New Jersey (1953).

[91] Hazen, H.L., "Theory of servomechanisms," *J. Franklin Inst.*, **218**, 279-331 (1934).

[92] Kalman, R.E., Falb, P.L., and Arbib, M.A., *Topics in Mathematical System Theory*, McGraw-Hill Company, New York (1969).

[93] Korovin, S.K. and Nosov, A.P., "Synthesis of robust feedback systems by the methods of asymptotic invariance and its application to uncertain delay plants," Trans. First Asian Control Conf. (ASCC), July 27-30, 1994, Japan, Tokyo, N FE-14-4.

[94] Lee, E.B. and Markus, L., *Foundation of Optimal Control Theory*, Wiley, New York (1967).

[95] Maxwell, D., *On Governors*, Royal Soci, London (1868).

[96] Nyquist, H., "Regeneration theory," *Bell Syst. Tech. J.*, **11**, 126 (1932).

[97] Routh, E.S., *A Treatise on the Stability of a Given State of Motion*, McMillan & C°, London (1877).

[98] Stodola, A., *Dampf und Gasturbinen*, Berlin (1922).

Index